The Second International Conference on Maintenance and Rehabilitation of Constructed Infrastructure Facilities (MAIREINFRA2)

The Second International Conference on Maintenance and Rehabilitation of Constructed Infrastructure Facilities (MAIREINFRA2)

Editor

Hosin (David) Lee

MDPI

Basel • Beijing • Wuhan • Barcelona • Belgrade • Novi Sad • Cluj • Manchester

Editor
Hosin (David) Lee
University of Iowa
Iowa City, IA
USA

Editorial Office
MDPI
St. Alban-Anlage 66
4052 Basel, Switzerland

This is a reprint of articles from the Proceedings published online in the open access journal *Engineering Proceedings* (ISSN 2673-4591) (available at: https://www.mdpi.com/2673-4591/36/1).

For citation purposes, cite each article independently as indicated on the article page online and as indicated below:

Lastname, A.A.; Lastname, B.B. Article Title. *Journal Name* **Year**, *Volume Number*, Page Range.

ISBN 978-3-7258-0733-8 (Hbk)
ISBN 978-3-7258-0734-5 (PDF)
doi.org/10.3390/books978-3-7258-0734-5

Contents

About the Editor

Hosin (David) Lee

Dr. Lee is a Professor in Civil and Environmental Engineering and a Director of Laboratory for Advanced Construction Technology (LACT), Iowa Technology Institute, University of Iowa. Dr. Lee is an internationally recognized expert in asphalt pavement recycling research and infrastructure asset management. He developed the pioneering "PicCrack" software that automatically analyzes cracks on highway pavements using AI technology. Dr. Lee currently serves as the Associate Editor of the MDPI *Infrastructures* Journal and on the Editorial Board of the ASCE *Journal of Infrastructure System*. He is a Member of the National Academy of Engineering of Korea. He has served as President of the International Society for Maintenance And Rehabilitation of Transport infrastructures (iSMARTi), Koren–American Scientists and Engineers Association (KSEA), KSCEA, and KOTAA; on the Board of Trustees of the Seoul Institute of Technology; and as Chairman of the ASCE Highway Pavements Committee. He has received life-time achievement awards from the Ministry of Science and ICT of Korea, KSEA, Korea Agency for Infrastructure Technology Advancement (KAIA), and iSMARTi, and outstanding contribution awards from the Asphalt Paving Association of Iowa (APAI), KOFST, KSRE, Utah Engineers Council, and the Utah section of the ASCE. He received the Otto Monsted Professorship and COWI Foundation Fellowship in Denmark.

Preface

The Second International Conference on Maintenance and Rehabilitation of Constructed Infrastructure Facilities (MAIREINFRA2: Chairman Hosin "David" Lee, Professor of University of Iowa, Associate Editor MDPI *Infrastructures* Journal, and Immediate Past President of iSMARTi) is organized by the International Society for Maintenance and Rehabilitation of Transport Infrastructures (iSMARTi). The sustainable and resilient maintenance and rehabilitation of infrastructure is becoming a key challenge.

The objective of this conference is to exchange technological innovations among researchers, government employees, consultants, and contractors for maintaining and rehabilitating sustainable and resilient infrastructures, which include roads, rails, bridges, and buildings. MAIREINFRA2 is for sustainable infrastructure engineers, smart city designers, disaster resilience professionals, intelligent transportation engineers, construction managers, maintenance engineers, consulting engineers, government employees, and academic researchers involved in designing and managing sustainable infrastructures worldwide. All papers accepted for the conference are published in the Open Access Scopus-Indexed MDPI *Engineering Proceedings* journal.

Hosin (David) Lee
Editor

engineering proceedings

Editorial

Preface of the Second International Conference on Maintenance and Rehabilitation of Constructed Infrastructure Facilities (MAIREINFRA2) †

Hosin (David) Lee

Civil and Environmental Engineering, University of Iowa, Iowa City, IA 52242, USA; hosin-lee@uiowa.edu

† All papers published in the volume are presented at the Second International Conference on Maintenance and Rehabilitation of Constructed Infrastructure Facilities, Honolulu, HI, USA, 16–19 August 2023.

Citation: Lee, H. Preface of the Second International Conference on Maintenance and Rehabilitation of Constructed Infrastructure Facilities (MAIREINFRA2). *Eng. Proc.* **2023**, *36*, 70. https://doi.org/10.3390/engproc2023036070

Published: 3 January 2024

1. Conference Overview

The Second International Conference on the Maintenance and Rehabilitation of Constructed Infrastructure Facilities (MAIREINFRA2; Chairman: Hosin "David" Lee, Professor of University of Iowa, Associate Editor of the MDPI journal *Infrastructures*, Immediate Past President of iSMARTi and the Founding Board of Trustees of the Seoul Institute of Technology), was organized by the International Society for Maintenance and Rehabilitation of Transport Infrastructures (iSMARTi) in Honolulu, Hawaii, USA, from 16 to 19 August 2023.

The MAIREINFRA2 is for sustainable infrastructure engineers, smart city designers, disaster resilience professionals, intelligent transportation engineers, construction managers, maintenance engineers, consulting engineers, government employees, and academic researchers involved in designing and managing sustainable infrastructures worldwide.

The objective of this series of conferences is to provide a forum for researchers, government employees, consultants, and contractors to exchange technological advancements and innovations with regard to maintaining and rehabilitating sustainable and resilient infrastructures, which include roads, bridges, rails, and buildings.

The sustainable and resilient maintenance and rehabilitation of constructed infrastructure facilities is the backbone of economic prosperity and public welfare. But the aspects of sustainability and resiliency challenge constructors and managers to respond creatively to a new paradigm shift in rehabilitating and maintaining constructed infrastructure facilities in the most environmentally friendly manner by lowering energy costs, enhancing safety, and minimizing air and water pollution. The three main themes of MAIREINFRA2 are (1) the maintenance and rehabilitation of pavements, (2) the automation/innovations in bridge/building construction, and (3) safety, disaster resilience and sustainability.

This conference features the following:

World-famous keynote speakers presenting in opening and plenary sessions and three technical tracks.

Over eighty papers from sixteen countries in three technical tracks: (1) asphalt pavements, (2) concrete pavements/bridges/buildings, and (3) safety resilience and sustainability.

Many networking opportunities, including Wednesday's welcoming reception at the Rainbow Suite and Patio; Thursday's lunch and welcoming dinner; and Friday's lunch and closing banquet on the sixth floor of the Mid-Pacific Conference Center located on top of the parking lot.

2. Conference Committee

- Organizing Committee:

  Taha Ahmed, Australian University—Kuwait, Kuwait
  Ioannis Brilakis, Univerity of Cambridge, UK
  Filippo Giustozzi, RMIT University, Australia

Kyong Ju Kim, Chung-Ang University, Korea
Seong-Min Kim, Kyung-Hee University, Korea
Tae-Hwan Kim, Yong-In University, Korea
Soo-Ahn Kwon, KICT, Korea
Mike LaViolette, HDR, USA
Hosin "David" Lee, University of Iowa, USA
Seonha Lee, Kongju University, Korea
Davide Lo Presti, University of Palermo, Italy
Joao Merighi, Latersolo, Ltda, Brazil
Byungkyu Moon, ARA, Inc., USA
Athanassios Nikolaides, Aristotle University of Thessaloniki, Greece
Ghim Ping Ong, National University of Singapore, Singapore
Brian Park, University of Virginia, USA
Paulo Pereira, University of Minho, Portugal
Krishna Prapoorna, Indian Institute of Technology Tirupati, India
Omar Smadi, Iowa State University, USA
Bo Song, Beijing Institute of Technology, China
Susan Tighe, McMaster University, Canada
Haifang Wen, Washington State University, USA
Jon Young, HAPI, USA

- Technical Committee:

Serji Amirkhanian, University of Alabama, USA
Adrián Ricardo Archilla, University of Hawaii, USA
Gabriella Buttitta, University of Palermo, Italy
Carlos Chang, Florida International University, USA
Chunhee Cho, University of Hawaii, USA
Yoon-Ho Cho, Chung Ang University, Korea
Rita Fortes, Instituto Federal de São Paulo, Brazil
Gaspare Giancontieri, University of Palermo, Italy
Sungdo Hwang, KICT, Korea
John Harvey, UC Davis, USA
Byung-Suk Kim, KICT, Korea
Hee-Jeong Kim, University of Arizona, USA
Sung-Hee Sonny Kim, University of Georgia, USA
Seung Woo Lee, Kangneung-Wonju National University, Korea
Jenny Liu, Missouri S&T, USA
Sue McNeil, UNSW, Australia
Young-Jun Moon, KOTI, Korea
Jorge Pais, University of Minho, Portugal
Jee Woong Park, UNLV, USA
Mansour Solaimanian, Penn State Univ, USA
Aravind Swamy, Indian Institute of Technology Delhi, India
Jeb Tingle, US Army ERDC, USA
David Woodward, Ulster University, UK
Zhanping You, Michigan Tech., USA
Kyong-Ku Yun, Kangwon National University, Korea
Xiong Zhang, Missouri S&T, USA

3. Technical Program

The technical program of MAIREINFRA2 consists of more than 80 speakers from 16 countries (Italy, UK, France, Portugal, Netherlands, Australia, New Zealand, Kuwait, India, Singapore, Japan, Korea, Canada, Brazil, Chile and USA).

- Track A: Asphalt Pavements and Maintenance.

- Track B: Concrete Pavements, Bridges and Buildings.
- Track C: Traffic Safety, Resilience and Sustainability.

Keynote Presentations presided by Imad Al-Qadi, University of Illinois at Urbana-Champaign:

John Harvey (Professor of Civil and Environmental Engineering at the University of California, Davis, and Director of the UC Pavement Research Center and the City and County Pavement Improvement Center); presentation title: "Improving Pavement Sustainability through Integrated Design, Construction, Asset Management, LCA, LCCA, and S-LCA".

Ed Sniffen (Director of Hawaii department of Transportation). He has served as the chair of the AASHTO Committee on Transportation System Security and Resilience). Presentation title: "Driving Innovation: The Influence of Forecasted Conditions and State Priorities on Project Selection, Design, and Material choice".

Session A.1 Asphalt Materials:

Session Chair: Davide Lo Presti, University of Palermo.

- A.1.1 Preliminary mechanical characterization of HMA mixtures with a high content of recycled materials. Giulia Tarsi and Cesare Sangiorgi, University of Bologna, Italy.
- A.1.2 Bio-oils as asphalt bitumen rejuvenators. Amir Tabaković, Netherlands Organization for Applied Research (TNO)/Delft University of Technology, Netherlands; Dave van Vliet, Kirsten Roetert Steenbruggen, and Greet Leeghwater, Netherlands Organization for Applied Research (TNO), Netherlands.
- A.1.3 Development of Plant-Mix Type Modified Mixture with Excellent Flexibility and Stress Relaxation Property for Ensuring High Resistance to Cracking. Nhat Thanh Tran and Masashige Aoki, Taisei Rotec Corporation, Japan.
- A.1.4 Design and Evaluation of Ultra-Thin Overlay with High Viscosity and High Elasticity. Yijia Chen, Zhi Liao and Lide Chen, Road Intellitech Co., China; Tao Ma, Southeast University, China; Susan Tighe and Li Ningyuan, McMaster University, Canada.

Session B.1 Bridge Construction and Evaluation:

Session Chair: Seok Hong Lee, SN Construction Co.

- B.1.1 Full Load Test for the Sheikh Jaber Al-Ahmad Al-Sabah Causeway Bridge (PSC Girder: 35M). Kwangsoo Kim, AI Safety Institute; Dooyong Cho, Chungnam National University, Korea; Raechul Lee and Sangcheol Lee, SQ Engineering; Joungyong Park, Korea Construction Disaster Prevention Research, Korea; Wonrak Jang, AI Safety Institute, Korea.
- B.1.2 Accelerated Bridge Construction on Maui's Hana Highway. Sean Oroho, HDR Inc., USA and Tom Kubicz, Federal Highway Administration, USA.
- B.1.3 Field Application of Hydro-Demolition and Dry-Mix Shotcrete for Repairing the Understructure of Bridge Deck. Kyong-Ku Yun, Seunghak Choi, Taeho Ha, Changseok Song, Mohammad Shakhawat Hossain, and Valerii Panov, Kangwon National University, Korea; Yonggon Kim, Daesang E&C, Korea.
- B.1.4 Accelerated Construction of Unbraced Network Arch Bridge Using SPMTs. Mike LaViolette, HDR., Inc., USA.

Session C.1 Asset and Risk Management:

Session Chair: Ji Yun Lee, Washington State University.

- C.1.1 Asset Management Decision Support Tools: Computational Complexity, Transparency and Realism. Babatunde Atolagbe, University of Delaware, USA; Sue McNeil, University of Delaware, USA, and University of New South Wales, Australia.
- C.1.2 Multi-Parametric Delineation Approach for Homogeneous Sectioning of Asphalt Pavements. Naga Siva Pavani Peraka, GMR Institute of Technology, India; Krishna Prapoorna Biligiri and Satyanarayana N. Kalidindi, Indian Institute of Technology Tirupati, India.

- C.1.3 A Framework for Smart Pavements in Canada, Pejoohan Tavassoti. Hassan Baaj, Moojan Ghafurian, University of Waterloo, Canada; Omran Maadani and Mohammad Shafiee, National Research Council Canada, Canada.
- C.1.4 Development and Implementation of a Multihazard Risk Management System for Road Networks: volcanic, seismic and hydrometeorological hazards in Chile. Alondar Chamorro, Pontifica Universidad Católica de Chile, Chile.

Invited Presentation:

Lori Kahikina (Executive Director and CEO, Honolulu Authority for Rapid Transportation (HART)) and Huy Huynh (Director of Core Systems, Honolulu Authority for Rapid Transportation (HART)); presentation title: Honolulu Rail Transit Project Update.

Session A.2 Asphalt Pavement Texture and Aging:

Session Chair: Haifang Wen, Washington State University.

- A.2.1 Re-evaluating the Risk of Using Higher Skid Resistance Aggregates. David Woodward, Phillip Millar, and Paul Sargent, Ulster University, UK.
- A.2.2 Toward the Determination of the Appropriate Capturing Resolution of Surface Textures in Relation to Pavement Friction. Malal Kane and Minh-Tan Do, Université Gustave Eiffel, France.
- A.2.3 Rheological and Aging Characteristics of Polymer-modified Asphalt with Addition of Sulfur. Ana Luiza Rodrigues, Caio Falcão and R. Chris Williams, Iowa State University, USA.
- A.2.4 Sun Damage on Roads: from UV Radiation to Bituminous Binders and the Protecting Effect of End-of-Life Tires. Marie Enfrin, RMIT University, Australia; Jaffer Bressan Borinelli, Johan Blom and Cedric Vuye, University of Antwerp, Australia; Filippo Giustozzi, RMIT University, Australia.

Session B.2 Sensing and Machine Learning for Structures:

Session Chair: Lu Gao, University of Houston.

- B.2.1 3D Printing Technique for Passive Wireless Strain Sensing. Joshua Dyogi, Xi Song, University of Hawaii at Manoa, USA; Sung-Hwan Jang, Hanyang University, Korea; Sang-Hyeok Nam, ENGSOFT Co., Korea; and Chunhee Cho, University of Hawaii at Manoa, USA.
- B.2.2 CNN-based Automatic Mobile Reporting System and Quantification for Concrete Crack Size of Precast Members of OSC Construction. Ali Akbar, James Mugo Njoroge, Seojoon Lee, Younghee Chang and Soonwook Kwon, Sungkyunkwan University, Korea.
- B.2.3 Prediction of Ultimate Bond Strength between UHPC and Titanium Alloy Bars using a Machine Learning Approach. Mahesh Acharya, Idaho State University, USA; Luis Bedriñana, Universidad de Ingenieria y Tecnologia, Peru; Jared Cantrell, Ankit Bhaukajee and Mustafa Mashal, Idaho State University, USA.
- B.2.4 Optimizing Pothole Detection in Pavements: A Comparative Analysis of Deep Learning Models. Tiago Tamagusko and Adelino Ferreira, University of Coimbra, Portugal.

Session C.2 Traffic Safety:

Session Chair: Ghim Ping Ong, National University of Singapore.

- C.2.1 Safety Assessment of Cooperative Platooning in Mixed Traffic. B. Brian Park, University of Virginia, USA; Hyejin Lee, Seoul National University, Korea; Ilsoo Yun, Ajou University, Korea; Jeehyung Park, The Korea Transport Institute, Korea.
- C.2.2 Comparison Between Two Different Deployment Types of Road-side Devices Reducing Incident-Related Potential Conflicts. Jae-Hyeong Lee and Jin-Tae Kim, Korea National University of Transportation, Korea.
- C.2.3 Analysis of LDWS Recognition Rate According the Aging of Road Marking. Soon Yong Park and Sung Bum Yun, Seoul Institute of Technology, Korea.
- C.2.4 Infrastructure Measures to Protect the Unrecognized Vulnerable Road User: Motorcyclists. Georgene M Geary, GGfGA Engineering, USA.

Session A.3 Asphalt Pavement Evaluation:
Session Chair: Orazio Baglieri, Politecnico di Torino.

- A.3.1 Study of Long-Term Field Performance of Chip Seal in Washington. Angelique Umutoniwase, Washington State DOT, USA; Haifang Wen and Kevin Littleton, Washington State University, USA.
- A.3.2 Evaluation of Longitudinal Irregularity in Airport Pavements and unpaved Runway. Livia Merighi, Claudia Pereira, and Jose Schiavon, Aeronautics Institute of Technology, Brazil.
- A.3.3 Laboratory Evaluation of Recycled Asphalt Pavement and Engineered Polymer Binder for Small Airfield Repairs. William D. Carruth, Webster C. Floyd, and Jeb S. Tingle, U.S. Army Engineer Research and Development Center, USA.
- A.3.4 Optimized Selection of Pavement Maintenance and Rehabilitation Techniques: A Comparative Life Cycle Assessment. Imad L. Al-Qadi and Qingwen Zhou, University of Illinois Urbana-Champaign, USA.

Session B.3 Innovative Bridges and Buildings:
Session Chair: Tom Kubicz, Federal Highway Administration.

- B.3.1 Pullout Behavior of Titanium Alloy Reinforcing Bars in Ultra-High Performance Concrete. Mahesh Acharya, Jared Cantrell, and Mustafa Mashal, Idaho State University, USA.
- B.3.2 Benchmarking Material Use Efficiency for Building Projects. Jiyong Choi, Myungjin Chae, and Namhun Lee, Central Connecticut State University, USA.
- B.3.3 ABC Components of the Commonwealth Avenue Superstructure Re-Placement Project. Charles Swanson, HDR, Inc., USA.

Session C.3 Traffic Data Analysis:
Session Chair: Taha Ahmed, Australian University of Kuwait.

- C.3.1 Driving speed analysis using real-time traffic light status information at signalized intersections. Eunjin Choi, Hyangmi Han, Ockhee Jeon, Seungcheol Lee, and Kwangyoung Ko, Korean Road Traffic Authority, Korea.
- C.3.2 Data-Driven Analysis for Road Traffic Condition Using Digital Tachograph Data. Sung Bum Yun and Soon Yong Park, Seoul Institute of Technology, Korea.
- C.3.3 A Study on Artificial Neural Network-Based Real-Time Traffic Signal Timing Design Model Utilizing Smart Intersection Data. Sang-Tae Oh and Jin-Tae Kim, Korea National University of Transportation, Korea.
- C.3.4 Deep learning and clustering-based analysis of text narratives for identification of traffic crash severity contributors. Cristian Arteaga and JeeWoong Park, University of Nevada Las Vegas, USA.

Feature Presentation:

Yongho Sohn (Pegasus/Lockheed Martin Professor, University of Central Florida); presentation title: Renaissance Engineering via Additive Manufacturing.

Keynote Presentations Presided by Sue McNeil, University of Delaware/University of New South Wales:

Ioannis Brilakis (Laing O'Rourke; Professor of Construction Engineering and the Director of the Construction Information Technology Laboratory at the University of Cambridge); presentation title: Digital Twinning for the Built Environment (co-presented with Tim Embley).

Susan Tighe (Provost and Vice-President (Academic) and a Professor of Civil Engineering at McMaster University, Canada, and a past President of the Canadian Society for Civil Engineering (CSCE)); presentation title: High Performance Materials and Management Systems to Support Resilient Pavement infrastructure.

Invited Presentations Presided byFilippo Giustozzi, RMIT University:

Massimo Losa (Full Professor of Road, Railway, and Airport Engineering; Chief of the University Road Research Laboratory; and Vice-Director of the Department of Civil and Industrial Engineering at the University of Pisa); presentation title: The bad story and positive effects of the collapse of an iconic Italian bridge: a critical literature review.

Jorge Pais (Associate Professor, University of Minho, Portugal, and co-chairman of the 10th International Conference on Maintenance and Rehabilitation of Transport Infrastructures (MAIREPAV10) held in Guimarães, Portugal, on 24–26 July 2024); presentation title: Pavement rehabilitation in the XXI century.

Session A.4 Asphalt Pavement Construction QA/QC:
Session Chair: Jon Young, Hawaii Asphalt Paving Industry.

- A.4.1. Unveiling the Benefits of Engineered Crumb Rubber for Asphalt Mixtures by Means of Performance-Related Characterization: Rutting Behavior. Usman Ghani, Silvia Milazzo, Gaspare Giancontieri, Chiara Mignini, Gabriella Buttitta, University of Palermo, Italy; Fan Gu, Changsha University of Science and Technology, China; Davide Lo Presti, University of Palermo, Italy.
- A.4.2 Systematic Evaluation of the Field Constructability and Performance of Asphalt Mixes Containing High Percentage Recycled Asphalt. Logan Cantrell, Granite Construction, USA; Haifang Wen, Washington State University, USA.
- A.4.3 Application of Road Compaction Quality Control System to Road Pavement Construction for Advanced Quality Control. Kei Sasaki and Hiroaki Aoki, Taisei Corporation, Japan; Masakazu Jomoto, Taisei Rotec Corporation, Japan; Yasuhiro Mori, Soil and Rock Engineering Corporation, Japan.
- A.4.4 Real-time Field Quality Management System for Asphalt Pavement Using Cloud. Kyu-Dong Jeong, Dong-Hyuk Kim, Jae-Won Kim and Soo-Ahn Kwon, KICT, Korea; Nam-Ho Kim, Korea University of Technology and Education, Korea; Sung-Do Hwang, KICT, Korea.

Session B.4 Concrete Pavement Repairs:
Session Chair: Mike LaViolette, HDR., Inc.

- B.4.1 Materials and Methods for Expedient Repairs of Concrete Pavements. Jeb S. Tingle, Charles E. Williams Jr., William D. Carruth, and Caitlin M. Tibbetts, U.S. Army Engineering Research and Development Center, USA.
- B.4.2 Prediction of Blow-up Potential due to Concrete Pavement Growth. Young Kyu Kim, Hui Rak Ahn, and Seung Woo Lee, Gangneung-Wonju National University, Korea.
- B.4.3 Construction and Design guidelines for Lightweight Cellular Concrete as Pavement Subbase. Abimbola Oyeyi, University of Waterloo, Canada; Frank Ni, University of Florida, USA; Susan Tighe, McMaster University, USA.
- B.4.4 A Data-Driven Approach for Fatigue Damage Prediction in Jointed Plain Concrete Pavement Subjected to Superloads. Yongsung Koh, Halil Ceylan, Sunghwan Kim, and In Ho Cho, Iowa State University, USA.

Session C.4 Resilience and Sustainability:
Session Chair: S. Sonny Kim, University of Georgia.

- C.4.1 Progress Toward More Resilient Infrastructures: Review of Recent Efforts. Amir Golalipour, Federal Highway Administration, USA.
- C.4.2 Quantifying and Reducing Uncertainty in Transportation System Resilience Assessment: A Dynamic Bayesian Network Approach. Vishnupriya Jonnalagadda and Ji Yun Lee, Washington State University, USA.
- C.4.3 Important Sustainability Determinants Meeting Sustainability Goals of California Infrastructure Construction Projects. Joseph J. Kim and Patricia McCarthy, California State University Long Beach, USA.
- C.4.4 Towards Positive Energy Districts: Defining a New Role for Sustainable Governance. Savis Gohari, Norwegian University of Science and Technology, Norway.

Poster Session—3D Laser, CFRP Dowel, AI detect Pothole, IoT QM, NOx Removal:

- P.1. Damage Detection and Monitoring of a Concrete Structure Using 3D Laser Scanning. Manik Das Adhikari, Gangneung-Wonju National University, Korea; Tae-Hwan

Kim, Yongin University, Korea; Sang-Guk Yum, Gangneung-Wonju National University, Korea; Joon-Yeong Kim, SQ Engineering Co., Korea.

- P.2. Large-scale Test Setup of Concrete Pavement Slabs Jointed by Carbon Fiber-Reinforced Polymer Dowel Bars as Load Transfer Devices. Taha Ahmed, Ahmad Saad and Abdulhadi Kazem, Australian University of Kuwait; Ali Radwan, International University of Kuwait; Ali AlMutairi and Sarah Ashkanani, Australian University of Kuwait.
- P.3 Investigation on the process of eliminating abnormal objects from the road for the creation of an AI program that can automatically detect potholes. Moonsup Lee, Taehoon Lee, Younghan Park and Seungyeon Han, KICT, Korea; Nuri Lee and Chulki Kim MOLIT, Korea.
- P.4 IoT(Internet of Things) Based Pavement Quality Management System Platform. Suwan Chung, Tae-wook Kang and Byungkon Kim, KICT, Korea.
- P.5 NOx Removal of Pervious Concrete Pavement Materials with TiO_2. Cheolwoo Park, Minsoo Cho, Dong Jun Kim, Ui Dae Park Yong Sik Kwon, Minkyu Ju and Seungwon Kim, Kangwon National University, Korea.

Session A.5 Asphalt Pavement Design and Recycling:
Session Chair: Omar Smadi, Iowa State University.

- A.5.1 Investigation of Long-Term Performance of Waste-Plastics Modified Asphalt Mixtures. Sin-Mei Lim, Gengren Hao, National University of Singapore, Singapore; Anggraini Zulfiki, Land Transport Authority of Singapore, Singapore; Ghim Ping Ong, National University of Singapore, Singapore.
- A.5.2 Performance Life using Mechanistic-empirical Analysis of Asphalt Mixtures in Arid Climatic Conditions-Case of Kuwait. Taha Ahmed and Aditya Singh, Australian University, Kuwait; Elie Hajj, University of Nevada, Reno, USA; Ahmad Saad, Australian University, Kuwait.
- A.5.3 Establishing Density Based Mix Design for Cold Recycled Asphalt Mixes. Mansour Solaimanian and Scott Milander, Pennsylvania State University, USA.
- A.5.4 Plastic Recycling in Asphalt Concrete Pavements: Preliminary Observations from Hawaii's Pilot Project. A. Ricardo Archilla, University of Hawaii at Manoa, USA.

Session B.5 Soil Stabilization:
Session Chair: David Woodward, Ulster University.

- B.5.1 Use of Ground Penetrating Radar to Detect Cement Content on Cement Stabilized Subgrade. Zack Hall and S. Sonny Kim, University of Georgia, USA.
- B.5.2 Machine Learning-based Slope Failure Prediction Model Considering Uncertainty of Prediction. Junhyuk Choi, POSTECH, Korea; Yongkyu Cho, Kangnam University, Korea; Yongjin Kim, Smartgeotech, Korea; Yongseong Kim, Bongjun Ji, Kangwon National University, Korea.
- B.5.3 Study of different stabilizers to dry sludge for use in confined landfill and ditches. Rita M. Fortes, post-graduate at Federal Institute of São Paulo, Brazil; A.S. Pinto, T.M. Gomes, Environmental Management of the Environmental Manaus; L. Rabelo, DD&L Consultores, Brazil; M. Dos Reis Paulista University, Brazil.
- B.5.4 Expeditionary Ground Rehabilitation for Military Vehicle Traffic. Haley Bell, Lulu Edwards, and John Rushing, U.S. Army Engineer Research and Development Center, USA.

Session C.5 Sustainable Pavements:
Session Chair: Halil Ceylan, Iowa State University

- C.5.1 Innovative Design of Paving Cold Mix and Cohesive Overlays for Sustainable Pavement Maintenance. Xiang Chen and Xiaohu Wang, Road Intellitech Co., China; Tao Ma, Southeast University, China; Susan Tighe, Li Ningyuan, McMaster University, Canada.
- C.5.2 Life Cycle Assessment of a Sustainable and Innovative Solution for Unpaved Rural Roads. Leonardo Urbano, Lucia Tsantilis, Pier Paolo Riviera and Orazio Baglieri, Politecnico di Torino, Italy; Ezio Santagata, Politecnico di Torino, Italy/Qatar University, Qatar.

- C.5.3 Consistent Foamed Asphalt Contents Needed for Cold In-place Recycled Pavement Layers in Practice. Hosin "David" Lee, University of Iowa; Byungkyu Moon, ARA Associates, USA; Ashley Buss, Iowa DOT, USA; Charles T. Jahren, Iowa State University, USA.
- C.5.4 Improving fatigue and rutting resistance of road pavements by using aramid fibers. George Pais, University of Minho, Portugal.

Session A.6 Pavement Maintenance and Management:
Session Chair: Mansour Solaimanian, Pennsylvania State University.

- A.6.1 Data-driven Approach to Decision-making for Pavement Preservation. Sara Arezoumand, Alireza Sassani, and Omar Smadi, Iowa State University, USA.
- A.6.2 Advancement of Pavement Management System for Efficient Management of National Highway in Korea. Seungyeon Han, Hyungmog You, Myeongill Kim and Moonsup Lee, KICT, Korea; Nuri Lee and Chulki Kim, Ministry of Land, Infrastructure and Transport, Korea.
- A.6.3 Automated Distress Detection, Classification and Measurement for Asphalt Urban Pavements Using YOLO. Paulina Gómez-Conti and Alelí Osorio-Lird, Federico Santa María Technical University, Chile.
- A.6.4 Combined Use of GPR and PMS Data for Composite Pavement Assessment. Tae-Soo Kim, Chul-Ki Jung, Young-Mi Yoon, Byeong-Seok Kwak, and Jung-Hun Lee, Roadkorea Inc., Korea.

Session B.6 Rail and Autonomous Vehicles:
Session Chair: B. Brian Park, Professor of University of Virginia.

- B.6.1 Is Maintaining a Train Network in New Zealand Worth the Cost? Eric Scheepbouwer and Daniel Van der Walt, University of Canterbury, New Zealand.
- B.6.2 Evaluating Remediation Techniques for Fouled Ballast on Army Installations. Charles E. Williams Jr. and Thomas J. Beasley, U.S. Army Engineering Research and Development Center, USA.
- B.6.3 Implementing Public Service Features in Autonomous Vehicles in Seoul, Hyerim Cho, SoonYong Park, Junchul Kim, and Seol Young Lee, Seoul Institute of Technology, Korea.
- B.6.4 Evaluating the demand for truck-only toll lanes in Southern California freeways with both owner-operator and company truck drivers. Jose Arroyo-Turcios and Joseph J. Kim, California State University Long Beach, USA.

Session C.6 Traffic Safety Features:
Session Chair: Adelino Ferreira, University of Coimbra.

- C.6.1 Study on Traffic Incident Management Boundary Based on GIS and Its Historical Travel Time Data. Donghyeop Kim and Jin-Tae Kim, Korea National University of Transportation, Korea.
- C.6.2 Simplified Deterioration Modeling for Highway Sign Support Systems. Myungjin Chae, Lucas Voghell and Jiyong Choi, Central Connecticut State University, USA.
- C.6.3 On-board evaluation of pavement wetness from water spray. Ebrahim Riahi, Wiyao Edjeou, Manuela Gennesseaux, Sebastien Buisson, Veronique Cerezo and Minh-Tan Do, Univ Gustave Eiffel, France.

Feature Presentation:

Krishna Prapoorna Biligiri (Associate Professor and Head of Civil and Environmental Engineering, Indian Institute of Technology, Tirupati; Chairman of Second International Conference on Smart Cities (ICSC2) in Tirupati, India, held from February 19 to 21, 2025); presentation title: Integrating Mechanistic Roadway Designs with Lifecycle Assessment: Moving Towards Achieving Sustainability in Roadway Technology & ICSC2.

Jorge Pais (Associate Professor, University of Minho, Portugal); presentation title: Presentation of MAIREPAV10, 24–26 July 2024.

4. Organizers and Sponsors

The MAIREINFRA2 is supported by the following organizations (Figures 1–3):

Figure 1. Co-organizers.

Figure 2. Sponsors.

Figure 3. Official endorsements.

Conflicts of Interest: The author declares no conflicts of interest.

engineering proceedings

MDPI

Editorial

Statement of Peer Review †

Hosin (David) Lee

Civil and Environmental Engineering, University of Iowa, Iowa City, IA 52242, USA; hosin-lee@uiowa.edu

† All papers published in the volume are presented at the Second International Conference on Maintenance and Rehabilitation of Constructed Infrastructure Facilities, Honolulu, HI, USA, 16–19 August 2023.

In submitting conference proceedings to *Engineering Proceedings*, the volume editors of the proceedings certify to the publisher that all papers published in this volume have been subjected to peer review administered by the volume editors. Reviews were conducted by expert referees to the professional and scientific standards expected of a proceedings journal.

- Type of peer review: single-blind;
- Conference submission management system: email;
- Number of submissions sent for review: 100 papers;
- Number of submissions accepted: 69 papers;
- Acceptance rate (number of submissions accepted/number of submissions received): 69%;
- Average number of reviews per paper: 1.5;
- Total number of reviewers involved: 50.

Conflicts of Interest: The authors declare no conflicts of interest.

Citation: Lee, H. Statement of Peer Review. *Eng. Proc.* **2023**, *36*, 71. https://doi.org/10.3390/engproc2023036071

Published: 3 January 2024

engineering proceedings

Proceeding Paper

Damage Detection and Monitoring of a Concrete Structure Using 3D Laser Scanning †

Manik Das Adhikari [1], Tae-Hwan Kim [2], Sang-Guk Yum [1,*] and Joon-Yeong Kim [3]

1 Department of Civil Engineering, Gangneung-Wonju National University, Gangneung 25457, Republic of Korea; rsgis.manik@gmail.com
2 Department of Security Service, Yongin University, Yongin 17092, Republic of Korea; twehwan@empas.com
3 Department of R&D Lab, SQ Engineering Co. Ltd., Sonpa-gu, Seoul 05818, Republic of Korea; kimjoonyeong@sqeng.co.kr
* Correspondence: skyeom0401@gwnu.ac.kr

† Presented at the Second International Conference on Maintenance and Rehabilitation of Constructed Infrastructure Facilities, Honolulu, HI, USA, 16–19 August 2023.

Abstract: Surface damage detection, geometry measurement and monitoring are important for assessing the condition and risk of concrete structures. Therefore, to effectively assess the damage to a concrete structure, a 3D laser scanner accurately estimates the damage within a short timeframe and with less cost than the traditional inspection approaches. This study presents a framework for automated surface damage detection and structural health monitoring of a concrete structure using a X7 laser scanner (Trimble, Westminster, CO, USA). The methodology includes the use of 3D laser scanning technology to capture the 3D geometry of the concrete structure, followed by a detailed analysis of the data to identify any areas of damage or crack. The isodata and object-based image analysis (OBIA) techniques were applied to a 2D image generated from 3D cloud points. Overall accuracy (>89.6) and kappa statistics (>0.83) of both classification techniques exhibit good agreement between the classified and reference image. The OBIA technique was shown to be more effective in detecting minor cracks (<5 mm) and damage on a concrete structure. It was observed that the proposed approach is effective at identifying and monitoring the structural health of a concrete structure. The ability to continuously monitor the structure in this manner allows for early detection of damage and can aid in the maintenance and repair of the structure. Furthermore, this approach can robustly perform structural health monitoring and damage estimation.

Keywords: laser scanning; point clouds; surface damage detection; image classification; monitoring; damage quantification

Citation: Adhikari, M.D.; Kim, T.-H.; Yum, S.-G.; Kim, J.-Y. Damage Detection and Monitoring of a Concrete Structure Using 3D Laser Scanning. *Eng. Proc.* **2023**, *36*, 1. https://doi.org/10.3390/engproc2023036001

Academic Editor: Hosin (David) Lee

Published: 28 June 2023

1. Introduction

Concrete structures are widely used in various applications, such as buildings, bridges, dams and highways. Over time, these structures may experience various types of damage, including cracking, spalling, and deformation, which can compromise their structural integrity and pose a risk to public safety. Therefore, the monitoring and maintenance of civil structures are critical to ensure the safety and long-term sustainability of infrastructure. Consequently, this study aims to develop an efficient and accurate method for detecting and monitoring damage in concrete structures. In recent years, 3D laser scanning technology has gained attention as a powerful tool for damage detection and monitoring of concrete structures [1–3]. This technology uses laser beams to capture the 3D geometry of the structure, creating a point cloud that can be processed and analyzed to detect any areas of damage or deformation [4]. Compared to traditional methods, such as visual inspection and manual measurements, 3D laser scanning offers several advantages, including high accuracy, precision, and efficiency [5]. Numerous researchers [2,3,6–10] used 3D laser scanning data to detect and monitor concrete structures' damages. Yoon et al. [6] developed an algorithm for detecting cracks in concrete tunnels based on 3D laser scanning data. Rabah et al. [8] used terrestrial laser scan data for automatic concrete crack detection and mapping. Cho et al. [2] used 3D laser scanning to monitor the deformation and cracks of a concrete structure. Another interesting application of 3D laser scanning for concrete structure monitoring was demonstrated by Ge et al. [10]. Therefore, it was observed that 3D laser scanning technology has proven to be a valuable tool for damage detection and monitoring of concrete structures.

The study aims to develop an efficient and accurate method for detecting and monitoring damage in concrete structures, contributing to developing more reliable and resilient infrastructure. Subsequently, we present a comprehensive study on the damage detection and monitoring of a concrete structure using 3D laser scanning technology. This technique transforms the point cloud into a 3D image, and the last is converted into a 2D intensity image for image processing. The intensity image is analyzed using isodata (unsupervised) and object-based image analysis (OBIA) techniques, and the damaged area is detected.

2. Materials and Methods

This study presents a framework for automated surface damage detection and structural health monitoring of a concrete structure using a 3D laser scanner. The proposed method comprises the following steps: (a) scanning the concrete structure to create a detailed point cloud; (b) pre-processing and analyzing the point cloud data to identify potential damage states such as cracks, voids, and deformations; and (c) evaluating the precision of the identified damage surfaces. The workflow of 3D scanner-based structural health monitoring of concrete structures is depicted in Figure 1.

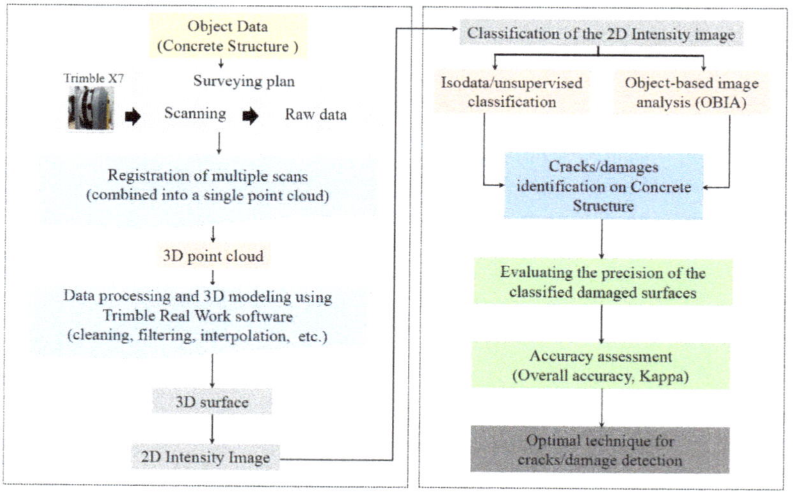

Figure 1. The workflow of 3D laser scanner-based structural health monitoring (data acquisition, processing, creation of 3D model, and damage detection) of a concrete structure.

We used a Trimble X7 laser scanner to collect the high-density 3D point cloud of a concrete structure. This instrument has a range of 0.6 m to 80 m, a precision of 2 mm and a maximum scanning speed of 500 kHz. In the present study, Trimble RealWorks 12.1 software (Trimble, Westminster, CO, USA)was used to pre-process point cloud data. The 3D surface model was then converted to the 2D intensity image for image processing purposes. The isodata and OBIA algorithm was used to classify the 2D intensity image [7,11]. The image classification was done using ERDAS Imagine 8.5 (ERDAS Inc., Norcross, GA., USA) and SAGA GIS 7.8.2 [12] software. Thereafter, the accuracy and kappa statistic were measured to evaluate the performance of classification models. These techniques are frequently used to determine a classification model's effectiveness [7]. The overall accuracy is a standard statistic that computes the proportion of correctly identified cases in relation to the total number of instances. On the other hand, the kappa statistic is a more sophisticated measure that considers the agreement between the actual observations and the predicted classifications while also considering the possibility of chance agreement [13]. Generally, a high overall accuracy and a high kappa statistic indicate good performance of the classification model and vice-versa. In the present study, we calculated the accuracy statistics based on the classified (i.e., isodata and OBIA) and reference image. The high-resolution RGB image collected by the same scanner was used as reference data. Subsequently, 550 random samples were collected from different parts of the high-resolution RGB image in order to compute the accuracy statistics for both classification models. The random samples contain information about the actual conditions of the concrete surface in the area covered by the classified image. After that, we developed a confusion matrix based on the classified and reference data in order to assess the overall accuracy and kappa statistic [14].

3. Results and Discussion

The study involved the use of a X7 3D laser scanner (Trimble, Westminster, CO, USA) to capture the 3D surface geometry of the concrete structure. The filtered cloud point was used to determine the region of interest, and the 3D surface was used to create the 2D intensity image. After that, the isodata and OBIA algorithm was used to classify the 2D intensity image into intact, damaged and cracked walls. Figure 2 depicts the classified image generated from 2D intensity data. It was observed that both classification techniques provide accurate and detailed information about the structure's condition, which can be used for further analysis and repair. To understand the precision of both classification techniques, we calculated the overall accuracy and kappa statistics based on the classified and reference image. The classification showed that the overall accuracy of the isodata and OBIA technique is 89.61% and 90.74%, respectively (Table 1). The calculated kappa value for the isodata and OBIA technique is 0.835 and 0.858, respectively, and exhibits good agreement of the classified and reference RGB images. The OBIA technique was shown to be more effective in detecting minor cracks (<5 mm) and damage on a concrete structure. The findings show that the Trimble X7 laser scanner has a high accuracy in capturing 3D cloud points, which is the sole cause of high accuracy in detecting the damage on the concrete wall. The use of 3D laser scanning can also reduce the time and cost required for inspection, as it eliminates the need for manual measurement.

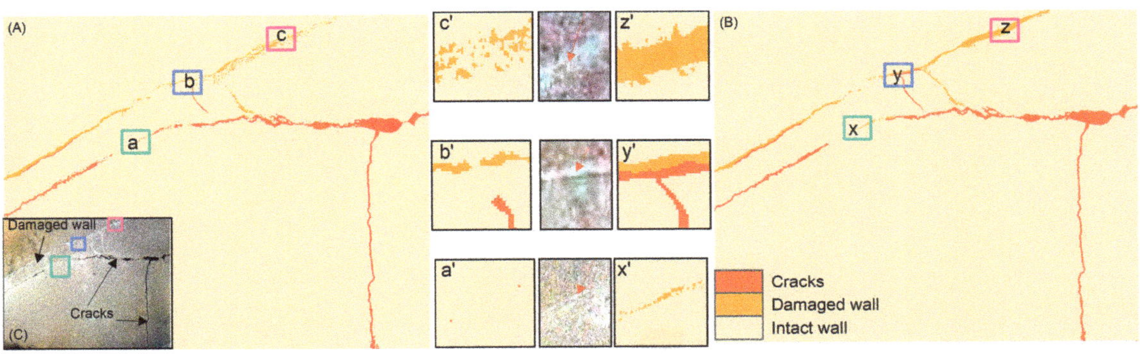

Figure 2. Cracks/damage detection of a concrete structure based on the (**A**) Isodata and (**B**) OBIA technique, whereas (**C**) represents the high-resolution RGB image. The zoom boxes represented by lowercase letters exhibit the detailed classification comparison of both techniques.

Table 1. Accuracy statistics of different classification techniques.

Classification Techniques		Cracks	Intact Wall	Damage Wall		Cracks	Intact Wall	Damage Wall	Overall Accuracy	Kappa
isodata	Producer's Accuracy	83.03	96.98	56.16	User's Accuracy	84.54	92.07	78.84	89.61	0.835
OBIA		90.69	97.63	45.31		87.31	92.60	80.55	90.74	0.858

4. Conclusions

The 3D laser scanning technology has been successfully used to detect and monitor damage in concrete structures. This research illustrated the step used to assess the efficiency of the Trimble X7 laser scanner in detecting cracks and damage on the concrete structure. The present methodology combined 3D cloud points and an image classification method to detect damage on the concrete structure. It was noted that the OBIA and isodata approaches show good agreement between the classified and reference images in terms of overall accuracy (>89.6) and kappa statistics (>0.83). Therefore, the proposed methodology can be suitable for monitoring an infrastructure's health. Furthermore, 3D laser scanning for damage detection and monitoring has several advantages over traditional inspection methods. This technique is non-destructive, meaning that it does not damage the inspected structure. This is particularly important in cases where the structure is still in use and cannot be shut down for inspection. In addition, the present technique provides accurate and detailed information about the structure's condition, which can be used for further investigation and restoration purposes. These findings have important implications for the field of civil engineering and suggest that 3D laser scanning will continue to play an important role in detecting and evaluating surface damage in concrete structures.

Author Contributions: Conceptualization, M.D.A., T.-H.K., S.-G.Y. and J.-Y.K.; methodology, M.D.A. and S.-G.Y.; software, M.D.A. and S.-G.Y.; validation, M.D.A. and S.-G.Y.; formal analysis, M.D.A. and S.-G.Y.; investigation, M.D.A., T.-H.K., S.-G.Y. and J.-Y.K.; resources, S.-G.Y.; data curation, M.D.A.; writing—original draft preparation, M.D.A., T.-H.K., S.-G.Y. and J.-Y.K.; writing—review and editing, M.D.A., T.-H.K., S.-G.Y. and J.-Y.K.; visualization, M.D.A. and S.-G.Y.; supervision, S.-G.Y.; project administration, S.-G.Y.; funding acquisition, S.-G.Y. All authors have read and agreed to the published version of the manuscript.

Funding: This work was supported by the National Research Foundation of Korea (NRF) grant funded by the Korea government (MSIT) NRF-2021R1C1C2003316).

Institutional Review Board Statement: Not applicable.

Informed Consent Statement: Not applicable.

Data Availability Statement: The datasets used and/or analyzed during the current study are available from the corresponding author on reasonable request.

Conflicts of Interest: The authors declare no conflict of interest.

References

1. Erkal, B.G.; Hajjar, J.F. Laser-based surface damage detection and quantification using predicted surface properties. *Autom. Constr.* **2017**, *83*, 285–302. [CrossRef]
2. Cho, S.; Park, S.; Cha, G.; Oh, T. Development of image processing for crack detection on concrete structures through terrestrial laser scanning associated with the octree structure. *Appl. Sci.* **2018**, *8*, 2373. [CrossRef]
3. Tzortzinis, G.; Ai, C.; Breña, S.F.; Gerasimidis, S. Using 3D laser scanning for estimating the capacity of corroded steel bridge girders: Experiments, computations and analytical solutions. *Eng. Struct.* **2022**, *265*, 114407. [CrossRef]
4. Wu, C.; Yuan, Y.; Tang, Y.; Tian, B. Application of terrestrial laser scanning (TLS) in the architecture, engineering and construction (AEC) industry. *Sensors* **2021**, *22*, 265. [CrossRef]
5. Mishra, M.; Lourenço, P.B.; Ramana, G.V. Structural health monitoring of civil engineering structures by using the internet of things: A review. *J. Build. Eng.* **2022**, *48*, 103954. [CrossRef]
6. Yoon, J.S.; Sagong, M.; Lee, J.S.; Lee, K.S. Feature extraction of a concrete tunnel liner from 3D laser scanning data. *NDT E Int.* **2009**, *42*, 97–105. [CrossRef]

7. Armesto-González, J.; Riveiro-Rodríguez, B.; González-Aguilera, D.; Rivas-Brea, M.T. Terrestrial laser scanning intensity data applied to damage detection for historical buildings. *J. Archaeol. Sci.* **2010**, *37*, 3037–3047. [CrossRef]
8. Rabah, M.; Elhattab, A.; Fayad, A. Automatic concrete cracks detection and mapping of terrestrial laser scan data. *NRIAG J. Astron. Geophys.* **2013**, *2*, 250–255. [CrossRef]
9. Valença, J.; Puente, I.; Júlio, E.N.B.S.; González-Jorge, H.; Arias-Sánchez, P. Assessment of cracks on concrete bridges using image processing supported by laser scanning survey. *Constr. Build. Mater.* **2017**, *146*, 668–678. [CrossRef]
10. Ge, Y.; Liu, J.; Zhang, X.; Tang, H.; Xia, X. Automated Detection and Characterization of Cracks on Concrete Using Laser Scanning. *J. Infrastruct. Syst.* **2023**, *29*, 04023005. [CrossRef]
11. Gui, R.; Xu, X.; Zhang, D.; Pu, F. Object-based crack detection and attribute extraction from laser-scanning 3D profile data. *IEEE Access* **2019**, *7*, 172728–172743. [CrossRef]
12. Conrad, O.; Bechtel, B.; Bock, M.; Dietrich, H.; Fischer, E.; Gerlitz, L.; Wehberg, J.; Wichmann, V.; Böhner, J. System for Automated Geoscientific Analyses (SAGA) v. 2.1.4. *Geosci. Model Dev.* **2015**, *8*, 1991–2007. [CrossRef]
13. Landis, J.R.; Koch, G.G. An application of hierarchical kappa-type statistics in the assessment of majority agreement among multiple observers. *Biometrics* **1977**, *33*, 363–374. [CrossRef] [PubMed]
14. Story, M.; Congalton, R.G. Accuracy assessment: A user's perspective. *Photogramm. Eng. Remote Sens.* **1986**, *52*, 397–399.

engineering proceedings

Proceeding Paper

A Data-Driven Approach for Fatigue Damage Prediction in Jointed Plain Concrete Pavement Subjected to Superloads †

Yongsung Koh [1,*], Halil Ceylan [1], Sunghwan Kim [2] and In Ho Cho [1]

1 Department of Civil, Construction and Environmental Engineering (CCEE), Iowa State University, Ames, IA 50011, USA; hceylan@iastate.edu (H.C.); icho@iastate.edu (I.H.C.)
2 Institute for Transportation, Iowa State University, Ames, IA 50011, USA; sunghwan@iastate.edu
* Correspondence: yskho27@iastate.edu; Tel.: +1-515-708-5512
† Presented at the Second International Conference on Maintenance and Rehabilitation of Constructed Infrastructure Facilities, Honolulu, HI, USA, 16–19 August 2023.

Abstract: The passage of superloads over the jointed plain concrete pavements (JPCPs) causes signification fatigue damage to the JPCPs. This mainly happens because of their non-standardized loading configurations and high gross vehicle and axle weights. Developing a high-accuracy prediction model for JPCP fatigue damage under superloads is strongly required to complement the mechanistic–empirical (ME) pavement design in aspects of its wide range of dimensions, including number, spacing, and loading of tires and axles. In this study, various data-driven models based on different theoretical approaches, including artificial neural network-based models, generalized additive models, and multiple linear regression models, were constructed using a well-established database derived from finite-element analysis results in order to predict the target response for JPCP fatigue damage when subjected to superloads. The prediction accuracies of these data-driven models were then evaluated to confirm their further applicability to the existing ME pavement design software.

Keywords: superload; jointed plain concrete pavement; fatigue cracking; finite element analysis; data-driven model

Citation: Koh, Y.; Ceylan, H.; Kim, S.; Cho, I.H. A Data-Driven Approach for Fatigue Damage Prediction in Jointed Plain Concrete Pavement Subjected to Superloads. *Eng. Proc.* **2023**, *36*, 2. https://doi.org/10.3390/engproc2023036002

Academic Editor: Hosin (David) Lee

Published: 28 June 2023

1. Introduction

Fatigue damage in jointed plain concrete pavement (JPCP) is one of the major damage types that traffic loads can induce. Particularly, fatigue damage is often caused by overweight trucks or superloads with gross vehicle or axle weights that exceed a state's permit limits. Unlike conventional vehicle types, superloads are distinguished from general vehicle types by their non-standardized loading configurations, including the number and spacing of tires and axles, for the purpose of transporting oversized and heavy payloads [1]. Previous studies have suggested using simple empirically based generalized power law theories to describe the nonlinear relationship between pavement damage and traffic loads, such as the fourth power law from the American Association of State Highway Officials Road Test [2], or using different power-law exponents dependent on different pavement structures and loading conditions from follow-up study [3]. As such, new mechanistic-based analysis approach that reflects the characteristics of superloads beyond the scope of general loading configuration is still required. AASHTOWare Pavement Mechanistic-Empirical (ME) Design software, which builds upon the ME pavement design guide [4], has provided ME-based JPCP fatigue damage calculation as a data analysis and back-calculation tool. It has done so by generating the horizontal tensile stresses in a Portland cement concrete (PCC) layer from a series of artificial neural network (ANN)-based models using a database of finite-element analysis (FEA) results produced by using ISLAB 2000 [5]. However, the database used to train the ANN models in this software did not include a wide range of dimensions with respect to superload loading configurations, and so the extrapolation of explanatory variables in terms of loading parameters may lead to inaccurate calculation of JPCP responses when subjected to superloads.

In this study, various data-driven models, including ANN-based models, generalized additive models (GAMs), and multiple-linear regression models, were constructed using a database of FEA results (i.e., the critical horizontal tensile stress at the bottom and top of the PCC slab when subjected to superloads) collected from ISLAB 2005. The prediction accuracy of these data-driven models was evaluated and compared to confirm their further applicability to the existing Pavement ME Design software.

2. Methodology

The FEA results, including the critical horizontal tensile stress in PCC slab depending on different JPCP structures and superloads, used as database to train and test the prediction models, were collected from the same author's previous study [1]. The experimental design for FEA and data-driven modeling is briefly summarized in Table 1. Note that each critical tensile stress of PCC slab was calculated from pre-determined critical loading locations differentiated by superload types and JPCP structures.

Table 1. Experimental matrix of JPCP structures for FEA and data-driven modeling [1].

Input Variables (Explanatory Variables)	Value	Unit
PCC layer thickness	200, 230, 250	mm
Composite modulus of subgrade reaction	14 (spring), 33 (summer), 33 (fall), 68 (winter)	MPa/m
Transverse joint spacing	4.5 for 200 mm and 230 mm PCC layer thickness 5 for 250 mm PCC layer thickness 6 for 200 mm, 230 mm, and 250 mm PCC layer thickness	m
Coefficient of thermal expansion	7.7, 9.4	$10^{-6}/°C$
Temperature gradients	−0.09, −0.04, 0, 0.04, 0.09	°C/mm
Elastic modulus of PCC layer	27,500	MPa
Traffic loads	IoH [1]: 18 types × 4 payload levels (100%, 75%, 50%, 0%) = 72 cases SHL [2]: 16 types × 4 payload levels (100%, 75%, 50%, 0%) = 64 cases	-

Note: Individual input variables applied to non-dimensional variables can have a wider range of values than those indicated in the table depending on other variables. [1] IoH = implements of husbandry; representative superload types for transporting agricultural products, slurry, or water, including grain carts, manure tankers, agricultural trailers and trucks. [2] SHL = superheavy load; representative superload types for transporting wind turbines or engine blocks, including drop-deck types, and single-row and dual-row modular types.

A database collected from FEA using ISLAB 2005 was then used to train and test surrogate data-driven models based on different theoretical approaches. (i) ANN-based models, (ii) GAMs, and (iii) multiple linear regression models were constructed in this study to compare and analyze each of their prediction accuracies. The data and statistical sets of data-driven models and a brief introduction to each approach are summarized and listed in Figure 1. Please refer to previous studies for a detailed theoretical background on each data-driven approach similarly applied to flexible pavement cases [6] and further application of GAM to earthquake engineering [7]. As can be seen from Figure 1, a total of 24 GAMs varying by three statistical distributions, two link functions, and two spline bases, and a total of 24 ANN models consisting of three different numbers of hidden layers and four different numbers of neurons per each layer were constructed for comparative evaluation.

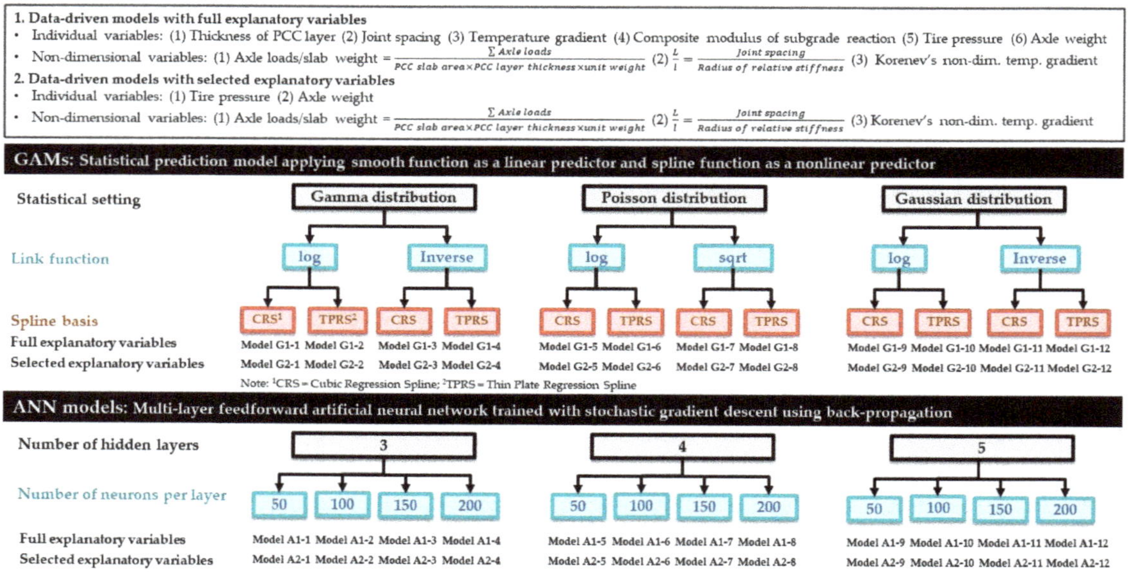

Figure 1. Data and statistical sets of data-driven models.

3. Prediction Accuracy of Data-Driven Models

All prediction models shown in Figure 1 were evaluated by splitting the original dataset into 70 percent and 30 percent portions to constitute a training set and a testing set. Figures 2 and 3, respectively, show the prediction accuracy of each model for the critical horizontal stress at the bottom and top of the PCC slab. It can be easily confirmed from the figures that ANN models outperform both GAMs and multiple regression models for all superload types when determining the best prediction model for each prediction approach and superload type. In the case of an agricultural trailer or truck, GAM exhibits a lower prediction accuracy for the critical horizontal stress at the top of the PCC slab than the multiple linear regression model due to the relatively small number of data points of the training dataset collected for those superload types.

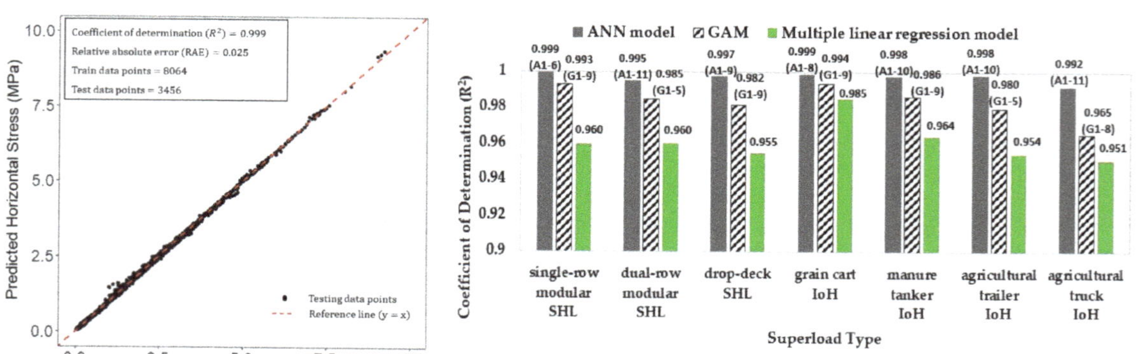

(a) (b)

Figure 2. Prediction accuracy evaluation in case of: (**a**) ANN model A1–6 predicting critical horizontal stress at the bottom of the PCC slab when subjected to single-row modular-type SHL; (**b**) models showing the best prediction accuracy for each prediction approach and superload type.

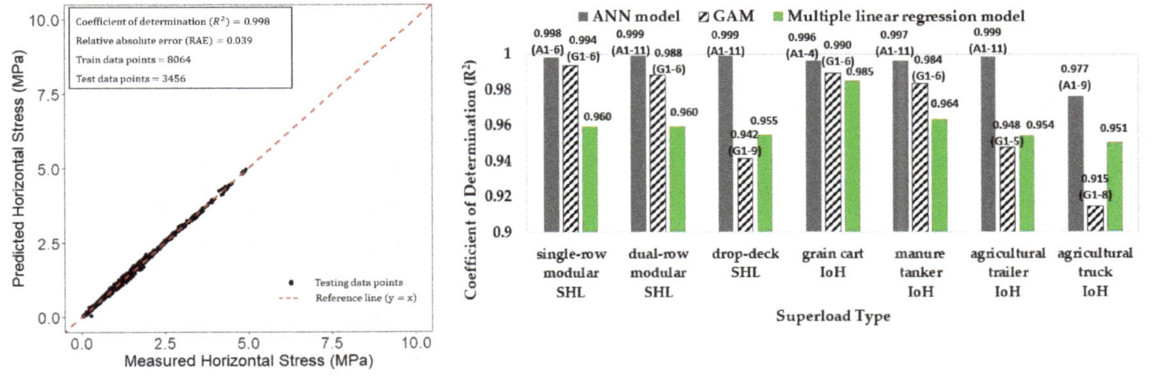

(a) (b)

Figure 3. Prediction accuracy evaluation in case of: (**a**) ANN model A1–6 predicting critical horizontal stress at the top of the PCC slab when subjected to single-row modular-type SHL; (**b**) models showing the best prediction accuracy for each prediction approach and superload type.

4. Conclusions

Overall, ANN-based prediction models exhibited better accuracy than GAMs or multiple regression models in predicting critical horizontal stress at both the bottom and top of the PCC layer in JPCPs under superload conditions. Further studies are needed to determine the prediction accuracy evaluation of other deep learning algorithms for superload issues.

Author Contributions: Conceptualization, H.C., S.K. and I.H.C.; data collection, Y.K.; analysis and interpretation of results, Y.K., H.C., S.K. and I.H.C.; draft manuscript preparation: Y.K., H.C., S.K. and I.H.C. All authors have read and agreed to the published version of the manuscript.

Funding: This research was funded by the Iowa Highway Research Board and the Iowa Department of Transportation (Grants TR-781).

Institutional Review Board Statement: Not applicable.

Informed Consent Statement: Not applicable.

Data Availability Statement: The data presented in this study are available on request.

Acknowledgments: The authors gratefully acknowledge Iowa Highway Research Board (IHRB) and Iowa Department of Transportation (Iowa DOT) for their support of this study. The project technical advisory committee members are gratefully acknowledged for their guidance. The authors would also like to express their sincere gratitude to other research team members from Iowa State University (ISU)'s PROSPER at InTrans for their assistance. The contents of this paper reflect the views of the authors who are responsible for the facts and accuracy of the data presented within, and do not necessarily reflect the official views and policies of the IHRB, Iowa DOT, or ISU.

Conflicts of Interest: The authors declare no conflict of interest.

References

1. Koh, Y.; Ceylan, H.; Kim, S.; Cho, I.H. Structural and Fatigue Analysis of Jointed Plain Concrete Pavement Top-Down and Bottom-Up Transverse Cracking Subjected to Superloads. *Transp. Res. Rec.* **2022**, *2676*, 76–93. [CrossRef]
2. Highway Research Board. *The AASHO Road Test, Report 5, Pavement Research, Special Report 61E*; National Academy of Sciences-National Research Council: Washington, DC, USA, 1962. Available online: https://onlinepubs.trb.org/onlinepubs/sr/sr61e.pdf (accessed on 24 January 2023).
3. Dawson, A.R. Rut Accumulation and Power Law Models for Low-Volume Pavements under Mixed Traffic. *Transp. Res. Rec.* **2008**, *2068*, 78–86. [CrossRef]

4. ERES Consultants. *Guide for Mechanistic–Empirical Design of New and Rehabilitated Pavement Structures*; NCHRP Project 1–37A; Final Report; Transportation Research Board of the National Academies: Washington, DC, USA, 2004.
5. Khazanovich, L.; Yu, H.T.; Rao, S.; Galasova, K.; Shat, E.; Jones, R. *ISLAB2000—Finite Element Analysis Program for Rigid and Composite Pavements. User's Guide*; ERES Consultants: Champaign, IL, USA, 2000.
6. Koh, Y.; Ceylan, H.; Kim, S.; Cho, I.H. Critical Responses of Flexible Pavements Under Superheavy Loads and Data-Driven Surrogate Model. *Int. J. Pavement Res. Technol.* **2022**, *16*, 513–543. [CrossRef]
7. Song, I.; Cho, I.H.; Wong, R.K. An Advanced Statistical Approach to Data-Driven Earthquake Engineering. *J. Earthq. Eng.* **2020**, *24*, 1245–1269. [CrossRef]

engineering proceedings

Proceeding Paper

Multi-Parametric Delineation Approach for Homogeneous Sectioning of Asphalt Pavements †

Naga Siva Pavani Peraka [1], Krishna Prapoorna Biligiri [2,*] and Satyanarayana N. Kalidindi [2]

1 Department of Civil Engineering, GMR Institute of Technology, Rajam 532127, India; nagasivapavani.peraka@gmail.com
2 Department of Civil and Environmental Engineering, Indian Institute of Technology Tirupati, Tirupati 517619, India; satyakn@iittp.ac.in
* Correspondence: bkp@iittp.ac.in; Tel.: +91-877-2503157
† Presented at the Second International Conference on Maintenance and Rehabilitation of Constructed Infrastructure Facilities, Honolulu, HI, USA, 16–19 August 2023.

Abstract: Maintenance of homogeneous road sections is one of the approaches to economizing the overall management of pavement systems. The objective of this study was to develop a multi-parameter-based delineation approach to segmenting the pavements into subsections in a way that considers multiple pavement characteristics. Deflection bowl parameters, pavement functional performance, surface layer modulus, and traffic were analyzed to develop a multi-parametric delineation index (MPDI), which was used in C-charts-based segmentation to obtain homogeneous sections. Importantly, the segmentation processes were automated using a deep neural network designed for rational implementation by practitioners. The devised approach was found to be efficient in segmenting the pavements, selecting the sections that are in direst need of maintenance, and necessitating prompt response from the agencies.

Keywords: homogeneous sectioning; deflection bowl parameters; delineation; pavement maintenance; C-charts; functional performance

Citation: Peraka, N.S.P.; Biligiri, K.P.; Kalidindi, S.N. Multi-Parametric Delineation Approach for Homogeneous Sectioning of Asphalt Pavements. *Eng. Proc.* **2023**, *36*, 3. https://doi.org/10.3390/engproc2023036003

Academic Editor: Hosin (David) Lee

Published: 28 June 2023

1. Introduction

Roadway maintenance and rehabilitation are critical to sustaining the economy of any nation. In 2021, the roadway network in the United States of America (USA) was ranked with a 'D' grade by the American Society of Civil Engineers (ASCE) [1]. According to the International Transport Forum, France spent EUR 2 billion on the maintenance of roadway infrastructure. This was found to be the highest amongst 63 participating countries, although the spending details were not available for India [2]. With the growing importance of preserving the existing pavement infrastructure, the Government of India initiated asset recycling processes to monitor and manage pavements in order to ensure that they cater to traffic needs during their remaining service lives [3]. In order to plan a maintenance intervention, it is essential to collect present pavement condition data and predict future performance. Additionally, it is essential to segment the pavements based on similar characteristics so as to establish an optimum level of maintenance intervention at the project level. The existing homogeneous sectioning methods include cumulative difference approach (CDA), absolute difference approach (ADA) [4–7], Bayesian algorithm [8], cumulative sum [9], quality control charts [10], and minimization of sum of squared error approach [11], all of which consider individual pavement performance parameter at a time for delineating the pavement sections.

In another study, Donev and Hoffmann [12] considered rutting, surface defects, and alligator approaches cracking in sectioning pavements based on similar characteristics for project-level maintenance applications. However, the outcomes were found to be suitable for short measurement series and were obtained with a methodology that may not be suitable for application elsewhere, especially in emerging economies [13]. Further, artificial intelligence (AI) techniques were used for pavement delineation, which reduced the tedious analysis process when using a single parameter for segmentation. However, the use of AI techniques for segmentation using multiple parameters was yet to be verified [14]. A recent study tested the performance of the C-charts method in segmenting the roadway sections that utilize two parameters: the international roughness index (IRI) and rutting [6]. It was found that the bi-parametric approach developed in the study was efficient in the segmentation process compared to the traditional methods. However, there is a need to explore the interaction of the other parameters such as deflection and traffic in the pavement segmentation process.

It is noteworthy that the previous studies only used one pavement condition parameter at a time when segmenting pavement sections. These studies repeated the procedure later for the remaining parameters to obtain the optimum homogenous segmentation. This was found to be monotonous, resulting in a rigorous analysis of the results, and tedious, possibly leading to erroneous results if the analyses were to be delayed. Therefore, there is a need to develop a multi-parametric-based sectioning approach, where multiple parameters can be considered simultaneously for use in sectioning. Thus, the objective of this research study was to develop a multi-parameter-based delineation approach (MPDA) to segment the pavements into subsections with similar features encompassing functional, structural, and traffic characteristics. It is envisioned that the developed approach would certainly reduce the analysis costs and duration and help the decision-making authorities to identify the optimum homogeneous sections at the project level.

2. Multi-Parametric Delineation Approach Framework

C-charts are among the statistical quality control methods widely used to monitor the defects in the production process, as well as for several engineering applications. The C-charts method was basically used for the data, which were obtained in a count-type fashion. Since homogeneous sectioning was performed for different pavement performance parameters that were measured at regular intervals, the data was presumed to be in a count-type arrangement. The results of the C-chart-based pavement homogeneous segmentation were better than those of the other segmentation methods [9]. Therefore, C-charts were used to perform homogeneous segmentation, with consideration given to multiple pavement characteristics. A series of tasks was performed to develop the MPDA for the homogeneous sectioning of roads.

- Metric formulation: a procedure was formulated to identify the parameters of homogeneous sectioning, pre-processing of data, and formulation of a multi-parametric delineation index (MPDI) as a function of identified parameters for sectioning and calculation of the control limits for MPDI, which included mean, standard deviation, and the upper control limit and lower control limit from the entire dataset.
- C-charts-based MPDA: a process that was adopted to construct C-charts for MPDI for each road section in the dataset. This could identify the outlier data and helped us to segment the sections between the outliers as homogeneous sections.
- Validation of the developed approach was undertaken as the final step.

2.1. Multi-Parametric Delineatoin Index

The necessity of segmenting the pavements that have the most similar characteristics formed the basis for the selection of multiple parameters for segmentation. The pavement deflection bowl parameters such as peak deflection, base layer index (BLI), middle layer index (MLI), and lower layer index (LLI), being representative of the structural integrity of the pavement system, were chosen. Additionally, a metric called unified pavement health index (UPHI) was utilized after its development by the authors. This value indicated the functional condition of the pavements based on their current distress levels on a scale of 0 to 100. In addition to these, higher traffic volumes that increase the rate of deterioration were also selected, along with the modulus corresponding to the structural capacity of the pavement system. The data from 26 road sections in the State of Andhra Pradesh, India, was used for this study, including: falling weight deflector (FWD) deflections, distresses, roughness, traffic measured in terms of annual average daily traffic, and existing pavement layer thickness details. The distresses were used to assess the present functional condition of the pavement in terms of UPHI, a functional performance measuring unit devised by the authors to rate the pavements on a scale of 1 to 100, where 100 points indicates a very good condition of the pavement and 1 signifies that the pavement needs to be reconstructed. Surface layer modulus was calculated using KGPBACK™ software using FWD deflections and existing pavement thickness data.

2.2. MPDI Formulation

A dimensionless parameter called MPDI was formulated with the normalized values of all the parameters, as show in Equation (1). Note that the normalized parameters were added to account for the individual contribution of each of the parameters during homogeneous sectioning.

$$MPDI_i = \frac{UPHI_i}{\mu_{UPHI}} + \frac{DO_i}{\mu_{DO}} + \frac{BLI_i}{\mu_{BLI}} + \frac{MLI_i}{\mu_{MLI}} + \frac{LLI_i}{\mu_{LLI}} + \frac{AADT_i}{\mu_{AADT}} + \frac{E_i}{\mu_E} \quad (1)$$

where i = datapoint; μ_{UPHI} = mean UPHI of the entire dataset (1781 data points); μ_{BLI} = mean BLI, µm; μ_{MLI} = mean MLI, µm; μ_{LLI} = mean LLI, µm; μ_{AADT} = mean AADT; and μ_E = mean back-calculated elasticity modulus of surface layer, MPa

The control limits identify an unexpected variation in the quality control process. In the past, researchers [15,16] have also used two standard deviations (in place of three standard deviations) to recognize the warning limits without compromising the quality of the established control limits and the process capability. Thus, in this study, the control limits of the MPDI accounted for the use of two standard deviations to rationally ascertain the homogeneous segments and the overall process stability.

3. Results

C-charts were constructed for MPDI for the whole dataset, covering 1,781 data points. The points that crossed the UCL and LCL were recorded as outliers. Section boundaries were introduced to the chart when the curve crossed either UCL or LCL. The data points that were present between any two section boundaries were regarded as "homogeneous segments". Based on the mean MPDI of the homogeneous segments, 34 homogeneous section types were defined, and maintenance strategies were suggested based on mean MPDI, whose details are available in Peraka [17]. The homogeneous sections of a road section and C-charts for the MPDI in the State of Andhra Pradesh between chainages of 22.41 and 28.71 km are shown in Figure 1. In the entire dataset of 26 road sections, a total of 389 homogeneous sections belonging to 34 homogeneous section categories were identified. Amongst all the sections, 57.58% of them were found to need corrective maintenance, while 25.96% would require minor rehabilitation. The minimum length of the homogeneous section was found to be 300 m (<500 m), which was appropriate for project-level maintenance applications, as also reported by Jannat et al. [18].

In order to validate the developed approach, the results were compared to the homogeneous sections obtained using CDA. There was such a section for each parameter considered in the formulation of MPDI. The results revealed that the MPDI approach was better for homogeneous segmentation. Finally, the data processing and basic computations involved in the calculation of MPDI, C-charts, and segmentation were automated using a regression-type feed-forward deep neural network approach.

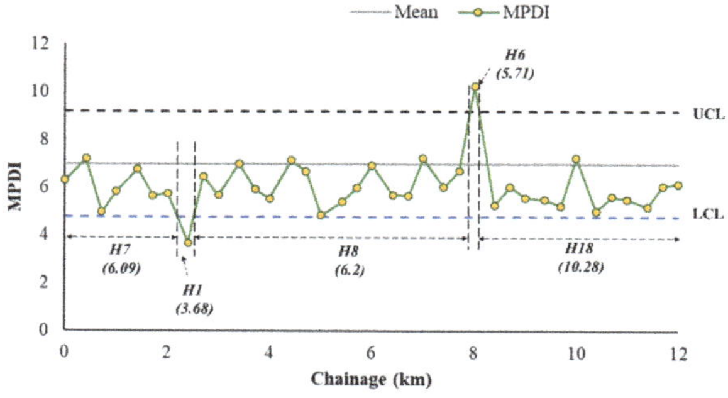

Figure 1. MPDA-based Homogeneous Sectioning of a Road Section in Andhra Pradesh State.

4. Conclusions and Recommendations

The delineation process developed in this study considered multiple parameters for the segmentation of asphalt pavements to accord maintenance activities. The major conclusions and recommendations are as follows:

- MPDI-based categorization for maintenance treatment selection: MPDI-based pavement maintenance selection scale was defined to provide insights into the roadway practitioners in order to select appropriate maintenance interventions for the designated homogeneous sections.
- Automation of delineation process: the DNN would serve as a one-stop solution for pavement segmentation in order to potentially help the practitioners in project-level maintenance applications.
- Recommendation: The multi-parametric delineation approach developed in this research study considered seven parameters in order to obtain the homogeneous roadway segments using a C-chart-based approach. However, other pavement characteristics must also be incorporated in future for better segmentation. These should be validated using the current method.

Author Contributions: Conceptualization, N.S.P.P. and K.P.B.; methodology, N.S.P.P.; software, N.S.P.P.; validation, N.S.P.P.; formal analysis, N.S.P.P.; investigation, N.S.P.P., K.P.B., S.N.K.; resources, N.S.P.P., K.P.B. and S.N.K.; data curation, N.S.P.P.; writing—original draft preparation, N.S.P.P.; writing—review and editing, N.S.P.P., K.P.B. and S.N.K.; visualization, N.S.P.P. All authors have read and agreed to the published version of the manuscript.

Funding: This research received no external funding.

Institutional Review Board Statement: Not applicable.

Informed Consent Statement: Not applicable.

Data Availability Statement: Data available on request due to restrictions e.g., privacy or ethical.

Conflicts of Interest: The authors declare no conflict of interest.

References

1. USA Report Card. *2017 ASCE Infrastructure Report Card-Roads*; ASCE: Reston, VA, USA, 2023.
2. International Transport Forum. Transport Infrastructure Investment and Maintenance Spending. 2023. Available online: https://www.oecd.org/ (accessed on 1 February 2023).
3. Invest India. Roadways in India-Road Industry, Network, Projects & FDI, Invest India-National Investment Promotion and Facilitation Agency. Available online: https://www.investindia.gov.in (accessed on 1 February 2023).
4. Misra, R.; Das, A. Identification of homogeneous sections from road data. *Int. J. Pavement Eng.* **2013**, *4*, 229–233. [CrossRef]
5. Thomas, F. Generating Homogeneous Road Sections Based on Surface Measurements: Available Methods. In Proceedings of the 2nd European Pavement and Asset Management Conference, Berlin, Germany, 21–23 March 2004; p. 48.
6. Eddula, S.V.; Peraka, N.S.P.; Biligiri, K.P. A Smart Bi-Parametric Approach for Homogeneous Delineation of Rural Roads. In Proceedings of the COSVARD 2020 International Conference, Guwahati, India, 7–8 December 2020.

7. Peraka, N.S.P.; Biligiri, K.P. Pavement asset management systems and technologies: A review. *Autom. Constr.* **2020**, *119*, 103336. [CrossRef]
8. Thomas, F. Statistical Approach to Road Segmentation. *J. Transp. Eng.* **2003**, *129*, 300–308. [CrossRef]
9. Tejeda, S.V.; Echaveguren, T. Proposal of a Segmentation Procedure for Skid Resistance Data. *Arab. J. Sci. Eng.* **2008**, *33*, 89–104.
10. El-Gendy, A.; Shalaby, A. Using Quality Control Charts to Segment Road Surface Condition Data. In Proceedings of the 7th International Conference on Managing Pavement Assets, Calgary, AB, Canada, 23–28 June 2008.
11. Cafiso, S.; Di Graziano, A. Definition of Homogenous Sections in Road Pavement Measurements. *Procedia Soc. Behav. Sci.* **2012**, *53*, 1069–1079. [CrossRef]
12. Donev, V.; Hoffmann, M. Optimisation of Pavement Maintenance and Rehabilitation Activities, Timing, and Work Zones for Short Survey Sections and Multiple Distress Types. *Int. J. Pavement Eng.* **2018**, *21*, 583–607. [CrossRef]
13. Donev, V.; Hoffmann, M.; Blab, R. Aggregation of Condition Survey Data in Pavement Management: Shortcomings of a Homogeneous Sections Approach and How to Avoid Them. *Struct. Infrastruct. Eng.* **2020**, *17*, 49–61. [CrossRef]
14. Abdelaty, A.; Attia, O.G.; Jeong, H.D.; Gelder, B.K. Dynamic Pavement Delineation and Visualization Approach Using Data Mining. *J. Comput. Civ. Eng.* **2018**, *32*, 04018019. [CrossRef]
15. Shewhart, W.A. *Statistical Method from the Viewpoint of Quality Control*; Dover Publications: Mineola, NY, USA, 2012.
16. Amin, S.G. Control Charts 101: A Guide to Healthcare Applications. *Qual. Manag. Health Care* **2001**, *9*, 1–27. [CrossRef] [PubMed]
17. Peraka, N.S.P. Development of Analytical Tools for Managing Asphalt Pavements: Distress Detection, Degree of Deterioration, and Delineation. Ph.D. Thesis, Department of Civil & Environmental Engineering, Indian Institute of Technology Tirupati, Tirupati, India, September 2021.
18. Jannat, G.E.; Henning, T.F.; Zhang, C.; Tighe, S.L.; Ningyuan, L. Road section length variability on pavement management decision making for Ontario, Canada, highway systems. *Transp. Res. Rec.* **2016**, *2589*, 87–96. [CrossRef]

engineering proceedings

MDPI

Proceeding Paper

Use of Ground-Penetrating Radar to Detect Cement Content in Cement-Stabilized Subgrade †

Zack Hall and S. Sonny Kim *

College of Engineering, The University of Georgia, Athens, GA 30605, USA; zch21221@uga.edu
* Correspondence: kims@uga.edu; Tel.: +1-706-542-9804

† Presented at the Second International Conference on Maintenance and Rehabilitation of Constructed Infrastructure Facilities, Honolulu, HI, USA, 16–19 August 2023.

Abstract: Cement stabilization has been successfully used to improve poor-quality subgrade soils by increasing the soil support to remedy these soils useful for pavement construction. Cement stabilization has the potential to reduce initial construction costs through improved subgrade stability in the pavement structure. Cement stabilization also provides greater long-term stability of the pavement structure and lower pavement life-cycle costs through reduced pavement maintenance. Unfortunately, flexible pavements over cement-stabilized subgrade are experiencing reflective cracking originating from the shrinkage cracks on top of cement-stabilized subgrade due to poor construction. In this study, ground-penetrating radar (GPR) was used to capture the inconsistent layer thickness of cement-stabilized subgrade and its cement content. The results show that GPR is capable of capturing different dielectric constants along with different percent cement contents in subgrade soils.

Keywords: soil cement; stabilization; ground-penetrating radar (GPR); cement-stabilized subgrade

Citation: Hall, Z.; Kim, S.S. Use of Ground-Penetrating Radar to Detect Cement Content in Cement-Stabilized Subgrade. *Eng. Proc.* **2023**, *36*, 4. https://doi.org/10.3390/engproc2023036004

Academic Editor: Hosin (David) Lee

Published: 28 June 2023

1. Introduction

The primary function of pavement structures is to protect the subgrade by reducing stresses and strains due to the loading to a tolerable level. Pavement structures are designed to decrease the stresses as they propagate through the layers above the subgrade. In the case of flexible pavements where close access to quarries is not feasible, the subgrade is stabilized to support and distribute the stresses.

Flexible pavements over stabilized subgrades are highly dependent on the subgrade elastic modulus. The compressive strains in the pavement structure greatly decrease as the subgrade modulus increases. When the compressive strains in the pavement become too high, pavement failures begin to occur. Likewise, the subgrade soil will experience deformation failure due to the high compressive forces. A weak soil underlaying the pavement structure can lead to accelerated pavement deterioration. Cement stabilization can be used to improve subgrade stability, leading to reduced initial construction costs and reduced repair costs [1]. GPR has been used to detect the thickness and cement content of the cement-stabilized layer [2]. GPR is widely used to evaluate the pavement structure using electromagnetic (EM) waves which find the dielectric constant, the primary material property obtained from GPR surveys. For homogenous layers, the speed of light is proportional to the EM waves passing through the layer. The relative permittivity, or dielectric constant, of a homogeneous medium connects the EM velocity in a substance to the speed of light in empty space [3].

$$\varepsilon_r = \left(\frac{c}{v}\right)^2, \quad (1)$$

where

ε_r = dielectric constant;

c = speed of light in free space of 3×10^8 m/s;
v = EM velocity in the material.

2. Research Approach

The dielectric constant can be used to characterize the microstructure of cement-based materials and, thus, was used as the method of strength prediction [4]. The percentage of capillary pores in a soil sample was directly proportional to the compressive strength of the soil. This relationship can be detected with the GPR due to the higher percentage of cement-stabilized soil samples possessing a greater water content, which led to an increased dielectric constant. In this study, mathematical models were developed to detect the different percentages of cement in cement-stabilized subgrade soils through the use of the relative dielectric constant.

3. Data Acquisition and Processing

3.1. Ground-Penetrating Radar (GPR) System

The GSSI 2 GHz air-coupled antenna was used for this study due to the wide range of possible applications in the field. The air-coupled antenna was ideal for pavement scanning, rather than the GSSI 400 MHz ground-coupled antenna, which must remain very close to the ground, as seen in Figure 1. The GSSI 2 GHz air-coupled antenna can be driven safely at normal road speeds without the need to worry about unlevel surfaces in the road. This model also more clearly captures the first layers of the pavement, as opposed to the GSSI 400 MHz ground-coupled antenna, due to the higher frequency not penetrating the road layers as deeply [5].

(a) (b)

Figure 1. (**a**) SIR30 Road scan system using a 2 GHz antenna; (**b**) UGA GPR Van system: (left) GSSI 2 GHz air-coupled antenna, (right) GSSI 400 MHz ground-coupled antenna.

3.2. Sample Preparation

The southern region of the state of Georgia does not have close access to quarries and, thus, relies on the use of cement stabilization for roadway construction. The GDOT uses plant-mixed cement and soil to provide an accurate mixture for cement-stabilized subgrades. Previous mix data show that the most common range of percentage of cement used is 5% to 7% [6]. These data are based on the road loading design requirements and are closely monitored by the mixing plant to meet the needs of the traffic.

Soil cement test specimens were prepared in a standard proctor mold and compacted to 100% of the maximum dry density to simulate site conditions in accordance with the American Society for Testing and Materials (ASTM) D1633-17 [7]. Varying soil mixes were provided by the GDOT to compare the data of different soil types. Three samples of 6% cement-stabilized soil were tested at different time intervals. The dielectric constant is known to decrease rapidly within the first 7 days due to the chemical properties of cement hydration [4]. Therefore, samples were scanned daily for the first week to monitor this chemical process.

4. Laboratory Test Results

A series of laboratory tests was conducted on soil samples of sand using the GPR setup. For the test, two 5 gallon buckets of sand were assembled and one was stabilized with 6% Type I Portland cement. Type I Portland cement was selected due to it being the type of cement used by the GDOT and 6% cement was used to provide a middle range of the cement percentages to be tested. The other bucket was not stabilized to provide a dielectric constant reading of the plain soil. The two buckets were then scanned at days 0, 7, 14, 21, and 28 using the GPR.

The preliminary soil cement test results show that the dielectric constant of the cement-stabilized soil decreased with time. This trend can be seen in Table 1. The decrease in the dielectric constant with time was due to the cement hydration process [4]. Water has a dielectric constant value of 81, which led to the large change in dielectric constant. It is important to note that the results shown in Table 1 were calculated using the amplitudes from the GPR scan. The two buckets were also subject to changing temperatures.

Table 1. Dielectric constant results.

% Cement	0 Days ε_r	7 Days ε_r	21 Days ε_r	28 Days ε_r
0	9.14	8.42	8.31	8.25
6	10.26	10.22	10.17	10.15

The same sand soil sample was also stabilized with 6% cement and prepared according to the sample preparation standards of ASTM D1633-17. The proctor mold-shaped cylinder was then scanned with the GPR every day for 7 days to detect the decrease in the dielectric constant. The soil sample was allowed to cure in a controlled environment at a constant temperature and humidity. The results of the scan are shown in Figure 2. The dielectric constant decreased the most within the first 2 days of the sample being constructed and then maintained a constant value.

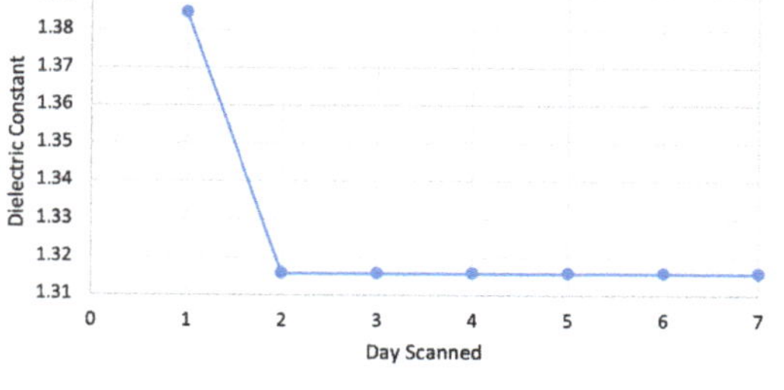

Figure 2. GPR dielectric constant scans for the first week.

5. Conclusions and Discussion

This research aims to identify a GPR method to detect the causes of failure in flexible pavements over cement-stabilized subgrades. This method of failure originates from the shrinkage cracks on top of cement-stabilized subgrade due to poor construction.

The application of this research is to provide a nondestructive testing method capable of predicting possible areas of roadways failures and to provide a method of asset management and quality control. This can be accomplished through the detection of cement content which is directly proportional to the compressive strength of the subgrade soil. The thickness of cement-stabilized subgrade also provides a method to ensure design field validation. Cement stabilization has been proven to provide greater long-term stability of the pavement structure and lower pavement life-cycle costs through reduced pavement maintenance.

The detection of the cement content in stabilized subgrades can be difficult to detect due to the cement hydration process being typically completed after 28 days. To analyze stabilized subgrades in the field, the electrical conductivity should be calculated along with the dielectric constant. By using both values, the percentage of cement stabilizer in the pore water may be detected.

The results in Figure 2 and Table 1 show that the size of the sample being scanned by the GPR affects the dielectric constant reading. This will be addressed by creating samples that cover the area of the GPR antenna. The cement hydration may also be different due to the first preliminary trial not being in a controlled environment.

Author Contributions: Formal analysis, investigation, visualization, Z.H. and S.S.K.; conceptualization, methodology, validation, S.S.K.; resources, data curation, writing—original draft preparation, Z.H. and S.S.K.; writing—review and editing, supervision, project administration, funding acquisition, S.S.K. All authors have read and agreed to the published version of the manuscript.

Funding: This research received no external funding.

Institutional Review Board Statement: Not applicable.

Informed Consent Statement: Not applicable.

Data Availability Statement: Some or all the data, models, or code that support the findings of this study are available from the corresponding author upon reasonable request.

Conflicts of Interest: The authors declare no conflict of interest.

References

1. Biswal, D.R.; Sahoo, U.C.; Dash, S.R. Durability and Shrinkage Studies of Cement Stabilsed Granular Lateritic Soils. *Int. J. Pavement Eng.* **2019**, *20*, 1451–1462. [CrossRef]
2. Saha, S. Characterization of Unbound and Stabilized Materials and Improved Consideration of Their Effects on Pavement Performance. Ph.D. Dissertation, Texas A&M University, College Station, TX, USA, 2019. Available online: https://oaktrust.library.tamu.edu/handle/1969.1/187924 (accessed on 18 January 2023).
3. Abdelmawla, A.; Kim, S.S. Application of Ground Penetrating Radar to Estimate Subgrade Soil Density. *Infrastructures* **2020**, *5*, 12. [CrossRef]
4. Shen, P.; Lu, L.; He, Y.; Wang, F.; Hu, S. Hydration Monitoring and Strength Prediction of Cement-Based Materials Based on the Dielectric Properties. *Constr. Build. Mater.* **2016**, *126*, 179–189. [CrossRef]
5. Solla, M.; Pérez-Gracia, V.; Fontul, S. A Review of GPR Application on Transport Infrastructures: Troubleshooting and Best Practices. *Remote Sens.* **2021**, *13*, 672. [CrossRef]
6. Bowen, R.; McMurry, R.; Pirkle, M. Standard Specifications Construction of Transportation Systems, 2021 ed. Available online: https://www.dot.ga.gov/PartnerSmart/Business/Source/specs/2021StandardSpecifications.pdf (accessed on 18 January 2023).
7. *ASTM D1633-17*; Standard Test Methods for Compressive Strength of Molded Soil-Cement Cylinders. ASTM International: West Conshohocken, PA, USA, 2017. Available online: http://www.astm.org (accessed on 10 October 2022).

engineering proceedings

MDPI

Proceeding Paper

Asset Management Decision Support Tools: Computational Complexity, Transparency, and Realism †

Babatunde Atolagbe [1] and Sue McNeil [1,2,*]

1 Department of Civil Engineering, University of Delaware, Newark, DE 19716, USA; atolagbe@udel.edu
2 Research Center for Integrated Transport Innovation, University of New South Wales, Kensington, NSW 2052, Australia
* Correspondence: smcneil@udel.edu; Tel.: +61-458-862-393

† Presented at the Second International Conference on Maintenance and Rehabilitation of Constructed Infrastructure Facilities, Honolulu, HI, USA, 16–19 August 2023.

Abstract: Asset management decision support tools determine which action (maintenance, rehabilitation, or reconstruction) is applied to each facility in a transportation network and when. Sophisticated tools recognize uncertainties and consider emerging priorities. However, these tools are often computationally complex and lack transparency, the models are difficult to evaluate, and the outputs challenging to validate. This paper explores computational complexity, transparency, and realism in transportation asset management decision support tools to better understand how to select the right tools for a particular context. The results provide direction for agencies when selecting decision support tools, and for researchers and tool developers working towards developing the right tool for an application.

Keywords: asset management; decision support; computational complexity; transparency; validation

Citation: Atolagbe, B.; McNeil, S. Asset Management Decision Support Tools: Computational Complexity, Transparency, and Realism. *Eng. Proc.* **2023**, *36*, 5. https://doi.org/10.3390/engproc2023036005

Academic Editor: Hosin (David) Lee

Published: 29 June 2023

1. Introduction

Asset management decision support tools are used by transportation agencies to determine which action (maintenance, rehabilitation, or reconstruction) is applied to each facility in a transportation network and when. These decisions are made considering existing and predicted condition and performance, asset life cycle costs, and user costs. Sophisticated tools recognize uncertainties and consider emerging priorities, such as resilience and sustainability. However, these tools are often computationally complex and lack transparency, the models are difficult to evaluate, and the outputs challenging to validate. This paper explores computational complexity, transparency, and realism in transportation asset management decision support tools to better understand how to select or develop the right tools for a particular context.

The paper reviews different types of decision support tools (ranking, prioritization, thresholds, and optimization) and the goals of state agencies in making decisions. Using a multi-asset roadway improvement scheduling tool [1] as a case study, the analysis compares the computational burdens, the parameters involved, and the range of outcomes for different scenarios.

The results demonstrate four issues in the selection of decision support tools in the context of state agencies in the United States. The first issue is the computational burden; running the decision support tool for a simple network requires several hours. The second issue is the sensitivity of the results to the input parameters. The results show the relative importance of different parameters. The third issue is the differences between simple and complex decision support tools and generalizing the circumstances in which to use one versus another. Some simple heuristics for selecting tools are identified. The fourth issue is the validation of the results. Strategies for qualitative validation are explored.

2. Background

Asset management processes have been adopted by state departments of transportation in the United States in response to aging infrastructure, traffic growth, unanticipated damage to infrastructure assets due to natural hazards, higher performance expectations of users, declining funding bases, and legislative requirements. Decision support tools are a key element of asset management.

The simplest tools rank and prioritize activities [2]. The more sophisticated and complex asset management tools use optimization to make decisions that select and prioritize activities. Formulations, which define the decision variables, objective function, and constraints, are classified as project or network level, single objective or multi-objective, and deterministic, heuristic, or other [3]. Another classification also considers whether the decision variables are discrete or continuous and whether uncertainty is considered or not. Optimization problems can also be classed as activity selection, scheduling, or both selection and scheduling. Invariably, optimization problems focus on a discrete set of locations and activities, a finite period (planning horizon), the condition of the asset, and the costs including agency and user costs. To simplify the solution process or reduce the size of the solution space, assumptions are usually made.

There is a large body of literature on decision support tools for asset management. Building from work in pavement management, the seminal work of Golabi, Kulkarni, and Way [4] is a foundation for advanced and more sophisticated optimization of maintenance and resurfacing decisions for pavements that recognized deterioration and uncertainty and then extended the work to bridges. Reviews of the state of the art provide context. Chen and Bai [3] review over 300 papers on optimization in asset management. Chen et al. [5] provide a review of optimization in transportation asset management for roads and bridges. Chen et al. [6] focus on multi-objective optimization for maintenance decisions. Other papers address the changing needs for asset management tools that embrace resilience, sustainability, and uncertainty [7–9]. Together, these papers provide a clear picture of the variety of approaches to the problem formulation and solution methods, both of which are tailored to a particular application. In this paper, we focus on understanding issues related to the computational complexity, transparency, and realism of tools.

3. Approach

The paper reviews the goals of state agencies in making decisions. The review is based on the Transportation Asset Management Plans submitted by each state department of transportation in 2019 to the Federal Highway Administration as required in the Moving Ahead for Progress in the 21st Century Act (MAP-21) [10].

Using a multi-asset roadway improvement scheduling tool [1] as a case study, the analysis compares the computational burdens, the parameters involved, and the range of outcomes for different scenarios. The tool is a bilevel program that prioritizes and schedules roadway improvement activities recognizing users' costs and disruption. The upper level involves a Markov decision process (MDP) to identify and prioritize potential roadway improvement actions. The lower level seeks to determine traffic flows based on a network user equilibrium solution across paths that is affected by capacities determined through actions determined at the upper level. The problem is solved using a reinforcement learning method.

4. Results

4.1. State Perspectives

A review of select state DOTs' Transportation Asset Management Plans revealed that most states aim to optimize their investments but do not optimize in the mathematical sense of the word [11]. At best, the states optimize investment in their bridge program or pavement program. Most conduct scenario analysis and explore alternative strategies. However, the TAMPs do recognize the value of optimization, the potential gains, and the importance of good models and reliable data.

Overall, the states are aiming to develop optimal plans that deliver the best serviceability given the budget constraints. Invariably, the implementations focus on independently reached optimization decisions for pavements and bridges based on a predefined set of scenarios. Essentially, the objective function is computed for each scenario that meets the constraints and the "optimal" solution chosen. Given the fact that there are many tradeoffs in terms of actions, timing, and location, it is possible for an optimal solution to be overlooked. However, given that scenarios are developed based on experience and data, the optimal solution selected is likely to be very desirable, and for the given problem and objective, either optimal or near optimal.

Thresholds for determining when to undertake a maintenance or improvement activity, decision trees, simulation, and scenario analysis are widely used in practice. While the strategies have proven to be effective for pavements and bridges independently, the need to consider cross-asset tradeoff and integrate more complex objectives, such as users' costs, disruption, and sustainability, adds to the complexity. On the other hand, the "black box syndrome" means that agencies are skeptical of the outputs from elaborate mathematical models. Wang and Pyle [12] recommend engaging the users, verification of results, and continued validation.

Another important gap in developing and implementing optimal decisions is the difficulty in assembling the required data. This is important because data collection is costly, and resources for maintaining and improving roads are scarce. Taking advantage of innovative data collection methods, more accurate and more timely data, and making better use of resources is important. Although agencies may not implement an "optimal" solution, exploring alternative solution methods provides insight into the factors that influence these decisions and will ultimately help agencies to deliver better transportation services.

4.2. Case Study

Using a simple network consisting of 10 nodes and 11 links with some redundancy, the case is used to demonstrate the computational complexity of the problem, sensitivity to input parameters, comparison of the solution using simple thresholding, and a brief discussion of the challenges involved in validating the results.

A single run of the simple case study on a Windows computer (16 GB RAM CPU with Intel Core i5 processor of 2.20 GHz speed) takes 3 h. Given this computational burden due to complex interactions among traffic, and activities, these types of problems do not encourage the exploration of alternatives or changes in the parameters. However, to explore the sensitivity of the solutions to changes in the parameters, we solved the problem for 1024 different scenarios using a designed experiment capturing changes in the deterioration rate, user cost, maintenance costs, discount factor, traffic factor, and observation accuracy for pavements and bridges. Using a Sobol global sensitivity analysis, we found that the solution was only sensitive to one parameter, the discount factor, suggesting that the time value of money is the most important parameter, but reasonable values for the other parameters should be selected.

A comparison of the results with optimal thresholds indicated that while the results are more efficient using the optimization methods, the differences are modest [11], suggesting that complex tools are not needed for uncongested networks or a network with a high degree of redundancy. Validation is also challenging. The results are optimal given the inputs. However, other parameters or omitted variables may influence the solution. The most common assessment is based on logic; outputs reflect the appropriate order of magnitude, and changes in outputs reflect changes in inputs in the right direction.

In summary, the results demonstrate four issues in the selection of decision support tools in the context of state agencies in the United States. The first issue is the computational burden; running the decision support tool for a simple network requires several hours. The second issue is the sensitivity of the results to the input parameters. The results show the relative importance of different parameters. The third issue is the differences between simple and complex decision support tools and generalizing the circumstances in which to use one versus another. Some simple heuristics for selecting tools are identified. The fourth issue is the validation of the results. Strategies for qualitative validation are explored.

5. Conclusions

Based on the case study results, the analysis suggests that the Sobol sensitivity analysis provides direction for agencies when selecting parameters for models, and that complex decision support tools are not always warranted. The results also serve as a reminder to researchers and tool developers of issues that must be considered in developing the right tool for an application. Further research can provide a more specific direction.

Author Contributions: All aspects and sections of the paper were developed by B.A. and S.M. All authors have read and agreed to the published version of the manuscript.

Funding: This work was sponsored by a grant from the Center for Integrated Asset Management for Multimodal Transportation Infrastructure Systems (CIAMTIS), a U.S. Department of Transportation University Transportation Center, under federal grant number 69A3551847103. The authors are grateful for the support.

Institutional Review Board Statement: Not applicable.

Informed Consent Statement: Not applicable.

Data Availability Statement: All data are available in referenced reports or papers.

Conflicts of Interest: The authors declare no conflict of interest.

References

1. Zhou, W.; Miller-Hooks, E.; Papakonstantinou, K.G.; Stoffels, S.; McNeil, S. A Reinforcement Learning Method for Multiasset Roadway Improvement Scheduling Considering Traffic Impacts. *J. Infrastruct. Syst.* **2022**, *28*, 04022033. [CrossRef]
2. Haas, R.; Hudson, W.R.; Falls, L.C. *Pavement Asset Management*; John Wiley & Sons, Inc.: Hoboken, NJ, USA, 2015.
3. Chen, L.; Bai, Q. Optimization in decision making in infrastructure asset management: A review. *Appl. Sci.* **2019**, *9*, 1380. [CrossRef]
4. Golabi, K.; Kulkarni, R.B.; Way, G.B. A statewide pavement management system. *Interfaces* **1982**, *12*, 5–21. [CrossRef]
5. Chen, Z.; Liang, Y.; Wu, Y.; Sun, L. Research on Comprehensive Multi-Infrastructure Optimization in Transportation Asset Management: The Case of Roads and Bridges. *Sustainability* **2019**, *11*, 4430. [CrossRef]
6. Chen, L.; Henning, T.F.P.; Raith, A.; Shamseldin, A.Y. Multiobjective Optimization for Maintenance Decision Making in Infrastructure Asset Management. *J. Manag. Eng.* **2015**, *31*, 04015015. [CrossRef]
7. Hessami, A.R.; Anderson, S.D.; Smith, R.E. Levels of Uncertainty in Infrastructure Asset Management. *Transp. Res. Rec.* **2021**, *2675*, 24–31. [CrossRef]
8. Liu, Y.; McNeil, S. Using resilience in risk-based asset management plans. *Transp. Res. Rec.* **2020**, *2674*, 178–192. [CrossRef]
9. Sinha, K.C.; Labi, S.; Agbelie, B.R. Transportation infrastructure asset management in the new millennium: Continuing issues, and emerging challenges and opportunities. *Transp. A Transp. Sci.* **2017**, *13*, 591–606. [CrossRef]
10. 112th Congress. Moving Ahead for Progress in the 21st Century Act. 2012. Available online: https://www.govinfo.gov/content/pkg/PLAW-114publ94/pdf/PLAW-114publ94.pdf (accessed on 15 July 2019).
11. Hooks, E.M.; McNeil, S.; Lattanzi, D.; Papakonstantinou, K.; Stoffels, S.; Zhou, W.; Kamranfar, P.; Saifullah, M.; Andriotis, C.; Withers, A. *Strategic Prioritization and Planning of Multi-Asset Transportation Infrastructure Maintenance, Rehabilitation, and Improvements: Phase 1–Prioritization through Optimization*; No. CIAM-UTC-REG5; Center for Integrated Asset Management for Multimodal Transportation Infrastructure Systems (CIAMTIS)(UTC): State College, PA, USA, 2021.
12. Wang, Z.; Pyle, T. Implementing a pavement management system: The Caltrans experience. *Int. J. Transp. Sci. Technol.* **2019**, *8*, 251–262. [CrossRef]

engineering proceedings

MDPI

Proceeding Paper

Machine Learning-Based Slope Failure Prediction Model Considering the Uncertainty of Prediction †

Junhyuk Choi [1], Yongkyu Cho [2], Yongjin Kim [3], Yongseong Kim [4] and Bongjun Ji [4,*]

1 Department of Industrial and Management Engineering, Pohang University of Science and Technology, Pohang 37673, Republic of Korea; cjh0102@postech.ac.kr
2 Department of Industrial Engineering, Kangnam University, Yongin 16979, Republic of Korea; yongkyu.cho@kangnam.ac.kr
3 Smartgeotech, Chuncheon 24341, Republic of Korea
4 Department of Regional Infrastructure Engineering, Kangwon National University, Chuncheon 24341, Republic of Korea
* Correspondence: bjji@kangwon.ac.kr
† Presented at the Second International Conference on Maintenance and Rehabilitation of Constructed Infrastructure Facilities, Honolulu, HI, USA, 16–19 August 2023.

Abstract: Slope failure is a severe natural disaster that can cause property damage and human costs. In order to develop a warning system for slope failure, various studies have been conducted, including research based on both physics-based models and machine learning-based models. While machine learning-based approaches have shown promise due to their ability to automatically extract hidden patterns in data, conventional machine learning models have their limitations. Specifically, while they can always provide a prediction value, they fail to provide information about the uncertainty of the prediction results. In this study, we developed a machine learning model that can predict the slope failure by training trends in time-series data. Our proposed model addresses the limitations of the conventional machine learning models by incorporating the Monte Carlo dropout to calculate the uncertainty during the prediction stage. The experimental results demonstrated that the proposed model significantly outperforms the conventional machine learning models in terms of both its prediction accuracy and the ability to estimate uncertainty. Furthermore, the model proposed in this study can support decision-makers by providing more accurate information than the conventional models.

Keywords: slope failure; machine learning; uncertainty

Citation: Choi, J.; Cho, Y.; Kim, Y.; Kim, Y.; Ji, B. Machine Learning-Based Slope Failure Prediction Model Considering the Uncertainty of Prediction. *Eng. Proc.* **2023**, *36*, 6. https://doi.org/10.3390/engproc2023036006

Academic Editor: Hosin (David) Lee

Published: 29 June 2023

1. Introduction

Slope failure is a severe natural disaster that can cause significant property damage and even the loss of life [1]. In order to develop a warning system for slope failure, various studies have been conducted in the past. One approach that has gained significant attention in recent years is machine learning [2]. Machine learning-based models have the potential to provide accurate predictions by automatically extracting the hidden patterns in data [3].

In contrast to machine learning, physics-based models are built on the principles of mechanics and physics [4]. These models typically require a detailed understanding of the geology and soil mechanics of the area in question. The major disadvantage of physics-based models is that obtaining an exact mathematical formula is often oversimplified and impractical, especially in complex geological settings [5]. These limitations lead to the formation of models that are not sufficiently precise or efficient for predicting the slope failure [6].

On the other hand, machine learning-based prediction models can be highly effective at predicting the slope failure by training on large datasets and identifying the hidden patterns in the data. However, traditional machine learning models often have their limitations. For example, they may need to consider the uncertainty associated with the prediction result, which can make it difficult for decision-makers to make informed decisions.

To address these limitations, we proposed a machine learning-based slope failure prediction model that considers the prediction result's uncertainty. Our model uses time-series data to train a machine learning algorithm and incorporates the Monte Carlo dropout to calculate the uncertainty of the prediction result [7]. In doing so, we aimed to provide decision-makers with more accurate and reliable information about the potential risks associated with a given slope.

The rest of this paper is organized as follows: Section 2 provides an overview of the materials and methods for slope failure prediction. Section 3 describes the results and discussion of the experiment. Finally, Section 4 concludes the paper and suggests some future works.

2. Materials and Methods

In this section, we describe the data collection and preprocessing, model architecture, training and validation, and model evaluation procedures that were used to develop a machine learning-based slope failure prediction model considering the uncertainty of prediction.

We used the displacement and remaining failure time data collected from a slope stability monitoring system installed on a slope located in Yeongdeok-gun, Gyeongsangbuk-do [8]. The monitoring system recorded the displacement of the slope at a frequency of day and the remaining failure time of the slope at regular intervals. The dataset consisted of 586 time-series data.

To prepare the displacement and remaining failure time data for use in our proposed model, we performed z-score normalization. This involved calculating the mean and standard deviation values of the data and subtracting the mean from each data point, then dividing the result by the standard deviation. This normalization step was necessary to perform to ensure that the data were on a similar scale, allowing our model to effectively identify patterns and relationships in the time-series data.

We utilized a long short-term memory (LSTM) model, a type of recurrent neural network (RNN) [9], to develop a machine learning-based slope failure prediction model. LSTMs are well-suited for sequence-to-sequence prediction tasks [10], as traditional RNNs often face the vanishing gradient problem where gradients become too small to learn long-term dependencies over many time steps. To address this issue, LSTMs incorporate a memory cell and gating mechanisms that regulate the flow of the information. The memory cell retains information over time, while the gating mechanisms control the amount of information retained, forgotten, or passed to the next time step. These mechanisms comprise three sigmoid functions and one tanh function, which compute the input gate, forget gate, output gate, and candidate memory vector, respectively.

LSTMs are widely used in natural language processing [11], speech recognition [12], and time-series prediction [13]. They are especially effective for tasks requiring long-term memory, such as language modeling, where the network must recall the context of a sentence over many words. In this study, we employed an LSTM network to capture the long-term dependencies present in the time-series data of displacement and remaining failure time.

The model architecture is composed of a stack of two LSTM layers, each with 64 units, followed by a dense layer with a single unit. To estimate the uncertainty, we applied the Monte Carlo dropout to the output layer. In further detail, dropout is a regularization technique that is commonly used in deep learning to prevent overfitting. It randomly sets some of the output activations of the layer to zero during training. During inference, dropout is applied by randomly dropping out neurons with a certain probability, and the prediction is then computed by taking the average of the outputs of the network over multiple dropout iterations. This approach allows us to estimate the model's uncertainty by measuring the variance of the predictions across the dropout iterations. Therefore, we applied the MC dropout to the output layer to obtain a more accurate estimate of the model's uncertainty.

We randomly split the collected data into a training data and a validation data set with a ratio of 70% and 30%, respectively. We used the Adam optimizer for training the model with a learning rate of 0.01. We used the mean squared error (MSE) loss function with an additional term that calculates the standard deviation of the MC dropout output as a measure of uncertainty. The model was trained for 100 epochs, and batch sizes of 32 were used. The validation loss was used to monitor the model's performance during training.

We assessed the performance of the prediction model using two popular metrics; the root mean squared error (RMSE) and the mean absolute error (MAE). To assess the model's ability to predict the uncertainty of the prediction, we generated 100 MC dropout samples for each test data point and estimated the uncertainty of the prediction. All computations were performed using the TensorFlow deep learning framework.

3. Results and Discussion

The results of our study indicate that the machine learning-based slope failure prediction model using a Bayesian LSTM with MC dropout can accurately predict the remaining failure time of a slope, while also estimating the uncertainty of the prediction. The Bayseian LSTM model showed a performance improvement of 3.5 in terms of MAE and 6.9 in terms of RMSE over the original model, respectively. The estimated uncertainty of the prediction allows decision-makers to have a more accurate and reliable assessment of the risks associated with slope instability. Furthermore, the examples of the predicted outputs presented in Figure 1 highlight the significance of estimating uncertainty. The large gap observed between the true and predicted values when the uncertainty of the output is significant emphasizes the importance of considering the uncertainty estimates in making decisions related to the slope stability.

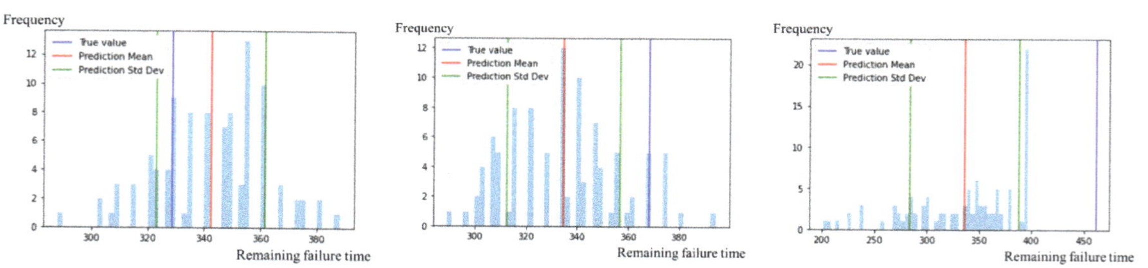

(a) (b) (c)

Figure 1. Examples of predicted output distribution which has actual values of; (**a**) 328, (**b**) 368, and (**c**) 460, respectively.

The use of machine learning for slope failure prediction has the potential to provide a more accurate and reliable warning system for slope failure. Compared to physics-based models, machine learning models can better capture the complex relationships between the various factors that contribute to slope failure, such as soil properties, rainfall, and slope geometry. However, the use of machine learning models can also introduce new sources of uncertainty which need to be considered.

In our study, we used a Bayesian LSTM with MC dropout to estimate the uncertainty of the prediction. By estimating the uncertainty, we can provide decision-makers with a more accurate assessment of the risks associated with the slope instability.

4. Conclusions

In conclusion, our study developed a machine learning-based slope failure prediction model that can accurately predict the remaining failure time of a slope, while also estimating the uncertainty of the prediction. This model has the potential to provide decision-makers with a more accurate and reliable warning system for slope failure. Future work should focus on validating the model on data from other monitoring systems and slopes, as well as exploring more efficient methods for estimating uncertainty.

Author Contributions: Conceptualization, J.C. and B.J.; methodology, B.J.; software, J.C.; validation, J.C., Y.C. and B.J.; formal analysis, J.C.; investigation, J.C. and Y.K. (Yongjin Kim); resources, Y.K. (Yongseong Kim); data curation, Y.C.; writing—original draft preparation, J.C.; writing—review and editing, Y.K. (Yongseong Kim) and B.J.; visualization, J.C.; supervision, B.J.; project administration, B.J. All authors have read and agreed to the published version of the manuscript.

Funding: First, this research was supported by "Ministry of the Interior and Safety" R&D program (RS-2022-00155667). Additionally this study was supported by 2022 Research Grant from Kangwon National University.

Institutional Review Board Statement: Not applicable.

Informed Consent Statement: Not applicable.

Data Availability Statement: The data presented in this study are available on request from the corresponding author. The data have been synthesized for the purpose of this study and are not publicly available.

Acknowledgments: Authors would like to acknowledge the significant support that has made this research possible.

Conflicts of Interest: The authors declare no conflict of interest.

References

1. Orense, R.P.; Shimoma, S.; Maeda, K.; Towhata, I. Instrumented model slope failure due to water seepage. *J. Nat. Disaster Sci.* **2004**, *26*, 15–26. [CrossRef]
2. Zhang, J.; Wang, Z.; Hu, J.; Xiao, S.; Shang, W. Bayesian machine learning-based method for prediction of slope failure time. *J. Rock Mech. Geotech. Eng.* **2022**, *14*, 1188–1199. [CrossRef]
3. Vasavi, S. Extracting hidden patterns within road accident data using machine learning techniques. In Proceedings of the Information and Communication Technology, Singapore, 12–13 December 2016.
4. White, J.A.; Singham, D.I. Slope stability assessment using stochastic rainfall simulation. *Procedia Comput. Sci.* **2012**, *9*, 699–706. [CrossRef]
5. Azmoon, B.; Biniyaz, A.; Liu, Z.; Sun, Y. Image-data-driven slope stability analysis for preventing landslides using deep learning. *IEEE Access* **2021**, *9*, 150623–150636. [CrossRef]
6. Ma, W.; Dong, J.; Wei, Z.; Peng, L.; Wu, Q.; Chen, C.; Xie, F. Landslide displacement prediction with gated recurrent unit and spatial-temporal correlation. *Front. Earth Sci.* **2022**, *10*, 1182. [CrossRef]
7. Kendall, A.; Badrinarayanan, V.; Cipolla, R. Bayesian segnet: Model uncertainty in deep convolutional encoder-decoder architectures for scene understanding. *arXiv* **2015**, arXiv:1511.02680.
8. Yoo, B.S. Study of Failure Analysis Methods Based on Real-Time Monitoring Data for Landslide Warning System. Doctoral Dissertation, Kumoh National Institute of Technology, Gumi-si, Republic of Korea, 2006.
9. Rumelhart, D.E.; Hinton, G.E.; Williams, R.J. Learning representations by back-propagating errors. *Nature* **1986**, *323*, 533–536. [CrossRef]
10. Hochreiter, S.; Schmidhuber, J. Long short-term memory. *Neural Comput.* **1997**, *9*, 1735–1780. [CrossRef] [PubMed]
11. Yao, L.; Guan, Y. An improved LSTM structure for natural language processing. In Proceedings of the IEEE International Conference of Safety Produce Informatization (IICSPI), Chongqing, China, 10–12 December 2018.
12. Graves, A.; Jaitly, N.; Mohamed, A.R. Hybrid speech recognition with deep bidirectional LSTM. In Proceedings of the IEEE Workshop on Automatic Speech Recognition and Understanding, Olomouc, Czech Republic, 8–12 December 2013.
13. Li, Y.; Zhu, Z.; Kong, D.; Han, H.; Zhao, Y. EA-LSTM: Evolutionary attention-based LSTM for time series prediction. *Knowl.-Based Syst.* **2019**, *181*, 104785. [CrossRef]

engineering proceedings

MDPI

Proceeding Paper

Construction and Design Guidelines for Lightweight Cellular Concrete as Pavement Subbase Material †

Abimbola Grace Oyeyi [1], Frank Mi-Way Ni [2,*] and Susan Tighe [3]

1 Department of Civil and Environmental Engineering, University of Waterloo, 200 University Avenue West, Waterloo, ON N2L 3G1, Canada; a2olaley@uwaterloo.ca
2 Department of Civil & Coastal Engineering, University of Florida, 1949 Stadium Rd., Gainesville, FL 32611, USA
3 Department of Civil Engineering, McMaster University, 1280 Main Street West, John Hodgins Engineering Building, Room 301, Hamilton, ON L8S 4L7, Canada; tighes1@mcmaster.ca
* Correspondence: marctise@ufl.edu

† Presented at the Second International Conference on Maintenance and Rehabilitation of Constructed Infrastructure Facilities, Honolulu, HI, USA, 16–19 August 2023.

Abstract: Lightweight cellular concrete (LCC) has gained attention in the pavement industry as a potential subbase material due to its workability, freeze–thaw resistance, and thermal insulation properties. Research has shown that LCC has sufficient strength to support pavement structures and reduce subgrade pressures. However, a successful application requires the consideration of construction provisions, such as equipment and quality control, and design parameters, such as strength requirements and structural coefficients. This paper provides recommendations for using LCC as a pavement subbase material, including when and how to design pavement with it.

Keywords: lightweight cellular concrete; construction guidelines; pavement design; subbase materials

Citation: Oyeyi, A.G.; Ni, F.M.-W.; Tighe, S. Construction and Design Guidelines for Lightweight Cellular Concrete as Pavement Subbase Material. *Eng. Proc.* **2023**, *36*, 7. https://doi.org/10.3390/engproc2023036007

Academic Editor: Hosin (David) Lee

Published: 30 June 2023

1. Introduction

Lightweight cellular concrete (LCC) or foamed concrete has gained attention in pavement industries due to its low density of 375 to 1600 kg/m^3 and homogenous structure of air bubbles throughout the mixture [1,2]. Studies have extensively examined its mechanical properties like compressive strength and elasticity modulus [3–6]. LCC has been successfully used as an alternative subbase to unbind granular material to solve the differential settlement on roadways in various locations in Canada and around the world [7–11]. LCC is proving to be an excellent alternative for pavement applications with improved bearing capacity [12,13]. However, there remains a requirement for a comprehensive assessment and set of recommendations for the incorporation of LCC into the design and construction of flexible pavements in Canada.

2. Investigation on Field Test Sections

In 2018 and 2021, two field test sections were constructed to evaluate the performance of LCC. The test sections involved three different densities (400, 475, and 600 kg/m^3) and thicknesses of LCC compared to the unbound granular subbase section. Different sensors were installed to monitor the pavement response and environmental conditions. The following sections summarize the findings from the construction process to post-construction performance tracking [14,15].

2.1. Construction Analysis

The results indicated in the past study that LCC subbase layers are a viable alternative to granular materials for reducing subgrade pressures by 78% during construction, with densities of 400 kg/m^3, 475 kg/m^3, and 600 kg/m^3 being effective [16]. When pouring LCC at lower temperatures (≤4 °C), insulating the layers after pouring may be necessary for proper curing. LCC subbase does not hinder drainage and can even improve it. However, construction methods, processes, and excessive vehicular traffic can impact performance. Careful design modifications such as LCC thickness, construction method, and training for construction crews are necessary to ensure good performance of LCC subbase pavements and limit the duration between LCC pour and asphalt paving operations.

2.2. Field Structural Evaluation

Regarding structural capacity, Oyeyi et al. documented that using LCC as a subbase material can significantly reduce subgrade pressure caused by traffic by three times compared to unbound granular subbase pavements (Figure 1) [15]. The pressure reduction depends on factors such as LCC layer thickness, pavement thickness, load magnitude, load distribution, tire pressure, and material characteristics. LCC subbase layers with a thickness of less than or equal to 250 mm could be used in areas with lower traffic volumes and smaller vehicles. Increased pavement depth reduces strains, and incorporating LCC subbase layers reduces strains more than unbound granular material. However, environmental conditions like temperature and moisture can affect the stresses and strains experienced at each layer (For instance, winter to have lower strain), and further investigation is necessary.

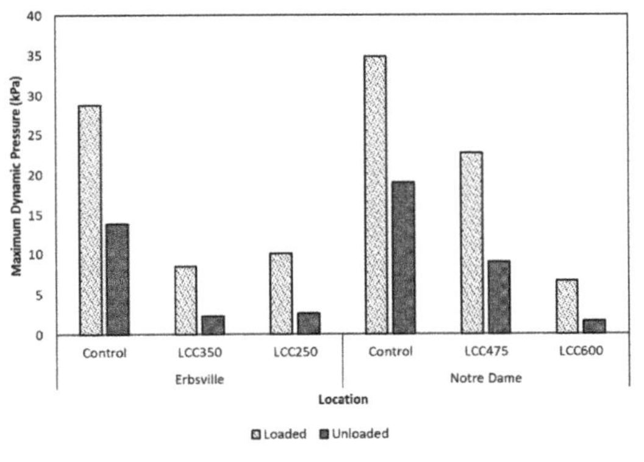

Figure 1. Maximum Dynamic Pressure Notre Dame and Erbsville [15].

3. Cost Analysis

Oyeyi et al. performed a life cycle assessment (LCA) for LCC pavements and compared them with the unbound granular subbase material (Granular A (GA) and B (GB)) pavements. Their results showed that up to 16% of CO_2 emission is reduced by substituting unbound granular material with LCC and reducing other harmful pollutants [16]. Based on the same maintenance and rehabilitation (M&R) schedule in [16], a preliminary life cycle cost assessment (LCCA) was performed. Figure 2 demonstrates the results [17]. This study considered only initial and M&R costs, but not the insulating capabilities of LCC which could contribute to the roadway longevity over frost susceptible subgrades.

The LCCA results showed that flexible pavement sections with granular A had between 10 and 13% and granular B between 4 and 6% lower total life cycle costs at the end of a 50-year analysis period compared with LCC subbases. The yearly cost commitment was also within the exact percentages. Initial construction contributed the most to all sections' total lifecycle costs. The initial construction costs for the control with granular A were 15%, 18%, and 21% lower than those with 400, 475, and 600 kg/m^3 LCC, respectively. These percentages were 7%, 10%, and 13% lower when using granular B. When only initial construction and maintenance LCCA phase costs were compared, the LCC sections yielded lower total life cycle costs, 400 kg/m^3 (6%), 475 kg/m^3 (5%), and 600 kg/m^3 (3%) than the control with granular A. Similarly, it was 12%, 11%, and 9%, respectively, less than the control with granular B. The costs increased with an increase in LCC density LCC. Total lifecycle cost was sensitive to changes in the cost of subbase material. Even with a 40 % LCC cost reduction, the granular A control section remained cheaper. However, a decrease in LCC cost by 40% caused the control with granular B to become more expensive than the LCC sections.

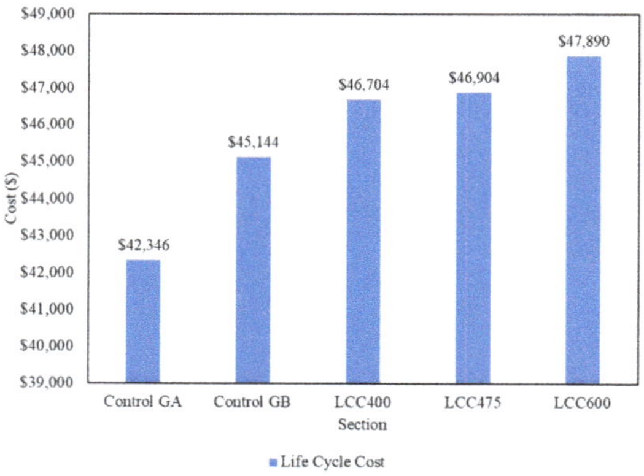

Figure 2. Total lifecycle cost of pavement design alternatives.

4. Recommendation of Use

LCC subbase can be a viable alternative to traditional subbase materials when weak subgrades need insulation. Thicknesses of at least 250 mm and densities between 400–600 kg/m^3 can improve performance and delay pavement failure.

Several methods can be used to design LCC subbase pavements, including granular base equivalent, MEPDG, and AASHTO 93. Linear elastic theory packages can also be used but may require adjustments. For example, WESLEA-generated strains and stresses could be reduced by a ratio of 0.2 and 0.65 for LCC between 400 and 600 kg/m^3. Other tools like WESLEA, KENLAYER, and PAVEXpress for better performance interpretation should supplement MEPDG analysis.

A structural coefficient of 0.22 is recommended for designing LCC pavements between 400–600 kg/m^3. The granular base equivalency strength coefficient is proposed to be 1.22 for 400 kg/m^3, 1.46 for 475 kg/m^3, and 1.91 for 600 kg/m^3. LCC subbase thickness can be reduced by 44–65% compared to unbound granular subbase.

In total, 600 kg/m^3 density LCC is recommended for major arterial roads, while minor arterial and collector roads can use 400 kg/m^3 and 475 kg/m^3 LCC. LCC subbase below 475 kg/m^3 should be at least 250 mm thick. Strain and stress patterns within the pavement structure should be considered when deciding on density for various road conditions and classes.

The timing for opening traffic after road construction is recommended after three days for 475 kg/m^3 density and 600 kg/m^3 density and seven days for 400 kg/m^3 density. The dry mix approach for LCC production is beneficial, and using slag or pozzolans can further reduce environmental damage.

Additionally, 475 kg/m^3 LCC is proposed over 400 kg/m^3 due to its stable pore structure, strength, and freeze–thaw resistance. Total lifecycle costs for 400 and 475 kg/m^3 LCC are comparable and less than 600 kg/m^3. Therefore, applying 475 kg/m^3 LCC is recommended due to its significant difference in performance from 400 kg/m^3 and relatively comparable performance to 600 kg/m^3.

5. Conclusions

This paper discusses the use of lightweight cellular concrete as an alternative pavement subbase material. Literature review, findings from field test sections, and results from LCA from past studies were reviewed. It presents preliminary lifecycle cost analysis (LCCA) information and provides recommendations for LCC use with flexible pavements. For future considerations, the freeze–thaw benefit of LCC should be considered in evaluation performance, an in-depth LCCA should be performed to consider social costs and insulation benefits associated with using LCC.

Author Contributions: Conceptualization, A.G.O., F.M.-W.N., and S.T.; methodology, A.G.O. and F.M.-W.N.; software, F.M.-W.N.; validation, A.G.O. and F.M.-W.N.; writing—original draft preparation, A.G.O.; writing—review and editing, F.M.-W.N.; supervision, S.T.; project administration, S.T. All authors have read and agreed to the published version of the manuscript.

Funding: This research was funded by Natural Sciences and Engineering Research Council of Canada (NSERC)and CEMATRIX, grant number funding reference CRDPJ 514908-2017.

Institutional Review Board Statement: Not applicable.

Informed Consent Statement: Not applicable.

Data Availability Statement: Not applicable.

Acknowledgments: The authors of this research gratefully acknowledge CEMATRIX (CANADA) Inc, the Region of Waterloo, Ontario, Canada, Natural Sciences and Engineering Research Council of Canada (NSERC), and Centre for Pavement and Transportation Technology (CPATT), the University of Waterloo for supporting this research.

Conflicts of Interest: The authors declare no conflict of interest.

References

1. Amran, M.; Onaizi, A.M.; Fediuk, R.; Danish, A.; Vatin, N.I.; Murali, G.; Abdelgader, H.S.; Mosaberpanah, M.A.; Cecchin, D.; Azevedo, A. An ultra-lightweight cellular concrete for geotechnical applications—A review. *Case Stud. Constr. Mater.* **2022**, *16*, e01096. [CrossRef]
2. Ozlutas, K. Behaviour of Ultra-Low Density Foamed Concrete. Doctoral Dissertation, University of Dundee, Dundee, UK, 2015.
3. Amran, Y.M.; Farzadnia, N.; Ali, A.A. Properties and applications of foamed concrete-a review. *Constr. Build. Mater.* **2015**, *101*, 990–1005. [CrossRef]
4. Jiang, J.; Lu, Z.; Niu, Y.; Li, J.; Zhang, Y. Study on the preparation and properties of high-porosity foamed concretes based on ordinary Portland cement. *Mater. Des.* **2016**, *92*, 949–959. [CrossRef]
5. Favaretto, P.; Hidalgo, G.E.N.; Sampaio, C.H.; Silva, R.D.A.; Lermen, R.T. Characterization and use of construction and demolition waste from south of Brazil in the production of foamed concrete blocks. *Appl. Sci.* **2017**, *7*, 1090. [CrossRef]
6. Fu, Y.; Wang, X.; Wang, L.; Li, Y. Foam concrete: A state-of-the-art and state-of-the-practice review. *Adv. Mater. Sci. Eng.* **2020**, *2020*, 6153602. [CrossRef]
7. Dolton, B.; Witchard, M.; Luzzi, D.; Smith, T.J. *Application of Lightweight Cellular Concrete to Reconstruction of Settlement Prone Roadways in Victoria*; GEOVANCOUVER: Vancouver, BC, Canada, 2016.
8. Griffiths, F.; Popik, M. *Pavement Evaluation-CEMATRIX Site Dixie Road, Caledon, Ontario*; Thurber Engineering Ltd.: Caledon, ON, Canada, 2013.
9. Maher, M.L.; Hagan, J.B. MAT-758: Constructability Benefits of the Use of Lightweight Foamed Concrete Fill (LFCF) in Pavement Applications. In Proceedings of the 2016 CSCE Annual Conference, London, ON, Canada, 1–4 June 2016.
10. Decký, M.; Drusa, M.; Zgútová, K.; Blaško, M.; Hájek, M.; Scherfel, W. Foam concrete as new material in road constructions. *Procedia Eng.* **2016**, *161*, 428–433. [CrossRef]
11. Drusa, M.; Decky, M. *Designing and Quality Control of Earth Structures on Transport Constructions*; Edis Uniza: Žilina-Vlčince, Slovakia, 2013; p. 522. ISBN 978-80-554-0823-1.
12. Ni, F.M.; Oyeyi, A.G.; Tighe, S. Structural capacity evaluation of lightweight cellular concrete for flexible pavement subbase. *Road Mater. Pavement Des.* **2021**, *23*, 2781–2797. [CrossRef]
13. Taylor, S.; Halsted, G. *Guide to Lightweight Cellular Concrete for Geotechnical Applications*; Portland Cement Association: Washington, DC, USA; National Concrete Pavement Technology Center at Iowa State University: Ames, IA, USA, 2021.
14. Oyeyi, A.G.; Ni, F.M.; Dolton, B.; Tighe, S. Investigating the Effect of Construction Activities on Lightweight Cellular Concrete Subbase Pavements. In Proceedings of the Transportation Association of Canada 2022 Conference and Exhibition-Changing Ways for our Changing Climate//Association des transports du Canada 2022 Congrès et Exposition-Approches Adaptées pour un Climat Changeant 2022, Edmonton, AB, Canada, 2–5 October 2022.
15. Oyeyi, A.G.; Ni, F.M.; Tighe, S. In-situ structural analysis of lightweight cellular concrete subbase flexible pavements. *Road Mater. Pavement Des.* **2023**, 1–17. [CrossRef]

16. Oyeyi, A.G.; Achebe, J.; Ni, F.M.; Tighe, S. Life cycle assessment of lightweight cellular concrete subbase pavements in Canada. *Int. J. Pavement Eng.* **2023**, *24*, 2168662. [CrossRef]
17. Oyeyi, A.G. Lightweight Cellular Concrete as Flexible Pavement Subbase Material: Field Performance and Sustainability Study Doctoral Dissertation, University of Waterloo, Waterloo, ON, Canada, 2022. Available online: http://hdl.handle.net/10012/18938 (accessed on 1 January 2023).

engineering proceedings

Proceeding Paper

NO_x Removal of Pervious Concrete Pavement Materials with TiO_2 †

Cheolwoo Park, Minsoo Cho, Dong-Jun Kim, Ui-Dae Park, Yong-Sik Kwon, Minkyu Ju and Seungwon Kim *

Department of Civil Engineering, Kangwon National University, 346 Joongang-ro, Samcheok-si 25913, Republic of Korea; tigerpark@kangwon.ac.kr (C.P.); olivars@naver.com (M.C.); kdj745@naver.com (D.-J.K.); euidea.park@halla.com (U.-D.P.); kwoun27@naver.com (Y.-S.K.); zenospy@nate.com (M.J.)
* Correspondence: seungwon.kim@kangwon.ac.kr; Tel.: +82-335706511
† Presented at the Second International Conference on Maintenance and Rehabilitation of Constructed Infrastructure Facilities, Honolulu, HI, USA, 16–19 August 2023.

Abstract: Various studies have been conducted on reducing NO_x emissions; titanium dioxide (TiO_2) is widely used to reduce NO_x in the air. This study proposes a method for exploiting the advantages of photocatalytic technology and water permeability to reduce NO_x emissions. The study comprises porosity, water permeability coefficient, compressive strength, and NO_x removal experiments. Based on the experiments, an optimum mix proportion is suggested. The results revealed that the NO_x removal effect is greater for variables with higher porosity. The removal is further enhanced by the use of siloxane, which hardens the surface of the TiO_2-incorporated cementitious materials in the mixture.

Keywords: NO_x removal; TiO_2; pervious concrete; pavement; permeability; porosity

1. Introduction

Over the past decade, the number of torrential rain events in Korea has increased approximately 1.5 times compared to the past, and the frequency of these events has increased six times [1]. Increases in phenomena such as heavy rainfall are closely related to climate change. Compared to the past, the annual average temperature reached its peak in 2016, confirming that global warming is continuing [2]. The main cause of these phenomena is air pollution caused by emissions of large amounts of substances such as fine dust, nitrogen oxides (NO_x), and carbon dioxide [3]. More than 50% of total NO_x emissions are caused by automobiles. Thus, reducing automobile usage can lead to reduced NO_x emissions; however, that is almost impossible to achieve because of increasing numbers of advance automobiles. Therefore, the development of structures such as a NOx absorbing infrastructure is necessary for reducing NO_x emissions. Various studies have been conducted worldwide to reduce NO_x emissions, and one of the most widely used materials in the construction field is titanium dioxide (TiO_2) [4]. TiO_2 is a photocatalytic material that can adsorb NO_x. To exploit this property, we sought to develop pervious concrete that can adsorb NO_x over a larger area.

Accordingly, as basic research for the development of photocatalytic pervious concrete to reduce NO_x, this study aims to evaluate the basic physical properties of pervious concrete, such as porosity, permeability coefficient, and compressive strength. Moreover, we intend to conduct experiments on the NO_x removal ratio to analyze the NO_x reduction effect according to two types of photocatalysts, namely, TiO_2 and spray-type.

Citation: Park, C.; Cho, M.; Kim, D.-J.; Park, U.-D.; Kwon, Y.-S.; Ju, M.; Kim, S. NO_x Removal of Pervious Concrete Pavement Materials with TiO_2 . *Eng. Proc.* **2023**, *36*, 8. https://doi.org/10.3390/engproc2023036008

Academic Editor: Hosin (David) Lee

Published: 30 June 2023

2. Material and Methods

2.1. Materials

Type I (equivalent to Type I) Ordinary Portland cement, coarse aggregates with a maximum size of 10 mm, TiO_2, isopropyl alcohol (IPA) solution with a specific gravity of 0.79 was used. Properties of aggregates and TiO_2 are listed in Tables 1 and 2, respectively.

Table 1. Physical properties of aggregates with a maximum size of 10 mm.

Aggregate Size	Specific Gravity	Absorption Rate	Fineness Modulus
10 mm	2.6	1.9%	5.9

Table 2. Properties of TiO_2 used.

Type	Specific Gravity	Content	Particle Size	Molecular Weight
anatase	4.0	98.5%	0.35–0.5 μm	77.9 g

2.2. Experimental Details

2.2.1. Mix Proportion

The mix proportions used in this study are listed in Table 3. TiO_2 was incorporated by substituting 5% and 10% of cement weight.

Table 3. Mix proportions used.

Variable	W	C	TiO_2	G	[1] S/P
OPC	108	360	-	1814	0.9
T5		342	18	1817	
T10		324	36	1820	

[1] S/P: superplasticizer added 0.25% binder weight to volume.

2.2.2. Porosity Measurement Method

The porosity of the pervious concrete was measured using the porosity test method suggested by the Concrete Research Committee of the Japan Concrete Institute (JCI). Equations (1) and (2) are used to measure the total and continuous porosities, respectively.

$$Total\ Porosity = \left(1 - \frac{W_2 - W_1}{V}\right) \times 100, \tag{1}$$

where, W_1 is the weight of the specimen in water; W_2 is the weight of the specimen in an absolutely dry state; and V is the specimen volume.

$$Continous\ Porosity = \left(1 - \frac{W_2 - W_1}{V}\right) \times 100, \tag{2}$$

2.2.3. Permeability Coefficient-Measurement Method

Because the permeability coefficient of pervious concrete is more than 105 times larger than that of ordinary concrete, it is impossible to measure the permeability coefficient using the permeability method for ordinary concrete. Therefore, we measured the permeability coefficient according to ASTM C 1701 "Standard Method for Infiltration Rate of In Place Pervious Concrete" using Equation (3).

$$I = \frac{K \times M}{D^2 \times t}, \tag{3}$$

where, I denotes the infiltrate in/h; M denotes mass of infiltrated water (lb); D denotes the inside diameter of the infiltration ring (in); t denotes the time required for the measured amount of water to infiltrate the concrete (s); and k = 126,870 (constant).

2.2.4. Compressive Strength Measurement Method

Using a Φ100 × 200 mm cylindrical mold, the compressive strength was measured after 28 days according to KS F 2405.

2.2.5. NO_x Removal Ratio Test Method

The NO_x removal ratio test was conducted according to KS L ISO 22197-1. The test was conducted by supplying a mixed gas with a certain concentration of nitric oxide and high-purity air at a certain ratio, while emitting ultraviolet light to activate TiO_2, which adsorbs NO_x when exposed to light.

2.2.6. TiO_2 vs. Spray-Type Photocatalyst Test Method

To compare the NO_x removal effect of pervious concrete using TiO_2, the specimens with dimensions of 100 mm × 100 mm × 400 mm, height, breadth and length, respectively, were used. A spray-type photocatalyst was sprayed on the pervious concrete to create a test specimen for comparison. A comparison of the NO_x removal ratios between the TiO_2-substituted and photocatalyst-sprayed specimens was conducted using the same process as the removal ratio test method described in Section 2.2.5.

3. Experimental Results and Analysis

3.1. Basic Property Evaluation

Table 4 presents the experimental results for evaluating the basic properties, including the porosity, permeability coefficient, and compressive strength. The continuous porosity of all the variables was approximately 7%, and the permeability coefficients were similar.

Table 4. Results of basic property evaluation experiments.

Variable	Compressive Strength (MPa)	Total Porosity (%)	Continuous Porosity (%)	Permeability Coefficient (cm/s)
OPC	17	9.30	7.60	1.25
T5	18.3	9.00	7.02	1.22
T10	18.0	8.90	6.89	1.20
T5-IPA	18.3	8.90	7.00	1.20
T10-IPA	18.4	8.80	6.84	1.19

3.2. NO_x Removal Ratio Test Results

Table 5 presents the NO_x removal ratio and total porosity results. For the OPC without photocatalysts, the NO_x removal ratio was 0.2%, indicating almost no removal effect. However, the removal ratios for the 5% and 10% TiO_2 substitution samples were 49% and 37%, respectively, indicating excellent NO_x removal. The removal ratio was expected to increase as the TiO_2 substitution rate increased; however, the experimental results were inconsistent with this expectation.

Table 5. NO_x removal ratio and total porosity.

Variable	Total Porosity (%)	Removal Ratio (%)
OPC	9.30	0.2
T5	9.00	49.0
T10	8.90	37.0
T5-IPA	8.90	35.1
T10-IPA	8.80	27.7

4. Conclusions

This study evaluated the basic properties of pervious concrete, including porosity, permeability coefficient, and compressive strength, and conducted experiments on the NO_x removal ratio by using TiO_2 and spray-type photocatalysts. The conclusions are as follows.

According to the NO_x removal ratio test results for pervious concrete with TiO_2, the 5% substitution variable showed a better removal ratio than the 10% substitution variable. Thus, we conclude that the best NOx removal ratio can be achieved with an appropriate mix proportion of TiO_2, rather than based on the amount of TiO_2. Therefore, further experimental research on the optimal mixing proportion of TiO_2 is necessary.

TiO_2 was applied to pervious concrete with a large air-exposure area to increase the NO_x removal ratio. However, pervious concrete formed a somewhat high porosity, which made it difficult to apply TiO_2 to concrete structures such as road pavements and parking lot decks. Therefore, additional evaluations of the durability characteristics should be conducted along with evaluations of the mechanical performance because of the high porosity.

Author Contributions: Conceptualization, C.P. and S.K.; methodology, C.P., M.C., D.-J.K., U.-D.P., Y.-S.K., M.J. and S.K.; formal analysis, C.P. and S.K.; writing—original draft preparation, C.P., M.C., D.-J.K., U.-D.P., Y.-S.K., M.J. and S.K.; writing—review and editing, C.P. and S.K. All authors have read and agreed to the published version of the manuscript.

Funding: This work was supported by the National Research Foundation of Korea (NRF) grant funded by the Korea government (MSIT) (No.2021R1A2C201409713).

Institutional Review Board Statement: Not applicable.

Informed Consent Statement: Not applicable.

Data Availability Statement: Not applicable.

Conflicts of Interest: The authors declare no conflict of interest.

References

1. Kim, Y.T.; Park, M.Y.; Kwon, H.H. Spatio-Temporal Summer Rainfall Pattern in 2020 from a Rainfall Frequency Perspective. *Korean Soc. Disaster Secur.* **2020**, *13*, 93–104.
2. *Korean Climate Change Assessment Report 2020*; Korea Meteorological Administration: Seoul, Korea, 2020; ISBN 978-89-954715-8-6.
3. Kwon, W.T. Current Status and Perspectives of Climate Change Sciences. *Asia-Pac. J. Atmos. Sci.* **2005**, *41*, 325–336.
4. Seo, J.H.; Yoon, H.N.; Kim, S.H.; Bae, S.J.; Jang, D.I.; Kil, T.G.; Park, S.M.; Lee, H.K. An Overview on the Physicochemical Properties and Photocatalytic Pollutant Removal Performances of TiO_2-incorporated Cementitious Composites. *Compos. Res.* **2020**, *33*, 68–75.

engineering proceedings

Proceeding Paper

Laboratory Evaluation of Recycled Asphalt Pavement and Engineered Polymer Binder for Small Airfield Repairs †

William D. Carruth *, Webster C. Floyd and Jeb S. Tingle

U.S. Army Engineer Research and Development Center, Vicksburg, MS 39180, USA; webster.c.floyd@usace.army.mil (W.C.F.); jeb.s.tingle@usace.army.mil (J.S.T.)

* Correspondence: william.d.carruth@usace.army.mil; Tel.: +1-601-634-2492

† Presented at the Second International Conference on Maintenance and Rehabilitation of Constructed Infrastructure Facilities, Honolulu, HI, USA, 16–19 August 2023.

Abstract: Conducting small asphalt repairs on airfields in remote locations can be technically and logistically challenging. An alternative to cold patch products is using an engineered polymer binder (EPB) mixed with recycled asphalt pavement (RAP). This paper presents the results of a laboratory evaluation of EPB with both wet and dry RAP. Compacted specimens were tested for rut resistance, indirect tensile strength (ITS), and Cantabro mass loss (ML). The results indicate that RAP mixed with EPB exhibited substantial rut resistance with ITS and ML similar to that of conventional dense-graded asphalt. Overall, the EPB and RAP blend appears to be a promising alternative for airfield repairs.

Keywords: airfields; repair; laboratory; RAP; asphalt; maintenance

Citation: Carruth, W.D.; Floyd, W.C.; Tingle, J.S. Laboratory Evaluation of Recycled Asphalt Pavement and Engineered Polymer Binder for Small Airfield Repairs. *Eng. Proc.* **2023**, *36*, 9. https://doi.org/10.3390/engproc2023036009

Academic Editor: Hosin (David) Lee

Published: 30 June 2023

1. Introduction

Conducting small-sized repairs on aged asphalt concrete on airfields in remote locations can be technically and logistically challenging, particularly if hot-mix asphalt (HMA) is utilized since it requires specialized equipment to produce. Cold patch products used on roadways are typically unable to support the high pass levels of aircraft with high wheel loads and tire pressures, particularly after only a few hours of curing [1,2]. Novel methods for producing small amounts of HMA quickly have been studied [3], including inductive HMA, which still requires specialized equipment, as well as a polymer-modified emulsion mixed with RAP that enables the repair product to be spray injected [4,5]. An alternative to HMA and cold patching is using an engineered polymer binder (EPB) mixed cold with recycled asphalt pavement (RAP) to produce a strong, durable repair with minimal equipment.

A recent research effort documented the use of 100% RAP millings stabilized with a specially designed polymer binder to produce a high-performing recycled asphalt mixture for roadway applications [6]. The proprietary EPB G5® is manufactured by Technisoil Inc. (Redding, CA, USA) and is liquid at room temperature, allowing it to be mixed with RAP millings, placed, and compacted, all without the addition of heat. Laboratory results indicated that the stabilized material showed high resistance to rutting and cracking [6]. A full-scale section consisting of RAP stabilized with G5® has also been constructed in Doha, Qatar, measuring 10 ft by 200 ft with traffic consisting of more than 1000 heavily loaded water trucks per day [7]. No construction related issues were reported, the material appeared to bond well, and visual surveys showed no pavement distress after 6 months of operations.

A mix design procedure was developed for determining the optimal Technisoil G5 content [6]. The procedure is a Marshall-based mix design procedure that begins with drying the RAP before compacting specimens at several binder contents for indirect tensile strength (ITS) and Marshall stability and flow testing. Once mixed, the specimens are compacted with 150 blows per side with a Marshall hammer before being cured and tested.

Anticipated usage in austere environments involves a scenario where only light compaction equipment may be available, so several changes to these procedures were implemented for the work described in this paper.

2. Materials and Methods

In order to evaluate RAP stabilized with EPB, RAP was first obtained from a local supplier and dried. The RAP source was a blend obtained from several asphalt milling operations in central Mississippi. The gradation of the RAP material was not obtained, but the gradation was likely similar to those provided in [8], who reported RAP gradations (without extraction) from several stockpiles in central Mississippi. The RAP material was screened over a 0.75 in sieve, and the oversize material was discarded before mixing with EPB. The EPB used was Technisoil G5®, the same product described previously.

EPB was added to the RAP and blended using a laboratory mixer capable of mixing materials in a 5 gal bucket before being batched and compacted in a gyratory compactor as shown in Figure 1. RAP was either used dry or moisture was added to achieve a RAP moisture content of 4% to represent "wet" RAP. In austere environments, RAP stockpiles could be exposed to moisture without a means of drying the material, so the effect of RAP moisture was important to this research effort. EPB binder contents of 3, 5, and 7% by dry weight of RAP were mixed with both dry and wet RAP. Specimens were compacted with 30 gyrations from a Superpave gyratory compactor. This 30 gyration compactive effort is notably less aggressive than that used by [6] and was selected to simulate light compaction equipment, as mentioned previously.

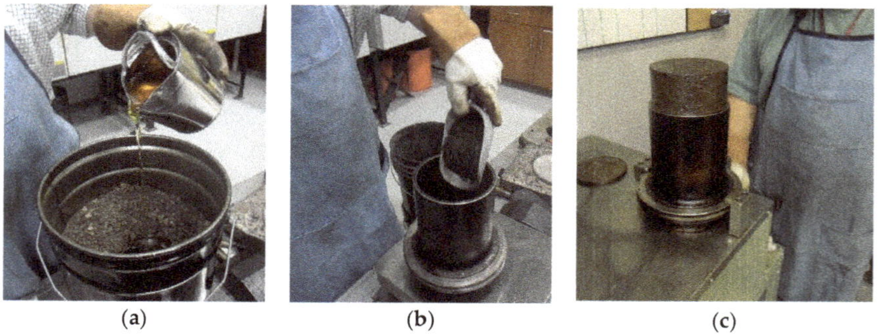

(a) (b) (c)

Figure 1. Batching, mixing, and compaction of RAP with EPB. (**a**) Addition of binder; (**b**) batching; (**c**) compacted specimen.

The weight of the mixed materials used to create each specimen was varied during preliminary testing so that compacted specimens would meet AASHTO T 340 dimensional criteria (approximately 75 by 150 mm) for evaluation via the asphalt pavement analyzer (APA). After compaction, specimens were cured in the mold in a 60 °C oven for 2 h, then extracted and placed back in the oven for an additional 22 h before being cured at room temperature for 24 h. Bulk density was measured via ASTM D6752, and average results are shown in Table 1.

After curing, all specimens were subjected to APA testing with an increased hose pressure (250 psi). However, after 8000 cycles, no measurable rutting was observed. Following APA testing, half of the specimens were subjected to Cantabro mass loss testing via ASTM D7064, and the other half were tested for indirect tensile strength (ITS) via ASTM D6931.

Table 1. Average G_{mb} results.

% Binder	% RAP Moisture	Avg G_{mb}
3	0	1.979
	4	2.079
5	0	2.038
	4	2.048
7	0	1.985
	4	1.986

3. Results and Discussion

As mentioned in the previous section, all specimens exhibited no measurable rutting after 8000 cycles of APA testing. Cantabro mass loss (ML) results are shown in Figure 2. For dry RAP, ML clearly decreased with increasing binder content. However, for wet RAP, ML increased when binder content increased from 3 to 5 percent, but then decreased at 7 percent. Historically, the Cantabro test was initially utilized primarily for open-graded friction coarse HMA mixtures, for which a maximum ML of 20% was suggested [9]. The results in Figure 2 were all well below 20% and were mostly similar to the dense-graded asphalt (DGA) Cantabro results [10–12].

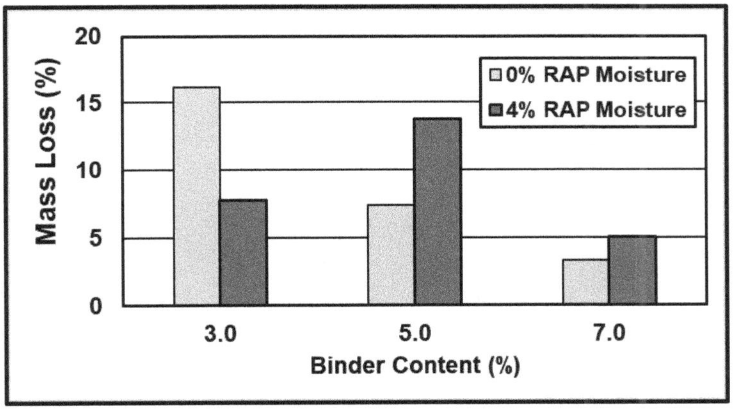

Figure 2. Average Cantabro mass loss results.

The ITS results are presented in Figure 3. The average ITS increased slightly with binder content for dry RAP, but ITS was similar for wet RAP at 3 and 5 percent binder contents. The ITS also decreased considerably at 7 percent binder content. Overall, with the exception of wet RAP at 7 percent binder content, the ITS results ranged from 149 to 226 psi. Other research groups reported a range of ITS for DGA at room temperature of approximately 150 to 200 psi, which is similar to the data presented in Figure 3 [13,14]. The ITS of a cold patch product averaged approximately 10 psi, which is much lower than the RAP–EPB mixture [1].

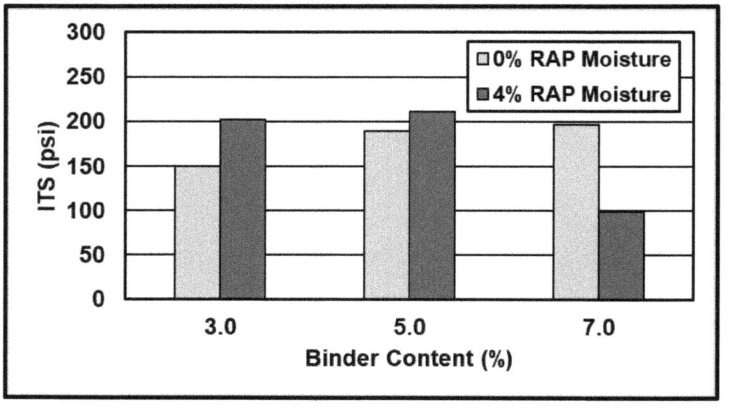

Figure 3. Average ITS results.

4. Conclusions and Recommendations

Overall, the results indicate that EPB mixed with RAP could be a suitable material for conducting small airfield repairs in locations where obtaining quality HMA from a plant is difficult. APA results indicate that once cured, this material should be able to resist rutting even when trafficked with high tire pressure aircraft. This initial investigation indicates that an EPB binder content of approximately 5% may be suitable for most RAP millings less than 0.75 in., with increased repair durability when dry RAP millings are utilized. Additional work is recommended to further investigate the effectiveness of the binder at various RAP moisture contents, as well as conducting a full-scale evaluation with simulated aircraft loads.

Author Contributions: Conceptualization, W.D.C. and W.C.F.; methodology, W.D.C. and W.C.F.; writing—original draft preparation, W.D.C.; writing—review and editing, W.D.C., W.C.F. and J.S.T.; supervision, J.S.T. All authors have read and agreed to the published version of the manuscript.

Funding: This research was funded by the U.S. Office of the Undersecretary of Defense for Research and Engineering.

Data Availability Statement: All data are contained within this article. Additional details can be obtained by contacting any of the authors.

Acknowledgments: The authors wish to acknowledge the U.S. Office of the Undersecretary of Defense for Research and Engineering for providing funding for this work, Tim McCaffrey for conducting laboratory testing, and Elie Hajj for providing technical advisement.

Conflicts of Interest: The authors declare no conflict of interest.

References

1. Rushing, J.F.; Cox, B.C.; Floyd, W.C. Rutting Performance of Cold-Applied Asphalt Repair Materials for Airfield Pavements. In *Airfield and Highway Pavements*; American Society of Civil Engineers: Philadelphia, PA, USA, 2017; pp. 185–195.
2. Wang, T.; Dra, Y.A.S.S.; Cai, X.; Cheng, Z.; Zhang, D.; Lin, Y.; Yu, H. Advanced cold patching materials (CPMs) for asphalt pavement pothole rehabilitation: State of the art. *J. Clean. Prod.* **2022**, *366*, 133001. [CrossRef]
3. Han, S.H.; Lee, S.Y.; Rhee, S.K.; Kwon, B.J. Development of a Spray-Injection Patching System and a Field Performance Evaluation of 100% RAP Asphalt Mixtures using a Rapid-Setting Polymer-Modified Asphalt Emulsion. *Int. J. Highw. Eng.* **2018**, *20*, 77–85. [CrossRef]
4. Mejias-Santiago, M.; Del Valle-Roldan, F.; Priddy, L.P. *Certification Tests on Cold Patch Asphalt Repair Materials for Use in Airfield Pavements*; ERDC/GSL TR-10-14; U.S. Army Engineer Research and Development Center: Vicksburg, MS, USA, 2010. [CrossRef]
5. Cox, B.C.; Floyd, W.C.; Rushing, J.F.; Rutland, C.A. Rapid Inductive Heating of Asphalt Concrete to Hot Mix Temperatures for All-Season Pothole Patching: Feasibility Study. *Transp. Res. Rec. J. Transp. Res. Board* **2019**, *2673*, 477–491. [CrossRef]
6. Hajj, E.Y.; Piratheepan, M.; Sebaaly, P.E. Performance Evaluation of a Polymer Binder Stabilized Aggregate Mixture: A Pilot Study. In *Airfield and Highway Pavements*; American Society of Civil Engineers: Philadelphia, PA, USA, 2017; pp. 406–415. [CrossRef]

7. Hajj, E.Y.; Piratheepan, M.; Sebaaly, P.E. Performance evaluation of a 100% recycled asphalt pavement mixture using a polymer binder: A pilot study. In *Bearing Capacity of Roads, Railways and Airfields*; CRC Press: London, UK, 2017; pp. 1217–1224.
8. Howard, I.L.; Cooley, L.A., Jr.; Doyle, J.D. *Laboratory Testing and Economic Analysis of High RAP Warm Mixed Asphalt*; Report FHWA/MS-DOT-RD-08-200; Mississippi Department of Transportation: Jackson, MS, USA, 2008; p. 104.
9. Watson, D.E.; Cooley Jr, L.A.; Moore, K.A.; Williams, K. *Laboratory Performance Testing of Open-Graded Friction Course Mixtures*; Transportation Research Record No.1891; Transportation Re-Search Board: Washington, DC, USA, 2004; pp. 40–47.
10. Mejías-Santiago, M.; Doyle, J.D.; Howard, I.L.; Brown, E.R. *Evaluation of Warm-Mix Asphalt Technologies for Use on Airfield Pavements*; ERDC/GSL TR-12-3; U.S. Army Engineer Research and Development Center: Vicksburg, MS, USA, 2013.
11. Doyle, J.D.; Howard, I.L. Characterization of Dense-Graded Asphalt with the Cantabro Test. *J. Test. Eval.* **2014**, *44*, 77–88. [CrossRef]
12. Cox, B.C.; Smith, B.T.; Howard, I.L.; James, R.S. State of Knowledge for Cantabro Testing of Dense Graded Asphalt. *J. Mater. Civ. Eng.* **2017**, *29*, 04017174. [CrossRef]
13. Doyle, J.D.; Robinson, W.J.; Stache, J.M.; Cox, B.C. Consideration of a Tack Coat Bond Strength Test to Minimize the Potential for Slippage Failures on Flexible Airfield Pavements. *Transp. Res. Rec. J. Transp. Res. Board* **2022**, *2676*, 374–388. [CrossRef]
14. Cox, B.C. *Cold In-Place Recycling Characterization Framework for Single or Multiple Component Binder Systems*; Mississippi State University: Starkville, MS, USA, 2015.

engineering proceedings

Proceeding Paper

Performance Life Using Mechanistic–Empirical Analysis of Asphalt Mixtures in Arid Climatic Conditions—Case of Kuwait †

Taha Ahmed [1,*], Aditya Singh [2,*], Elie Hajj [2] and Ahmad Saad [1]

1 Civil Engineering Department, College of Engineering, Australian University, Kuwait City 13015, Kuwait; a.said@au.edu.kw
2 Civil and Environmental Engineering Department, College of Engineering, University of Nevada, Reno, NV 89557, USA; elieh@unr.edu
* Correspondence: t.ahmed@au.edu.kw (T.A.); adityasingh@nevada.unr.edu (A.S.)
† Presented at the Second International Conference on Maintenance and Rehabilitation of Constructed Infrastructure Facilities, Honolulu, HI, USA, 16–19 August 2023.

Abstract: The extreme arid climatic conditions and poor asphalt mix design characteristics have further accelerated the rate of deterioration of in-service asphalt pavements in Kuwait. Pavement distresses, such as raveling, rutting, and fatigue cracks, worsen in severe climatic and loading conditions, such as high temperatures, elevated humidity levels, and high traffic loads on the pavement surfaces. In this study, a life performance evaluation using mechanistic–empirical analysis of a new modified Superpave mix design was undertaken. The performance life of the modified Superpave asphalt mixture was evaluated, and the results showed that the new modifications to the mix improved the rutting and fatigue cracking resistances of the asphalt mixture for the unconditioned state. However, fatigue cracking resistance under the moisture-conditioned state still needs further improvement for the newly modified Superpave asphalt mixture.

Keywords: arid climate; asphalt mixture; pavement distresses; Superpave; mechanistic empirical

Citation: Ahmed, T.; Singh, A.; Hajj, E.; Saad, A. Performance Life Using Mechanistic–Empirical Analysis of Asphalt Mixtures in Arid Climatic Conditions—Case of Kuwait. *Eng. Proc.* **2023**, *36*, 10. https://doi.org/10.3390/engproc2023036010

Academic Editor: Hosin (David) Lee

Published: 30 June 2023

1. Introduction

The proper design and maintenance of asphalt pavement will extend its service life and help it to maintain a satisfactory level of service. However, due to in-service conditions, asphalt pavements can be damaged by certain types of distresses. Pavement distresses, such as raveling, rutting, and fatigue cracking, worsen in severe climatic and loading conditions, such as high temperatures, elevated humidity levels, and high traffic loads on the pavement. The extreme arid climatic conditions of Kuwait and poor asphalt mix design characteristics have further accelerated the rate of deterioration of in-service asphalt pavements [1]. To combat this issue, Kuwait's Ministry of Public Works (MPW) adopted the Superpave asphalt mix design from the Qatari Construction Specifications, which considers both climate and traffic, unlike the Marshall mix design approach. However, the long-term performance of this newly adopted Superpave asphalt mix design must be evaluated to determine its long-term resistance to common pavement distresses [2].

Mechanistic–empirical (ME) analysis is based on the theory of mechanics and predicts the responses of the pavement to the application of loads. The empirically derived models employ these pavement responses to forecast distress in terms of cracking and rutting. ME analysis can forecast asphalt pavement's long-term performance over its design period. As a result, the objective of this study was to use a ME analysis to assess the resistance of the Superpave asphalt mixture against prevalent pavement distresses, including rutting and fatigue cracking. This ME study assisted in comprehending how the modified material properties impact pavement performance over time and if further modifications are required to the newly approved mix design procedure.

2. Methods and Materials

In this research study, the local raw materials were acquired from Kuwait to develop Superpave asphalt mix design. The raw material types and the desired number of samples were selected according to the General Specifications for Kuwaiti Roads and Highways (MPW) [3] and the Qatar Construction Specifications [4]. Only one aggregate source and one asphalt binder grade were used for this research study. The selected aggregate was Gabroo aggregate, and the evaluated properties met the ASTM standard requirements. Figure 1 depicts the final aggregate gradation obtained for the Superpave mix design. Similarly, the asphalt binder properties were examined, and the final performance grade obtained following AASHTO M320 was PG 76-22 [2].

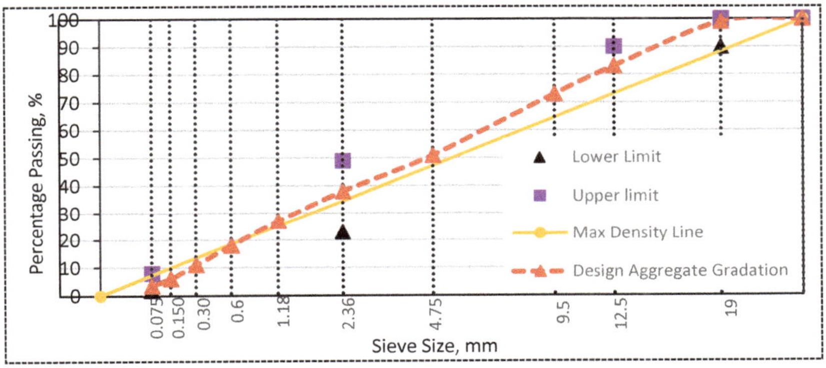

Figure 1. Verified Aggregate Gradation.

After evaluating the raw materials, the Superpave mix was performed as per AASHTO M 323, where the optimum binder content selected was 4.1% (TWM) at 4% air voids meeting the required volumetrics shown in Table 1.

Table 1. Superpave Mix Design Volumetric Requirements.

Property	Average Result	QCS Specifications
Design air void, %	4.0	4.0
Pb, %	4.1	-
VMA, %	13.21	13
VFA, %	70	50–75
DP	1.12	0.75–1.20
G_{mm} @ N_{ini}	87	Max. 88
G_{mm}	2.648	-
Air voids at N_{max}, %	2.7	Min. 2.0

After finalizing the mix design, performance testing such as dynamic modulus, cyclic fatigue, and stress sweep rutting tests were conducted for the Superpave asphalt mixture at the unconditioned state (UC) and moisture-conditioned at three freeze–thaw cycles (3-C). The engineering properties measured from these tests were used as the input parameters in the mechanistic–empirical analysis to predict the long-term pavement performance over 20 years.

Mechanistic–Empirical Analysis

The FlexPave™ software was used to perform mechanistic–empirical analysis of the Superpave asphalt mixture at the UC state and after the 3-C state. FlexPave™ is a modified linear viscoelastic pavement design for critical distresses (LVECD) based on the three-dimensional (3D) viscoelastic finite element technique (FEM) [5].

3. Results and Discussion

The required input parameters for the ME analysis are climate data, traffic data, material properties, pavement structure, and performance criteria. Local climate information for Kuwait was taken from the MEERA-2 global climatic database. Currently, the MPW is updating its traffic data collection system, and the most recent traffic data is not available. Therefore, the traffic parameters used in the analysis were the default values, as shown in Table 2. In Kuwait, there are currently no performance standards for rutting and fatigue cracking. Hence, the performance criteria used by researchers in neighboring countries with similar arid climate regions were considered.

Table 2. Input Parameters for Mechanistic–Empirical Analysis.

Single Axle Load	80 kN
Tire pressure	120 psi
Traffic levels	30 MESALS
Speed	97 km/h
Design Life	20 years
Cracking Limit	20%
Rutting Limit	0.75 inch (1.9 cm)

In this study, the thickness of the pavement was determined using the three methods listed in Table 3. Method A employs a similar pavement structure built in Kuwait before 2012, but Method B employs the AASHTO Guide to the Design of Pavement Structures, 1993 (referred to as AASHTO 93). Method C employs a pavement structure similar to Method B, but the asphalt (AC) layer thickness is increased until the rutting and cracking performance criteria are met. The AC layer was assumed to be a single layer, and the analysis was conducted for the Superpave asphalt mixture at the UC state and after the 3-C, respectively. Figure 2 shows the pavement structure used in the ME analysis for each method.

Table 3. Pavement Design Thickness for All Three Methods.

Design Method	AC Layer Thickness
Typical sections (A)	6.3 in. (16 cm)
AASHTO 93 (B)	7.9 in. (20 cm)
FlexPave^{TM} (C)	8.7 in. (22 cm)

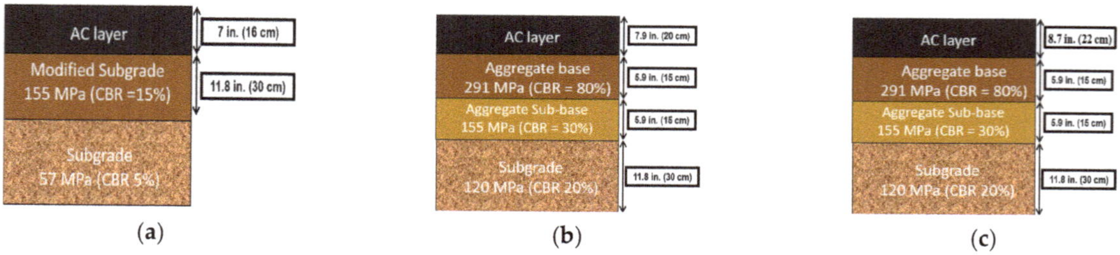

Figure 2. Pavement structure used in ME analysis. (**a**) Typical sections constructed before 2012 in Kuwait, (**b**) AASHTO 93 pavement structure, and (**c**) FlexPave^{TM} pavement structure.

After inputting the above-mentioned parameters, the analysis was run using the FlexPave^{TM} software. An example of the results obtained for the cracking and rutting for design method A is shown in Figure 3. Similarly, the analysis was performed for the other two design methods, and the final obtained results are discussed in Table 4.

As illustrated in Figure 3 and Table 4, pavement sections constructed before 2012 in Kuwait were inadequately designed for 30 MESALS of traffic. The pavement constructed using Method B had a longer fatigue and rutting life, but it still required improvement to withstand the 20-year design traffic of 30 MESALS. Method C increased the asphalt pavement thickness until the pavement met the performance parameters for the Superpave asphalt mixture at the UC state. However, the pavement could only endure 5.5 years after the 3-C state indicating a durability issue in the mix design.

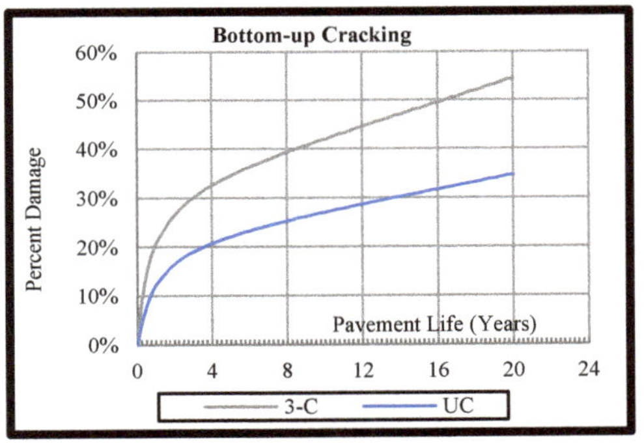

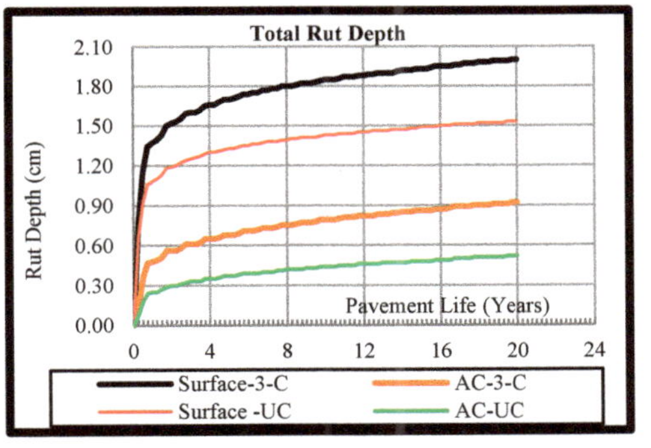

Figure 3. Example of FlexPave™ results for design method A.

Table 4. Pavement Design Life of the Superpave Asphalt Mixture for 30 MESALS Traffic.

Design Method	Fatigue Life (UC)	Rutting Life (UC)	Fatigue Life (3-C)	Rutting Life (3-C)
Typical sections (A)	3.75 years	20 years	1 year	12.75 years
AASHTO 93 (B)	18.25 years	20 years	4.25 years	20 years
FlexPave™ (C)	20 years	20 years	5.5 years	20 years

4. Conclusions and Recommendations

This study focused on the process and importance of conducting mechanistic–empirical analysis for the newly adopted Superpave asphalt mixture. Results showed that the newly adopted Superpave asphalt mixture enhanced the rutting and fatigue cracking resistances at the unconditioned state. However, fatigue cracking resistance under the moisture-conditioned state still needs further improvement for the newly modified Superpave asphalt mixture.

Author Contributions: T.A. and A.S. (Aditya Singh): Conceptualization, data curation, supervision, writing—original draft, software, and data analysis. E.H. and A.S. (Ahmad Saad): Conceptualization, investigation, methodology, data curation, writing—review and editing. All authors have read and agreed to the published version of the manuscript.

Funding: This research study was jointly funded by the Australian University-Kuwait and Kuwait Foundation for the Advancement of Sciences under project code CR20-15EV-01.

Institutional Review Board Statement: Not applicable.

Informed Consent Statement: Not applicable.

Data Availability Statement: Data are available from the corresponding authors upon request.

Acknowledgments: The authors would like to acknowledge the great help and support provided by the Australian University-Kuwait.

Conflicts of Interest: The authors declare no conflict of interest.

References

1. Ahmed, T.; Hajj, E.; Warrag, A.; Piratheepan, M. Postmortem evaluation of accelerated rate of raveling of in-service asphalt pavements in arid climatic conditions-case of Kuwait. *Case Stud. Constr. Mater.* **2021**, *14*, e00533. [CrossRef]
2. Singh, A. Comprehensive Evaluation of Asphalt Mixtures in Arid Climatic Conditions. Master's Thesis, University of Nevada, Reno, NV, USA, 2022.
3. Hannah, A.; Prime, J.; Hassan, K.; Attia, M. *Asphalt Wearing Course Pavement Investigation: Investigation into Pavement Raveling*; Report No. RPN3225; Ministry of Public Works, Roads Administration: Kuwait City, Kuwait, 2015.
4. Public Works Authority (Ashghal). *Qatar Construction Specifications*; Standard QCS.: Doha, Qatar, 2014.
5. Pivetta, F. Development of Software Programs for Performance Related Specifications of Asphalt Concrete Pavements. Master's Thesis, North Carolina State University, Raleigh, NC, USA, 2021.

engineering proceedings

MDPI

Proceeding Paper

Optimizing Pothole Detection in Pavements: A Comparative Analysis of Deep Learning Models †

Tiago Tamagusko and Adelino Ferreira *

Research Centre for Territory, Transports and Environment (CITTA), Department of Civil Engineering, University of Coimbra, 3030-788 Coimbra, Portugal; tamagusko@mail.com

* Correspondence: adelino@dec.uc.pt

† Presented at the Second International Conference on Maintenance and Rehabilitation of Constructed Infrastructure Facilities, Honolulu, HI, USA, 16–19 August 2023.

Abstract: Advancements in computer vision applications have led to improved object detection (OD) in terms of accuracy and processing time, enabling real-time solutions across various fields. In pavement engineering, detecting visual defects such as potholes, cracking, and rutting is of particular interest. This study aims to evaluate YOLO models on a dataset of 665 road pavement images labeled with potholes for OD. Pre-trained deep learning models were customized for pothole detection using transfer learning techniques. The assessed models include You Only Look Once (YOLO) versions 3, 4, and 5. It was found that YOLOv4 achieves the highest mean average precision (mAP), while its shortened version, YOLOv4-tiny, offers the best-reduced inference time, making it ideal for mobile applications. Furthermore, the YOLOv5s model demonstrates potential, attaining good results and standing out for its ease of implementation and scalability.

Keywords: computer vision; object detection; pothole; road pavements; YOLO; deep learning

1. Introduction

This paper investigates state-of-the-art computer vision (CV) techniques in detecting pavement potholes, comparing the performance of various deep learning (DL) models. Object detection (OD) methods, which identify and locate objects in images or videos, have evolved from traditional image processing techniques, such as the Viola-Jones Detector [1] and Histogram of Oriented Gradients, to DL implementations [2,3]. These DL implementations have demonstrated better performance, particularly in complex scenarios, due to their supervised learning approach and the availability data. Hence, the community effort to create massive datasets such as MS COCO [4], PASCAL [5], and IMAGENET [6], has helped the field to evolve. Still, computation power, mainly with GPUs, rapidly increases year by year [7].

One-stage and two-stage detectors are the main categories of DL applications for OD. Two-stage detectors typically exhibit higher accuracy but are slower, while one-stage detectors are faster and more suitable for real-time applications. This article focuses on one of the most famous families of one-stage detectors: You Only Look Once (YOLO) [8]. The YOLO algorithm is a fast and accurate object detection model. It divides input images into grids for simultaneous object detection and classification. Despite lower average precision than some competitors, YOLO's detection makes it ideal for low-latency applications.

This article is organized into four sections, with a brief review of the background, a description of the data and methods used, a presentation of the results, and a conclusion with future recommendations. By comparing the performance of YOLO models, this research aims to determine the most effective method to detect potholes in road pavements, contributing to a safer and well-maintained infrastructure.

Citation: Tamagusko, T.; Ferreira, A. Optimizing Pothole Detection in Pavements: A Comparative Analysis of Deep Learning Models. *Eng. Proc.* **2023**, *36*, 11. https://doi.org/10.3390/engproc2023036011

Academic Editor: Hosin (David) Lee

Published: 30 June 2023

2. Data and Methods

This study aims to identify the best model for pothole detection using YOLO-based implementations. Six deep learning models were compared, including YOLOv3-tiny [9], YOLOv3 [9], YOLOv4-tiny [10], YOLOv4 [10], YOLOv5s [11], and YOLOv5x [11]. All models were pre-trained on the Common Objects in Context (COCO) dataset, and transfer learning was used, so the developed models use the base of previous models adapted to pothole detection.

A dataset created by Rahman Atikur [12] containing 665 road pavement images with labeled potholes was used, with a 70/20/10 split for training, validation, and testing. An example of labeled images can be seen in Figure 1.

Figure 1. Example image of potholes with labels.

This experiment used a computer specifically assembled to perform high-demand computing tasks with the following specifications:

- CPU: AMD Ryzen 9 5950x
- Memory: 32GB DDR4 3000MHz RAM
- GPU: NVIDIA GeForce RTX 3090
- NVIDIA Driver: 510.68.02
- CUDA: 11.6
- OS: Arch Linux, Kernel 5.17.5-arch1-1

Models run on Python version 3.8.13 and Pytorch 1.10.2. Furthermore, instructions for installation are in the repositories of YOLOv3, YOLOv4, and YOLOv5. In addition, the customized models were trained using Pytorch and Darknet frameworks, specifically YOLOv3 and YOLOv5 for Pytorch and YOLOv4 for Darknet. The base models used in this study are as follows: YOLOv3-tiny (Pytorch, yolov3-tiny.pt), YOLOv3 (Pytorch, yolov3.pt), YOLOv4-tiny (Darknet, yolov4-tiny.conv.29), YOLOv4 (Darknet, yolov4.conv.137), YOLOv5s (Pytorch, yolov5s.pt), and YOLOv5x (Pytorch, yolov5x.pt).

Likewise, the same training hyperparameters were used for all models, namely:

- Batch size: 16
- Epochs: 3000
- Image size: 416
- Patience: 100 (for Pytorch base models)

No data augmentation technique was used, but the default parameters for each model were maintained.

3. Results

The models were compared based on their mean average precision (mAP) and time to infer an image (Figure 2). The goal is to find a highly precise model for detecting and locating potholes while maintaining a short inference time. YOLOv4, YOLOv4-tiny, and YOLOv5s models stood out as the best options.

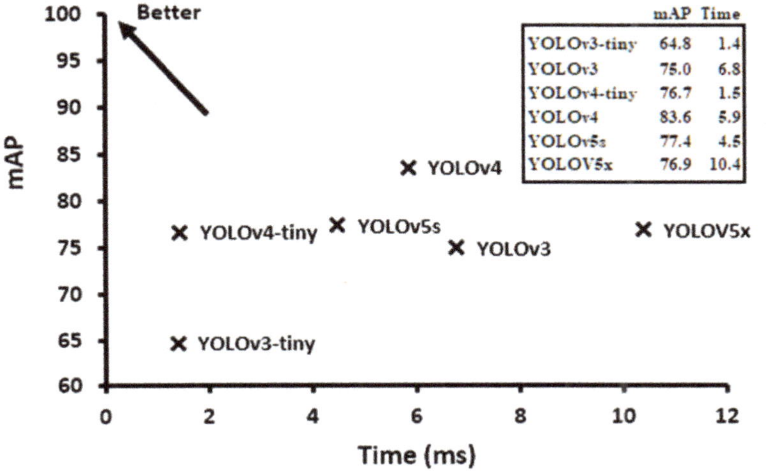

Figure 2. Mean average precision vs. time to infer one image.

YOLOv4 demonstrated greater confidence in predicting small potholes compared to YOLOv5-based models. The detected objects were similar in YOLOv4 versions, and their confidence levels were better than those of YOLOv5 models.

The detailed results are shown in Table 1, with YOLOv4 and YOLOv4-tiny presenting excellent results in terms of mAP, model size, and detection time. In addition, it is believed that mAP could be improved by improving label quality, increasing training data, tuning hyperparameters, and using data augmentation. More precise labeling, such as polygonal segmentation, could also help improve the results.

Table 1. Comparison of YOLO models.

Model	mAP @0.50	Size (MB)	mAP @0.50	Training (s)	Inference (ms)
YOLOv3-tiny	64.8%	16.6	64.8%	779.4	1.4
YOLOv3	75.0%	117.7	75.0%	1002.3	6.8
YOLOv4-tiny	76.7%	22.4	76.7%	249.2	1.5
YOLOv4	83.2%	244.2	83.2%	1254.1	5.9
YOLOv5s	77.4%	13.6	77.4%	512.6	4.5
YOLOv5x	76.9%	165.0	76.9%	1430.7	10.4

Lastly, limitations of this study include the small dataset of 665 images, limited quality of the labels, lack of hyperparameter tuning, and no direct testing of the algorithm's performance in real time. Furthermore, only YOLO implementations were evaluated.

4. Conclusions

The best result obtained for pothole detection in the dataset used was with YOLOv4, reaching a mAP of 83.2%. Still, the implementation with YOLOv4-tiny presents good potential for mobile applications or devices with less computational power. However, training a custom model with YOLOv4 and its usability turns out to be more complex with the use of the Darknet framework. This becomes an obstacle to putting the model into production and the solution's scalability. On the other hand, version 5 could have better results with some tuning. However, its Pytorch-based implementation is a plus. Consequently, it is recommended to keep the YOLOv4, YOLOv4-tiny, and YOLOv5s models in mind, depending on the application.

As a future research direction, the goal is to expand these custom models to detect more classes, such as alligator cracking, block cracking, longitudinal or transverse cracking, slippage cracks, and rutting. Additionally, more attention will be given to the data, which will be expanded and revised. The ultimate goal is to develop a real-time model capable of detecting various visual defects in road pavements, improving the management of road assets, reducing costs, and improving road safety.

Author Contributions: Conceptualization, T.T. and A.F.; methodology, T.T.; software, T.T.; validation, T.T., and A.F.; formal analysis, T.T.; investigation, T.T.; resources, T.T.; data curation, T.T.; writing—original draft preparation, T.T.; writing—review and editing, T.T. and A.F.; visualization, T.T.; supervision, A.F.; project administration, A.F.; funding acquisition, A.F. All authors have read and agreed to the published version of the manuscript.

Funding: The author Tiago Tamagusko is grateful to the Portuguese Foundation for Science and Technology for the PhD Grant 2020.09565.BD. This research was funded by the Research Center for Territory, Transports and Environment—CITTA (UIDP/04427/2020).

Institutional Review Board Statement: Not applicable.

Informed Consent Statement: Not applicable.

Data Availability Statement: Data and models are available at github.com/tamagusko/pothole-detection.

Acknowledgments: The authors would like to thank the support of the Research Centre for Territory, Transports, and Environment CITTA (UIDP/04427/2020) and also ACIV for the presentation of this paper in the conference.

Conflicts of Interest: The authors declare no conflict of interest.

References

1. Viola, P.; Jones, M. Robust Real-Time Face Detection. *Int. J. Comput. Vis.* **2001**, *57*, 137–154. [CrossRef]
2. Mittal, U.; Srivastava, S.; Chawla, P. Review of Different Techniques for Object Detection Using Deep Learning. In Proceedings of the Third International Conference on Advanced Informatics for Computing Research, Shimla, India, 15–16 June 2019; Association for Computing Machinery: New York, NY, USA, 2019.
3. Xiao, Y.; Tian, Z.; Yu, J.; Zhang, Y.; Liu, S.; Du, S.; Lan, X. A review of object detection based on deep learning. *Multimed. Tools Appl.* **2020**, *79*, 23729–23791. [CrossRef]
4. Lin, T.; Maire, M.; Belongie, S.; Hays, J.; Perona, P.; Ramanan, D.; Doll, P. Microsoft COCO: Common Objects in Context. *Eur. Conf. Comput. Vis.* **2014**, *8693*, 740–755.
5. Everingham, M.; Van Gool, L.; Williams, C.K.I.; Winn, J.; Zisserman, A. The pascal visual object classes (VOC) challenge. *Int. J. Comput. Vis.* **2010**, *88*, 303–338. [CrossRef]
6. Fei-Fei, L.; Deng, J.; Li, K. ImageNet: Constructing a large-scale image database. *J. Vis.* **2010**, *9*, 1037. [CrossRef]
7. Pal, S.K.; Pramanik, A.; Maiti, J.; Mitra, P. Deep learning in multi-object detection and tracking: State of the art. *Appl. Intell.* **2021**, *51*, 6400–6429. [CrossRef]
8. Redmon, J.; Divvala, S.; Girshick, R.; Farhadi, A. You Only Look Once: Unified, Real-Time Object Detection. In Proceedings of the 2016 IEEE Conference on Computer Vision and Pattern Recognition (CVPR), Las Vegas, NV, USA, 27–30 June 2016; Volume 1, pp. 779–788. [CrossRef]
9. Redmon, J.; Farhadi, A. YOLOv3: An Incremental Improvement. *arXiv* **2018**, arXiv:1804.02767. [CrossRef]
10. Bochkovskiy, A.; Wang, C.-Y.; Liao, H.-Y.M. YOLOv4: Optimal Speed and Accuracy of Object Detection. *arXiv* **2020**, arXiv:2004.10934. [CrossRef]
11. Jocher, G. "YOLOv5", Ultralytics. Available online: https://github.com/ultralytics/yolov5 (accessed on 23 July 2020).
12. Atikur, R. Annotated Potholes Image Dataset. Available online: https://www.kaggle.com/chitholian/annotated-potholes-dataset (accessed on 15 April 2022).

engineering proceedings

MDPI

Proceeding Paper

Life Cycle Assessment of a Sustainable and Innovative Solution for Unpaved Rural Roads †

Leonardo Urbano [1], Lucia Tsantilis [1,*], Pier Paolo Riviera [1], Orazio Baglieri [1] and Ezio Santagata [1,2]

1 Department of Environment, Land and Infrastructure Engineering, Politecnico di Torino, Corso Duca degli Abruzzi 24, 10129 Torino, Italy; leonardo.urbano@polito.it (L.U.); pierpaolo.riviera@polito.it (P.P.R.); orazio.baglieri@polito.it (O.B.); ezio.santagata@polito.it or ezio.santagata@qu.edu.qa (E.S.)

2 Department of Civil and Architectural Engineering, Qatar University, Doha 2713, Qatar

* Correspondence: lucia.tsantilis@polito.it

† Presented at the Second International Conference on Maintenance and Rehabilitation of Constructed Infrastructure Facilities, Honolulu, HI, USA, 16–19 August 2023.

Abstract: The use of recycled aggregates, including waste materials and by-products, has attracted increasing interest in the last decades as a sustainable and cost-effective solution for the construction and maintenance of road pavements, due to the reduction of excavation operations and depletion of natural resources. Life cycle assessment (LCA) represents a valuable methodology for the evaluation of the environmental sustainability of technologies involving the use of such materials. This paper deals with the LCA of an innovative emulsion-based cold recycled mixture specifically conceived to be used as a sustainable solution for the surface finishing of unpaved rural roads. Two scenarios entailing the employment of recycled or virgin materials were analyzed with the assessment of global warming potential (GWP), energy requirement, and water consumption. Results obtained confirmed that the scenario entailing the use of recycled materials represents the most sustainable and environmentally friendly solution.

Keywords: life cycle assessment; sustainability; reclaimed asphalt; mineral sludge; unpaved roads; rural roads

1. Introduction

The growing concerns regarding environmental issues and climate change have led the pavement construction industry to move towards increasingly sustainable solutions to preserve natural resources [1]. In this context, the recycling of by-products in paving mixtures represents a valuable option since it allows for the replacement of large volumes of virgin aggregates, thus contributing to reducing the depletion of raw materials [2–4]. The recycling of by-products appears to be especially suitable for rural roads, for which lower performance is generally required compared to high-volume highways and arterials [5].

Reclaimed asphalt pavement (RAP), obtained from the milling of existing asphalt pavements, and mineral sludge (MS), derived from industrial washing of natural aggregates in crushing plants, may be effectively used as alternative aggregates in the construction of pavement layers due to their large availability and problems faced in their stockpiling and/or disposal [6–8]. In fact, a previous research study conducted by the authors showed that RAP and MS can be successfully employed in large quantities in the production of emulsion-based cold recycled mixtures (CRM) for the surface finishing of unpaved rural roads [9].

In this paper, the environmental benefits related to the use of CRM were assessed using the life cycle assessment (LCA) methodology. To this purpose, the global warming potential (GWP), energy requirement, and water consumption related to the production, construction, and maintenance stages of the innovative mixture were analyzed and compared to a reference in which the use of only virgin materials is considered.

Citation: Urbano, L.; Tsantilis, L.; Riviera, P.P.; Baglieri, O.; Santagata, E. Life Cycle Assessment of a Sustainable and Innovative Solution for Unpaved Rural Roads. *Eng. Proc.* **2023**, *36*, 12. https://doi.org/10.3390/engproc2023036012

Academic Editor: Hosin (David) Lee

Published: 30 June 2023

2. LCA: Scenarios and Results

Figure 1 shows the cross sections (4.50 m width and 30 cm thick) corresponding to the two different scenarios compared based on LCA.

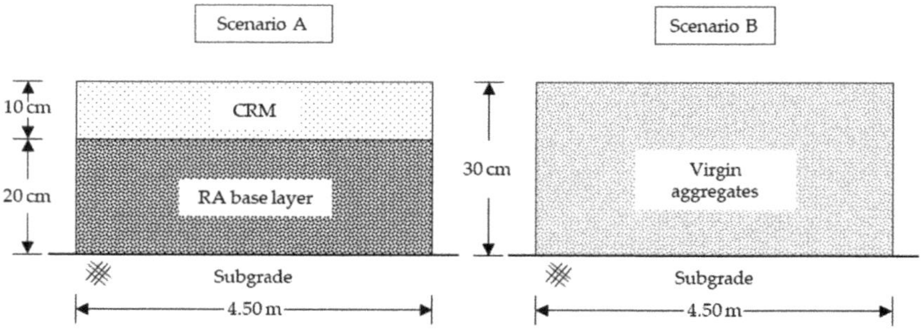

Figure 1. Cross-sections corresponding to the considered scenarios.

Scenario A refers to the innovative proposed solution consisting of a 20 cm thick RA base layer covered by a 10 cm thick CRM finishing surface layer. Scenario B refers to the typical solution currently adopted in northern Italy for low-volume unpaved rural roads, consisting of a single 30 cm thick crushed virgin aggregates layer.

The LCA was performed using the PaLATE tool, properly customized in order to account for the reference geographic context analyzed in the present study. The system boundaries included materials production, construction, and maintenance operations, according to a cradle-to-gate approach. The in-service phase was kept out of the analysis, due to its negligible impact in comparative terms and uncertainty in the forecast of traffic spectra. The end-of-life phase was also excluded from the system boundaries since all materials have the potential to be fully reclaimed without requiring landfill disposal. The functional unit consists of 1 km of road with a service life of 10 years.

The life cycle inventory was based on both primary and secondary data. Primary data were collected through interviews with local companies, while secondary data were collected from the available literature [10–13]. The inventory of equipment and trucks employed in the production, construction, transportation, and maintenance activities (Table 1) was based on primary data, while emission factors were obtained from the literature.

Table 1. Equipment and trucks considered in the analysis.

Equipment/Trucks	Activity
Crushing Plant	Crushing of virgin aggregates
Mobile Crushing Device	Crushing of RA
Wheel Loader + Blender	Production of CRM
Dump Trucks	Transportation of granular material
Tanker Trucks	Transportation of fluid material
Paver and Roller	Pavement construction

The analyzed impact categories were global warming potential (GWP), energy requirement, and water consumption.

The results of the analysis obtained for the two scenarios are given in Table 2. It can be highlighted that the adoption of the technical solution proposed for scenario A led to an overall reduction of the environmental burdens, which was particularly relevant when considering water consumption and GWP.

A comparison between these two solutions is also provided in a graphical form in Figure 2, where the relative impacts in terms of GWP, energy requirement, and water consumption values are expressed as a percentage of those found for reference scenario B.

Moreover, for each normalized impact, the relative contributions of materials production (MP), transportation (T), and construction/maintenance (C&M) phases are also provided.

Table 2. GWP, energy requirement and water consumption for scenarios A and B.

Impact	Scenario A	Scenario B
GWP (Mg of CO_{2eq})	18.7	24.0
Energy (MJ)	253,169	263,651
Water (m^3)	327	6701

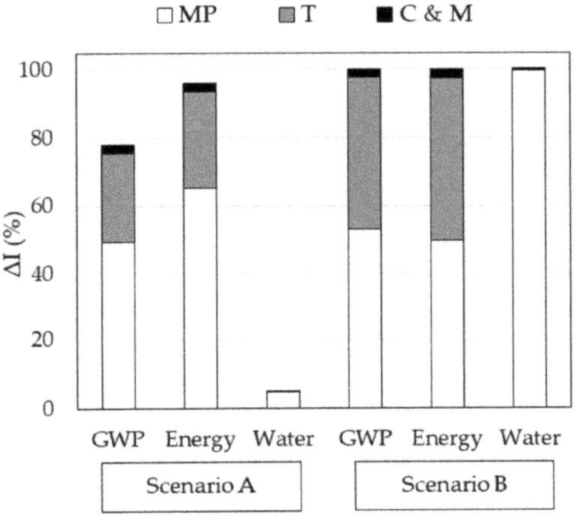

Figure 2. Normalized impacts (ΔI) of GWP, energy, and water expressed as a percentage of the results of reference scenario B, with relative contributions of materials production (MP), transportation (T), and construction /maintenance (C&M) phases.

It is interesting to observe that pavement construction and maintenance activities played a marginal role, as proven by contributions that were found to be always lower than 3%, regardless of the analyzed scenarios. The highest impacts were found to be those related to materials production. In particular, based on the comparison of the two solutions in terms of GWP and energy requirement, it was demonstrated that savings related to the use of a less amount of virgin aggregates were almost or totally offset by the innovative technical solution. This is due to the environmental burdens caused by the industrial production of the emulsion, the treatments of RAP, and the in-field preparation of CRM. On the other hand, relevant reductions in terms of water consumption were yielded by the dramatic decrease in the use of virgin aggregates. When focused on transportation, it can be noticed that the technical solution proposed in scenario A produced a reduction of more than 40% in the GWP and energy impact categories. Such a significant saving can be attributed to the use of in-place recycled materials that required fewer transportation distances with respect to solutions based on the use of virgin aggregates, as well as by the lower maintenance activities that are needed when bitumen-bound materials are used for wearing courses.

3. Concluding Remarks

In the present study, the environmental performances of two technical solutions for low-volume unpaved rural roads were evaluated and compared using LCA.

It was observed that the most critical stages are those related to materials production and transportation, while negligible impacts derive from construction and maintenance operations.

The outcomes of this LCA demonstrated the benefits, in terms of environmental impacts, associated with the considered innovative solution. Based on the assessment of GWP, energy requirement and water consumption related to materials production, transportation, construction, and maintenance operations, the proposed solution (which entails the use of large quantities of RAP and MS) outperformed the reference solution currently adopted for unpaved rural roads in northern Italy.

Further studies are certainly needed to extend the analysis to other impact categories and to widen the system boundaries.

Author Contributions: Conceptualization, L.T., P.P.R. and L.U.; methodology, L.T. and L.U.; formal analysis, L.T., P.P.R. and L.U.; investigation, L.T., P.P.R. and L.U.; resources, E.S. and O.B.; data curation, L.U.; writing—original draft preparation, L.T., P.P.R. and L.U.; writing—review and editing, L.T., P.P.R., O.B. and E.S.; supervision, E.S. and O.B. All authors have read and agreed to the published version of the manuscript.

Funding: This research received no external funding.

Institutional Review Board Statement: Not applicable.

Informed Consent Statement: Not applicable.

Data Availability Statement: Data available under request.

Conflicts of Interest: The authors declare no conflict of interest.

References

1. Riviera, P.P.; Bellopede, R.; Marini, P.; Bassani, M. Performance-based re-use of tunnel muck as granular material for subgrade and sub-base formation in road construction. *Tunn. Undergr. Space Technol.* **2014**, *40*, 160–173. [CrossRef]
2. Balaguera, A.; Carvajal, G.I.; Albertí, J.; Fullana-I-Palmer, P. Life cycle assessment of road construction alternative materials: A literature review. *Resour. Conserv. Recycl.* **2018**, *132*, 37–48. [CrossRef]
3. Plati, C. Sustainability factors in pavement materials, design, and preservation strategies: A literature review. *Constr. Build. Mater.* **2019**, *211*, 539–555. [CrossRef]
4. Salehi, S.; Arashpour, M.; Kodikara, J.; Guppy, R. Sustainable pavement construction: A systematic literature review of environmental and economic analysis of recycled materials. *J. Clean. Prod.* **2021**, *313*, 127936. [CrossRef]
5. Gomes, G.J.C.; Magalhães, A.J.; Rocha, F.L.L.; Fonseca, A. A sustainability-oriented framework for the application of industrial byproducts to the base layers of low-volume roads. *J. Clean. Prod.* **2021**, *295*, 126440. [CrossRef]
6. European Asphalt Pavement Association (EAPA). *Asphalt in Figures 2019*; European Asphalt Pavement Association: Brussels, Belgium, 2020.
7. Choorackal, E.; Riviera, P.P.; Santagata, E. Mix design and mechanical characterization of self-compacting cement-bound mixtures for paving applications. *Constr. Build. Mater.* **2019**, *229*, 116894. [CrossRef]
8. Hudson, W.; Little, D.; Razmi, A.; Anderson, A.; Weissmann, A. *An Investigation of the Status of by-Product Fines in the USA, International Center for Aggregates Research*; Research Report ICAR-101-1; International Center for Aggregates Research: Austin, TX, USA, 1997.
9. Urbano, L.; Dalmazzo, D.; Riviera, P.P.; Baglieri, O.; Santagata, E. A Sustainable Cold-Recycled Solution for the Surface Finishing of Unpaved Rural Roads. *Materials* **2022**, *15*, 3920. [CrossRef]
10. Blengini, G.A.; Garbarino, E. Resources and waste management in Turin (Italy): The role of recycled aggregates in the sustainable supply mix. *J. Clean. Prod.* **2010**, *18*, 1021–1030. [CrossRef]
11. Eurobitume. *Life Cycle Inventory: Bitumen*; European Bitumen Association: Brussels, Belgium, 2012; ISBN 2-930160-26-8.
12. European Environment Agency. *EMEP/EEA Air Pollutant Emission Inventory Guidebook 2019 Technical Guidance to Prepare National Emission Inventories*; EEA Report No 13/2019; European Environment Agency: Copenhagen, Denmark, 2019; ISSN 1977-8449.
13. Ecoinvent. *Life Cycle Inventories of Production Systems*; Swiss Centre for Life Cycle Inventories: Zürich, Switzerland, 2007.

engineering proceedings

MDPI

Proceeding Paper

Pullout Behavior of Titanium Alloy Reinforcing Bars in Ultra-High Performance Concrete †

Mahesh Acharya *, Jared Cantrell and Mustafa Mashal

Department of Civil & Environmental Engineering, Idaho State University, Pocatello, ID 83209, USA; cantjare@isu.edu (J.C.); mashmust@isu.edu (M.M.)

* Correspondence: achamahe@isu.edu; Tel.: +1-(980)-339-0764

† Presented at the Second International Conference on Maintenance and Rehabilitation of Constructed Infrastructure Facilities, Honolulu, HI, USA, 16–19 August 2023.

Abstract: This paper presents the novel concept of titanium alloy reinforced ultra-high performance concrete (TARUHPC) that can be utilized in bridges. This research highlights the advantages associated with titanium alloy bars (TiABs) and ultra-high performance concrete (UHPC). To validate the concept, pullout tests were implemented to assess the pullout behavior of TiABs with concrete and UHPC. Twelve beam samples (normal concrete and UHPC) were prepared using #6 (ø 0.75 inch) TiABs with different embedment lengths. TARUHPC specimens performed exceedingly better, with an average ultimate force of about 29 kips and a resulting shear stress of 1.82 ksi.

Keywords: UHPC; titanium alloy bars; pullout; durability; concrete structures; bond behavior

Citation: Acharya, M.; Cantrell, J.; Mashal, M. Pullout Behavior of Titanium Alloy Reinforcing Bars in Ultra-High Performance Concrete. *Eng. Proc.* **2023**, *36*, 13. https://doi.org/10.3390/engproc2023036013

Academic Editor: Hosin (David) Lee

Published: 3 July 2023

1. Introduction

Recently many advanced novel materials have been introduced into civil infrastructures in order to improve structural performance and durability. Some of these advanced materials that have a huge potential to benefit civil and critical infrastructure include ultra-high performance concrete (UHPC) and titanium alloy bars (TiABs).

1.1. Ultra-High Performance Concrete

Ultra-high performance concrete (UHPC), which is a new class of concrete falling under high-performance fiber-reinforced cement-based composite (HPFRCC) materials, was developed decades ago. A recent study has shown that the research efforts on HPFRCC seismic performance over the past two decades have been increasing every year [1]. The research areas on HPFRCC materials include but are not limited to material testing on beams, columns, structural walls, beam-column joints, and more. UHPC is also considered for rehabilitating bridges and bridge components across the world [2]. UHPC has been utilized in civil infrastructure lately because of its exceptional advantages. It is an advanced cementitious composite material with exceptionally high strength and high durability compared to conventional forms of concrete. The steel fibers in the UHPC concrete restrain the development of cracks and can result in added ductility in structures. UHPC has advantages in many aspects compared to conventional or high-strength concrete, for example, higher compressive and tensile strength, higher elastic modulus, and lower porosity [3].

UHPC has demonstrated its superiority in certain cases, for example, resulting in less maintenance and significantly longer service life due to its better mechanical properties, and the bond performance of the reinforcements embedded in UHPC is critically important for the safety of the UHPC structures. Several investigations are being carried out by researchers across the globe to study the bond behavior of deformed steel bars and UHPC. Graybeal [3], Fehling et al. [4], Holschemacher et al. [5], Saleem et al. [6], Jungwirth and Muttoni [7], and Aryal [8] are among those to study the bond development of steel bar reinforcement embedded in UHPC. It has been found that the UHPC specimens are capable of developing higher bond strengths in bar pullout specimens and are largely dependent on bar spacing, concrete cover, development length, and bar size. It has been evident that compared with normal-strength concrete, the particular material properties of UHPC will inevitably lead to changes in bond properties between rebar and UHPC. The mechanical interlocking between rebar and UHPC is improved due to the material's ultra-high compressive strength and elastic modulus. Bond strength and peak slip have been found to be the key parameters to study the bond behavior between rebar and UHPC.

1.2. Titanium Alloy Bars (TiABs)

Titanium alloy bars (TiABs) are an advanced material that is mostly used in the aerospace industry and has been gaining popularity for applications in concrete structures. TiABs (Ti6Al4V) offer higher strength, superior fatigue performance, high strength-to-weight ratio, lighter weight, lower modulus of elasticity, higher modulus of resilience, reduced rebar congestion, smaller inelastic residual deformation, and excellent corrosion resistance compared to conventional rebars [9]. Presently, TiABs are used to retrofit bridges in the civil engineering industry. They are strong, durable, and naturally resistant to rust and corrosion; however, they cannot be cast like aluminum or iron. This makes TiABs an expensive material compared to steel, stainless steel, aluminum, etc. Several research institutions have carried out studies of TiABs for concrete structures applications, for example, Idaho State University (mechanical properties testing of TiABs [9], bond testing and splicing of TiABs [10], bridge columns reinforced with TiABs [11], cap beam reinforced with TiABs [12]) and Oregon State University (retrofit square column with TiABs [13], retrofit deficient reinforced concrete girders [14]).

The Oregon Department of Transportation (ODOT) and Texas Department of Transportation (TxDOT) have been among the very first to use TiABs for retrofitting bridges in the United States. They have utilized near-surface-mounted (NSM) techniques with TiABs in retrofitting the bridges. TiABs provided increased strength across cracks. Although TiABs are more expensive compared to normal steel rebars, they offer two and a half times higher yield stress compared to grade 60 ksi steel rebars; this results in having less TiABs [15]. Past applications have shown that the use of TiABs provides higher strength and durability at a lower cost and less construction time [14,16]. The use of TiABs for retrofitting bridges offers cost savings for labor and materials, causes less traffic disruption, and provides a durable and accelerated retrofitting process that is competitive to other conventional materials over the service life of a structure. Several DOTs in the United States have adopted TiABs as a retrofitting material for bridges, and the American Association of State Highway and Transportation Officials (AASHTO) recently released a new publication for the use of TiABs [17].

2. Experimental Program

The paper presents the construction and pullout testing of titanium alloy reinforced ultra-high performance concrete (TARUHPC) and compares the performance with titanium alloy reinforced normal concrete (TARNC). A proprietary UHPC that has been common in North America, Ductal JS1000 produced by Lafarge, was selected for the experimental study [18]. The ingredients include dark grey premix (pre-blended cement, sand, ground quartz, silica fume), liquid admixture (high-range water reducer), steel fibers (0.008 in diameter, 0.5 in long), and water. It is claimed by the producer that this type of UHPC provides superior performance in terms of abrasion and chemical resistance, freeze-thaw, carbonation, and chloride penetration. This type of UHPC should be batched in accordance with the Lafarge's Ductal Batching Procedure [19]. Most importantly, high-shear mixers (Figure 1b) are recommended by the producer to properly and efficiently mix the UHPC. UHPC is self-consolidating and should not be vibrated to avoid the segregation of steel fibers. Similarly, a 5000 psi concrete mix with locally available materials (Portland cement, fly ash, fine aggregates, coarse aggregates, and water) was used as the normal concrete to prepare the samples for pullout test.

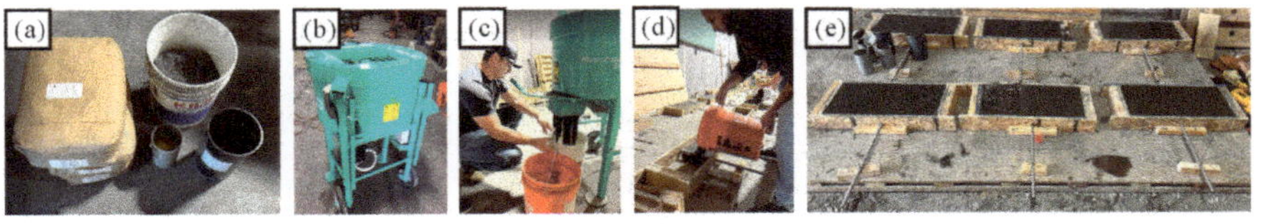

Figure 1. Construction of TARUHPC specimen: (**a**) batching, (**b**) high-shear mixture, (**c**) discharging UHPC, (**d**) pouring UHPC, (**e**) after pour.

2.1. Test Specimen Preparation

A total of 12 beam samples (six samples TARUHPC and six samples TARNC) of size (22 in × 10 in × 4 in) with different embedment lengths (5 in, 7 in, and 9 in) were prepared and casted for the pullout test. The construction process for TARNC specimens follows the traditional process. However, the preparation and construction process of TARUHPC specimens is slightly different, which is presented in Figure 1.

2.2. Test Setup

The specimens were tested in tension using a displacement-controlled servo-valve-hydraulic actuator. The actuator was supported by a reaction frame. The specimen was supported on two heavy steel blocks. The specimen was then tied to a strong floor using threaded rods bolted down to the strong floor. To tie the TiABs to the actuator head, they were attached to the coupler (ø 2.785 in) using fifteen 0.5 in threaded bolts. The coupler was then tied with the actuator head using four 0.75 in threaded rods and a 1 in thick plate. The loading rate of 0.02 in/min was used to pull the specimen.

2.3. Testing Results

Testing was carried out on a total of 12 specimens, and the summary of pullout testing is presented in Table 1.

Table 1. Pullout testing results.

	TARUHPC			TARNC		
Embedment Length, l (inch)	5	7	9	5	7	9
Average Ultimate Force, P (kip)	26.2	30.0	30.6	5.4	4.9	4.7
Average Slip at Ultimate Force (inch)	0.036	0.045	0.044	0.0025	0.0019	0.0027
* Shear Stress, τ (ksi)	2.22	1.82	1.44	0.46	0.30	0.22

* Note: Shear Stress ($\tau = P/\pi dl$); d = diameter of bar; 1 in = 25.4 mm; 1 kip = 4.45 kN; 1 ksi = 6.89 MPa.

3. Discussion

The next generation novel concept titanium alloy reinforced ultra-high performance concrete (TARUHPC), which is a combination of titanium alloy bars (TiABs) and ultra-high performance concrete (UHPC), has great potential for civil infrastructures. Pullout testing results showed improved performance of the TARUHPC compared to the TARNC, as shown in Table 1. For the test setup provided, the TARNC had a premature failure with an average ultimate force of about 5 kips and a resulting shear stress of only 0.33 ksi. However, TARUHPC performed exceedingly better, with an average ultimate force of about 29 kips and a resulting shear stress of 1.83 ksi. To ensure that TARUHPC can be structurally efficient to be used in civil infrastructures, small-scale experiments on structural elements, such as beams and columns, should be carried out. There is presently little to no data on testing of the TARUHPC on structural elements. Researchers at Idaho State University are currently working on an effort to test the TARUHPC concept on structural elements and are even utilizing various artificial intelligence (AI) tools to evaluate the performance of the TARUHPC concept.

Author Contributions: Conceptualization: M.M.; methodology: M.M., J.C. and M.A.; validation: M.M., M.A. and J.C.; formal analysis: M.M., M.A. and J.C.; investigation: M.M., M.A. and J.C.; resources: M.M. and J.C.; data curation: M.A.; writing—original draft preparation: M.A.; writing—review and editing; M.M., M.A. and J.C.; project administration: M.M.; funding acquisition: M.M. All authors have read and agreed to the published version of the manuscript.

Funding: This research was funded by Idaho State University Center for Advanced Energy Studies Seed Grant.

Institutional Review Board Statement: Not applicable.

Informed Consent Statement: Not applicable.

Data Availability Statement: The data is available upon request to the corresponding author.

Acknowledgments: This work was supported by State of Idaho appropriated funding for the Center for Advanced Energy Studies (CAES).

Conflicts of Interest: The authors declare no conflict of interest. The funders had no role in the design of the study; in the collection, analyses, or interpretation of data; in the writing of the manuscript; or in the decision to publish the results.

References

1. Shao, Y.; Nguyen, W.; Bandelt, M.J.; Ostertag, C.P.; Billington, S.L. Seismic Performance of High-Performance Fiber Reinforced Cement-Based Composite Structural Members: A Review. *ASCE J. Struct. Eng.* **2022**, *148*, 03122004. [CrossRef]
2. Kim, H.; Moon, B.; Hu, X.; Lee, H.; Ryu, G.-S.; Koh, K.-T.; Joh, C.; Kim, B.-S.; Keierleber, B. Construction and Performance Monitoring of Innovative Ultra-High-Performance Concrete Bridge. *Infrastructures* **2021**, *6*, 121. [CrossRef]
3. Graybeal, B. *Bond Behavior of Reinforcing Steel in Ultra High Performance Concrete*; Federal Highway Administration, FHWA-HRT-14-089; Office of Infrastructure Research and Development: Washington, DC, USA, 2014; p. 12.
4. Fehling, E.; Lorenz, P.; Leutbecher, T. Experimental Investigations on Anchorage of Rebars in UHPC. In *Proceedings of Hipermat 2012 3rd International Symposium on UHPC and Nanotechnology for High Performance Construction Materials*; Schmidt, M., Fehling, E., Eds.; Kassel University Press: Kassel, Germany, 2012.
5. Holschemacher, K.; Weibe, D.; Klotz, S. Bond of reinforcement in ultra high strength concrete. In *Proceedings of the International Symposium on Ultra High Performance Concrete*; Schmidt, M., Fehling, E., Geisenhanslüke, C., Eds.; Kassel University Press: Kassel, Germany, 2004; pp. 375–387.
6. Saleem, M.A.; Mirmiran, A.; Xia, J.; Mackie, K. Development length of high-strength steel rebar in ultrahigh performance concrete. *J. Mater. Civil Eng.* **2013**, *25*, 991–998. [CrossRef]
7. Jungwirth, J.; Muttoni, A. Structural Behavior of Tension Members in UHPC. In *Proceedings of the International Symposium on Ultra High Performance Concrete*; Kassel University Press: Kassel, Germany, 2004.
8. Aryal, B. Pullout Behavior of Steel Reinforcing Bars in Ultra-High Performance Concrete for Field-Cast Connection of Bridge Precast Decks. Master's Thesis, Idaho State University, Pocatello, ID, USA, 2022.
9. Khadka, R.; Mashal, M.; Cantrell, J. Experimental Investigation on Mechanical Properties of Titanium Alloy Bars: Comparison with High-Strength Steel. *Spec. Publ. Am. Concr. Inst.* **2020**, *341*, 160–187.
10. Acharya, M.; Khadka, R.; Mashal, M. Preliminary Bond Testing and Splicing of Titanium Alloy Bars. *Transportation Research Record* **2022**, *2676*, 410–427. [CrossRef]
11. Acharya, M.; Mashal, M. Titanium Alloy Bars for Construction of Resilient and Durable Concrete Structures. In Proceedings of the 12th National Conference on Earthquake Engineering, Salt Lake City, UT, USA, 27 June 27–1 July 2022.
12. Thapa, A.; Mashal, M.; Acharya, M. Large-Scale Flexural Testing of Concrete Beams Reinforced with Conventional Steel and Titanium Alloy Bars. In Proceedings of the IABSE Symposium: Challenges for Existing and Oncoming Structures, Prague, Czech Republic, 25–27 May 2022; pp. 272–276.
13. Shrestha, S. Seismic Retrofit of Square Reinforced Concrete Bridge Columns using Titanium Alloy Bars. Ph.D. Thesis, Civil Engineering, Oregon State University, Corvallis, OR, USA, 2019.
14. Higgins, C.; Amneus, D.; Barker, L. Methods for Strengthening Reinforced Concrete Bridge Girders Containing Poorly Detailed Flexural Steel Using Near-Surface Mounted Metallics. In *Final Report to Oregon Dept. of Transportation*; Report No. FHWA-OR-RD-16-02; Oregon Department of Transportation: Salem, OR, USA, 2015.
15. *Transportation News*; Texas Department of Transportation: Austin, TX, USA, 2020; Volume 44, Issue 3, May/June 2020; pp. 16–17.

16. Adkins, J.; George, W. Titanium Finds a Home in Civil Engineering. *Concr. Int.* **2017**, *39*, 51–55.
17. AASHTO. *Guide for Design and Construction of Near-Surface Mounted Titanium Alloy Bars for Strengthening Concrete Structures*; American Association of State Highway and Transportation Officials: Washington, DC, USA, 2020.
18. Stéphane Rigaud; Phillippe Fonollosa; Gilles Chanvillard. Concrete Composition. Lafarge, Assignee. U.S. Patent 8303708 B2, 6 November 2012.
19. Lafarge North America: Product Data Sheet JS1000, Chrome-Extension. Available online: https://www.ductal.com/en (accessed on 1 May 2020).

engineering proceedings

Proceeding Paper

Establishing Density-Based Mix Design for Cold Recycled Asphalt Mixes †

Mansour Solaimanian * and Scott Milander

Larson Transportation Institute, Pennsylvania State University, University Park, PA 16802, USA
* Correspondence: msol@psu.edu

† Presented at the Second International Conference on Maintenance and Rehabilitation of Constructed Infrastructure Facilities, Honolulu, HI, USA, 16–19 August 2023.

Abstract: Design of emulsified cold asphalt mixes using reclaimed asphalt pavement requires optimization of the emulsion and moisture content to achieve either maximum density or a minimum strength level of the mix, and in some cases both. Municipalities and local governments most often lack advanced testing equipment to design the cold mixes based on strength or stiffness results. This study was conducted with the objective of developing a simple mix design process for emulsion-based cold mixes using simple laboratory equipment such as proctor molds which are commonly used for optimization of moisture content for soil compaction.

Keywords: asphalt; cold mix; reclaimed asphalt; emulsion; strength; proctor; unit weight; RAP

Citation: Solaimanian, M.; Milander, S. Establishing Density-Based Mix Design for Cold Recycled Asphalt Mixes. *Eng. Proc.* **2023**, *36*, 14. https://doi.org/10.3390/engproc2023036014

Academic Editor: Hosin (David) Lee

Published: 3 July 2023

1. Introduction

The usage of RAP in the United States has increased over the years; for example, its use grew from 15.6 percent in 2009 to 21.31 percent in 2020, with an estimated 87 million tons used in 2020 [1]. The use of RAP in the construction of pavements is economical and helps save resources and mitigate associated environmental impacts [2,3]. While most RAP is used in hot-mix and warm-mix asphalt, states have also been utilizing it in cold mixes to some extent. Processes such as cold recycling (CR) and full-depth reclamation (FDR) make use of 100 percent of the previously existing asphalt pavement material in the recycling process to produce new and improved pavement base layers. The CR process can be classified into two types: Cold In-Place Recycling (CIR) and Cold Central Plant Recycling (CCPR). Regardless of the cold recycling technique, the compacted mat needs to undergo a curing process to gain strength. The curing in the field is specified by the amount of water content present in the CR layer. The maximum allowable amount of water in the CR layer before an overlay placement should be limited to 1.0 to 1.5 percent [4].

Various approaches have been taken for the design of cold mixes and optimization of the needed emulsion and water content. The most rational design protocols include some measure of the mix engineering properties, such as stiffness or strength. An example of such can be found in the design process developed by Solaimanian et al. [5]. However, there is a lack of adequate testing equipment or manpower at the local level in some municipalities and counties to develop a mix design based on mechanical testing. This research was undertaken to address this problem through a simple design system. The objective was to develop general guidelines for use by maintenance and local forces in determining the necessary emulsion and water content to be blended with RAP in cold recycling work without the need for an elaborate mix design system.

2. Materials

Five sources of RAP were included in the study. Initial characterization of RAP included determination of the binder content, black rock gradation, and extracted aggregate gradation. Only the RAP passing a 3/4-in (19-mm) sieve was used. The RAP was fractionated on the #4 sieve (4.75 mm) to provide two stockpiles of coarse RAP (retained on the #4 sieve) and fine RAP (passing the #4 sieve). The average binder content of the RAP among the five sources varied in the range of 5.2 to 7.8%. The only virgin aggregate used in this work was an AASHTO #57. A cationic, slow-setting asphalt emulsion (CSS-1 h) was used for all specimens.

3. Procedures, Methods, and Tests

3.1. Curing and Preparation of Loose Mix

The air-dried RAP material was mixed with water first. Before mixing with the wetted RAP, the selected emulsion was heated in a 60 °C (140 °F) oven for 30 min and stirred to ensure homogeneity. The RAP–emulsion blend was cured in a 40 °C (104 °F) chamber for 30 min immediately after blending and before compaction. Two water contents (2% and 3%) and three emulsion contents (3%, 4%, and 5%) were considered, delivering a total of 90 Proctor-compacted specimens when including the replicates and different tests.

3.2. Compaction and Curing of Compacted Specimens

Compacted specimens were prepared two ways: (1) based on AASHTO T 180 Method C, and (2) using a Superpave gyratory compactor (SGC). Specimens were then cured at 104 °F (40 °C) for 72 h. Typical examples of compacted specimens are presented in Figure 1.

(a) (b)

Figure 1. Compacted 100% RAP emulsified cold mix specimens: (**a**) Proctor and (**b**) SGC.

3.3. Establishing Relationship between Dry Unit Weight and Emulsion Content

The dry unit weight of each emulsion content was simply determined based on the weight–volume relationship of the compacted specimens. Unit weights were plotted against the emulsion content at different water contents as well as total fluid content.

3.4. Measuring Strength of Proctor and SGC Specimens

The strength of the Proctor specimens was measured following ASTM D1633 [5] but at a displacement rate of 50 mm/min. The indirect tensile strength (IDT) of the SGC specimens was measured according to ASTM D6931 [6]. All testing was conducted at 77 °F (25 °C). The test assembly for both types of tests is presented in Figure 2.

(a) (b)

Figure 2. Specimen setup for (**a**) compression test and (**b**) indirect tension test.

4. Results and Discussion

4.1. Relationship between Dry Unit Weight and Emulsion Content

Figure 3a shows an example of the dry unit weight after extrusion from the Protor mold versus the emulsion content for one of the five sources. It can be seen that it is possible to develop a density–fluid content curve for these RAP mixes in a similar manner to the common methods used for soils. This was successfully performed for three of the five RAP sources. For the other two sources, it appeared that the range of water and emulsion content had to be extended. One should be consistent in using the dry unit weight (or density) either after extrusion or after curing. While there is a strong correlation between these two unit weights, the cured dry unit weight is larger because of the volume change (shrinkage), as shown in Figure 3b.

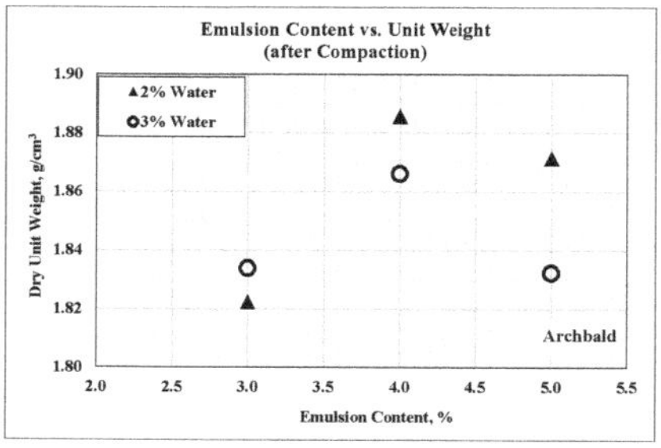

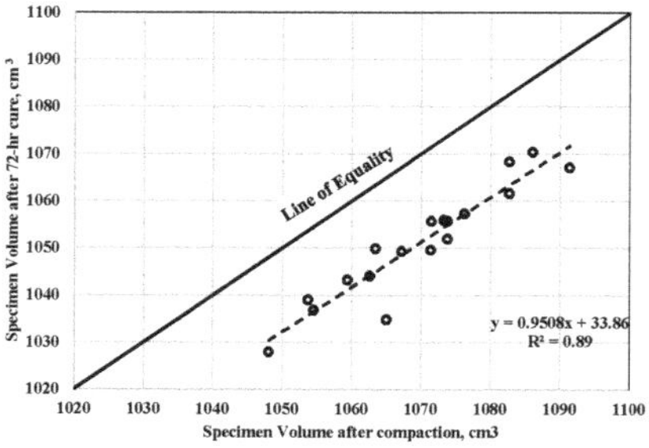

(a) (b)

Figure 3. (**a**) Relationship between unit weight after curing and unit weight after compaction, and (**b**) specimen volume after compaction versus volume after curing.

4.2. Strength of Proctor-Compacted and SGC Specimens

Figure 4a presents an example plot of how the compressive strength varies as a function of emulsion content for one of the RAP sources. The compressive strength is also strongly correlated with the indirect tensile strength, as shown in Figure 4b. The compressive strength varied in the range of 200 to 400 psi for these cold mixes, almost five times larger than the strength obtained from indirect tensile testing. In a previous study [7], the authors established the minimum required indirect tensile strength as 45 psi. Based on the correlation of ID strength with compressive strength, a minimum compressive strength of 170 psi is recommended for emulsified RAP cold mixes.

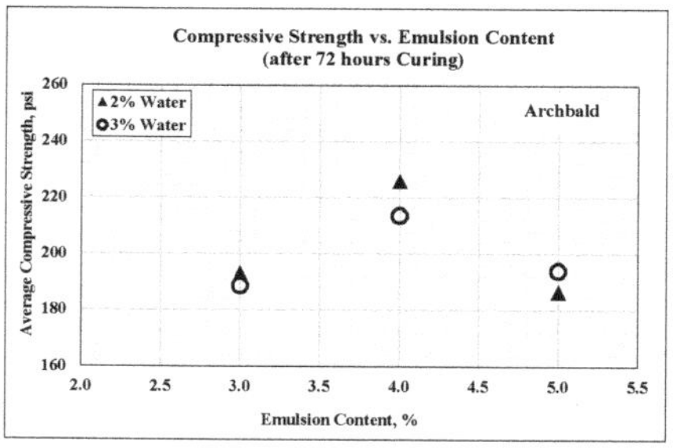

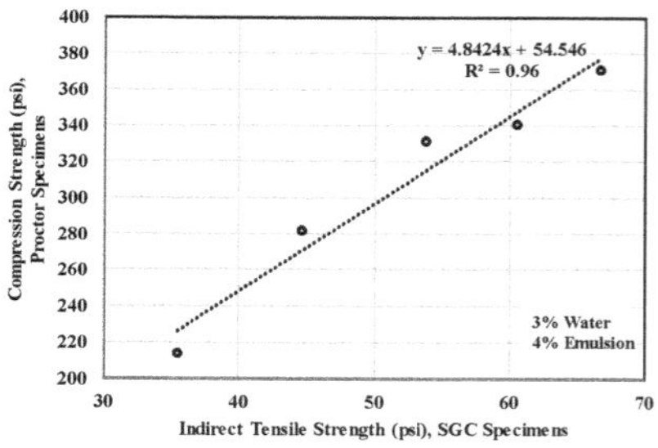

(a) (b)

Figure 4. (**a**) A typical strain–stress curve from the compressive strength test, and (**b**) the relationship between the compressive strength and indirect tensile strength of the cold RAP mixes.

One of the most important findings from this research was that, in general, the strength values were more sensitive to changes in emulsion content compared to the unit weight values. It was also observed that RAP mixes with 3% water content consistently delivered higher strength than those with 2% water content. Finally, the effect of emulsion content must be distinguished from the effect of water content, since different proportions of the two variables deliver different results when the total fluid content is fixed. For example, for a fixed total fluid content of 6%, strength values are higher for 3% emulsion and 3% water compared with 2% water and 4% emulsion (Figure 4a). However, at a total fluid content of 7%, higher strength is gained from 3% water and 4% emulsion compared with 2% water and 5% emulsion (Figure 4a).

The results presented in this paper were based on testing emulsified cold mixes prepared with 100 percent RAP. For one of the RAP sources, a series of specimens were prepared at 3% water and 4% emulsion content with addition of 10% AASHTO #57 aggregate to improve the coarse aggregate content. The results from this testing did not indicate any further improvements in strength despite the additional coarse aggregate.

5. Conclusions

The results indicated that in general, an optimum emulsion content could be established based on the maximum dry unit weight of the compacted RAP mixes, somewhat similar to the common approach used to optimize the water content for soil compaction. A design based on the Proctor mold density results does not require strength testing and extended curing. This simplified design process requires compaction of specimens according to AASHTO T 180 Method C with slight modifications to better suit the emulsified cold-mix design.

Author Contributions: Conceptualization, M.S.; methodology, M.S.; investigation, M.S. and S.M.; validation, M.S. and S.M.; writing—original draft preparation, M.S. and S.M.; writing—review and editing, M.S. All authors have read and agreed to the published version of the manuscript.

Funding: This research was funded by Pennsylvania Department of Transportation.

Institutional Review Board Statement: Not applicable.

Informed Consent Statement: Not applicable.

Data Availability Statement: Not applicable.

Acknowledgments: The authors greatly appreciate the financial support for this research project that was provided by the Pennsylvania Department of Transportation.

Conflicts of Interest: The authors declare no conflict of interest.

References

1. Williams, B.A.; Willis, J.R.; Ross, T. *Asphalt Pavement Industry Survey on Recycled Materials and Warm-Mix Asphalt Usage 2020*; Information Series 138, Technical Report no. IS138(11e); National Asphalt Pavement Association: Greenbelt, MD, USA, 2021.
2. Horvath, A. *Life-Cycle Environmental and Economic Assessment of Using Recycled Materials for Asphalt Pavements*; University of California Transportation Center: Berkley, CA, USA, 2003.
3. Huang, Y.; Bird, R.; Heidrich, O. Development of a Life-cycle Assessment Tool for Construction and Maintenance of Asphalt Pavements. *J. Clean. Prod.* **2009**, *17*, 283–296. [CrossRef]
4. Kim, Y.; Im, S.; Lee, H.D. Impacts of Curing Time and Moisture Content on Engineering Properties of Cold In-Place Recycling Mixtures Using Foamed or Emulsified Asphalt. *J. Mater. Civ. Eng.* **2011**, *23*, 542–553. [CrossRef]
5. *ASTM D1633*; Standard Test Methods for Compressive Strength of Molded Soil-Cement Cylinders. American Society for Testing and Materials: West Conshohocken, PA, USA, 2018.
6. *ASTM D6931*; Standard Test Method for Indirect Tensile (IDT) Strength of Asphalt Materials. American Society for Testing and Materials: West Conshohocken, PA, USA, 2017.
7. Solaimanian, M.; Chen, X.; Milander, S.M.; Kheiry, P. *Polymer Modified Cold Recycled Asphalt Evaluation and Methodology*; Final Report, FHWA-PA-2015-008-PSU WO 03A; The Pennsylvania State University, Larson Transportation Institute: University Park, PA, USA, 2015.

engineering proceedings

MDPI

Proceeding Paper

CNN-Based Automatic Mobile Reporting System and Quantification for the Concrete Crack Size of the Precast Members of OSC Construction †

Ali Akbar [1], James Mugo Njoroge [2], Seojoon Lee [1], Younghee Chang [1] and Soonwook Kwon [3,*]

1 Department of Global Smart City, College of Engineering, Sungkyunkwan University, Suwon 16419, Republic of Korea; greyice11@g.skku.edu (A.A.); sjlee3003@skku.edu (S.L.); yhyhchang@g.skku.edu (Y.C.)
2 Department of Civil, Architectural and Environmental System Engineering, College of Engineering, Sungkyunkwan University, Suwon 16419, Republic of Korea; mugojames254@g.skku.edu
3 School of Civil, Architectural Engineering and Landscape Architecture, College of Engineering, Sungkyunkwan University, Suwon 16419, Republic of Korea
* Correspondence: swkwon@skku.edu
† Presented at the Second International Conference on Maintenance and Rehabilitation of Constructed Infrastructure Facilities, Honolulu, HI, USA, 16–19 August 2023.

Abstract: Civil infrastructure over the years has experienced a dominant reliance on concrete material compared to other construction materials. Human inspection is the main mode of inspection for such structures, including concrete columns, which has been proven to be inaccurate and time-consuming. Convolutional neural networks (CNNs) are a substitute for such problems for both detection and quantification. However, storing the results and visualizing them at a later stage has always been a challenge. Additionally, integration of the concrete crack deep learning model to a mobile platform is an area that has received less attention. This study focuses on segmenting the concrete crack sections using the latest state-of-the-art (YOLOv7) neural network, which is then used to obtain the quantification data about the length and width of the detected crack using image binarization, and finally the results are published using a reporting system integrated to a mobile platform using a web and IoT system. The published report uses a checklist from the quantification results to grade the crack as well as its structure. The results show a mAP of 0.85, while the quantification results show a 10.82% absolute error, respectively. The reporting system takes a combined average of 5940 ms to store the data inside a database, which is then published through a mobile device. It has been demonstrated through this study that an automatic mobile reporting system is feasible to be used on buildings for maintenance, which can be further applied across other sectors of construction for monitoring and repair purposes.

Keywords: crack quantification; CNN; precast concrete (PC) member; database management system; off-site construction (OSC)

Citation: Akbar, A.; Njoroge, J.M.; Lee, S.; Chang, Y.; Kwon, S. CNN-Based Automatic Mobile Reporting System and Quantification for the Concrete Crack Size of the Precast Members of OSC Construction. *Eng. Proc.* **2023**, *36*, 15. https://doi.org/10.3390/engproc2023036015

Academic Editor: Hosin (David) Lee

Published: 3 July 2023

1. Introduction

Concrete as a construction material is heavily relied upon in the construction industry. In building construction projects there is a sixty percent usage of concrete as a material on average compared to other construction materials [1]. In civil engineering construction, concrete plays a critical role in the structural integrity of the structure. Hence, the safety of the majority of these civil engineering structures is determined by the quality of the concrete [2]. However, concrete structures are subject to defects over time, with one major defect being concrete cracks. The structural health of a concrete structure is mostly determined by the size of the concrete cracks. Large concrete cracks tend to expose the concealed structural steel members undermining the structure's overall integrity. The most dominant mode of monitoring concrete cracks is manual inspection, which on the contrary has been cited to be time-consuming and at the same time prone to errors [3].

The proposed system in this study uses a convolutional neural network, which is a computer vision algorithm to detect the location of these cracks and segment the crack regions. The segmented crack regions are then processed to a binarized format to obtain the crack length and width in a pixel format. Sinha (2006) briefed about the development of a statistical filter for the detection and segmentation of cracks in concrete pipes [4]. Yang (2018) used a system to identify cracks using the pixel segmentation technique [5]. Lee (2010) proposed a crack measurement technique on the length, width, and crack orientation based on the artificial neural network [6]. Kim's (2019) system was based on a mask RCNN model to measure the concrete crack width but not the concrete crack length [7]. Yuan (2021) used a convolutional neural network to measure the crack length and monitor the crack propagation, however the system did not measure the crack width [3]. Licun Yu (2022) detected concrete cracks using bounding boxes and measured the width of the cracks leaving crack length measurements [8]. Detection and segmentation have been discussed in the past, but presenting these results and using it for automatic reporting has rarely been discussed for the structures in these previous studies. Sohn (2004) put forward a new crack monitoring tool, CCMS, with a graphical user interface for crack analysis and detecting the arising changes in these cracks using images [9]. Shuang-rui (2015) detected and segmented the crack width using an android device application that had a user interface for saving the results [10]. Lee (2022) gave a deep learning-based system for the quality inspection of cracks on the concrete members [11]. Zhang (2022) proposed a bridge surface crack detection method and displayed it using a mobile device and a server without measurement parameters [12], while Li (2023) devised a system for the identification of the damage in bridge structures and displayed the evaluation results using commercial software [13]. However, all the reviewed studies were not based on a mobile system or reporting system of the quantified concrete cracks.

2. Methodology

In the fields of computer vision and construction, the instance segmentation technique has recently been employed. The capability of the model to perform real-time segmentation using neural networks with high accuracy makes it efficient for detecting the required content.

The model development began with the collection of data images containing concrete cracks. In this process, a total of 200 images were collected. Then, to make the model more robust in differentiating between the concrete cracks and concrete crack noise the images were cropped to only those sections that contained concrete cracks. The data was increased from 200 images to 1000 images through five image augmentation techniques: 50% horizontal flip, gaussian blurring, image recoloring, image translation, and image rotation. Data obtained through image augmentation and the original data were resized to a uniform size of $224 \times 224 \times 3$, where these dimensions indicate the length, height, and channels of the image data, respectively. The image data with their corresponding annotated files were passed through the instance segmentation, and a model was obtained which could make inferences to the new concrete crack data. The hyperparameters of the trained model involved a batch size of 16,400 epochs, an input image size of 224×224 corresponding to the length and width, an initial learning rate of 0.01, and one class (Crack). The internal structure of the model is composed of 104 convolutional neural network layers with 5 max-pooling layers using a stride of 1 and the concatenation layers which were used for accurate feature extraction. The overall model contained the input, the backbone for feature extraction, the head for refining the feature maps, and the final segmentation, as shown in Figure 1. The segmentation was performed in the final convolutional layers 102–104.

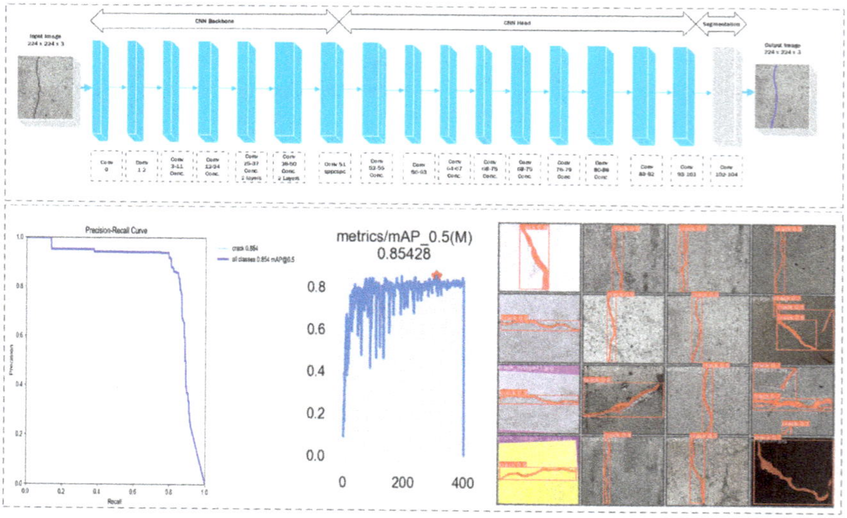

Figure 1. Convolutional neural network architecture and accuracy result for the devised system.

The system's architecture, as shown in Figure 2, includes a Flask API platform, which is connected to a mobile device and a computer server. Upon an HTTP request, the captured image from the mobile device is segmented onto the server, which stores it and then masks it using the image binarization technique, after which the nonzero mask pixels are counted along with the image size parameters to calculate the length, which is then divided by the total pixel values and halved to obtain the value of the width and classify it according to the criteria proposed [12].

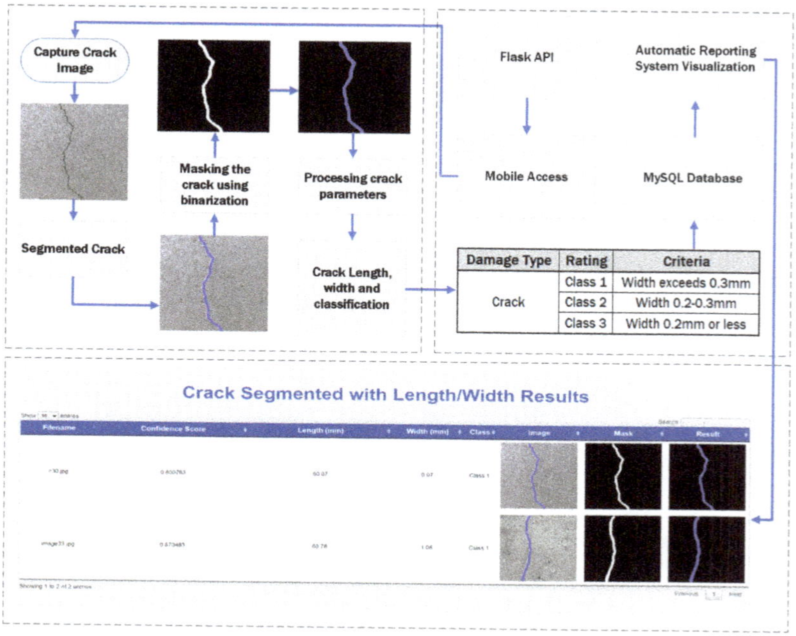

Figure 2. System framework for the segmentation and measurement of cracks with a GUI.

3. Conclusions

From the results shown in Table 1, as the width of the crack decreases the absolute error of the proposed system also increases. However, for the crack length, the measured length and the actual length of the crack show a minimal error, with the highest absolute error obtained from the test being twelve percent. While maintaining a fixed distance of 10 cm from the concrete crack object to the camera device, the system was able to quantify the length of the crack with an average of 90% percent accuracy and the width with 88% accuracy, respectively, as stated in Table 1. The hand measurement of the crack length was performed with a ruler, while for the width, a vernier caliper was used to accurately measure the width of up to 0.1 mm.

Table 1. Relative error percentages for the length and width of the proposed reporting system.

Crack	Proposed System Result	Hand Measurement	Error (%)
Length (mm)	49.66	52	4.5
	51.07	58	11.95
	52.60	60	12.33
Width (mm)	0.48	0.39	23.08
	0.73	0.67	8.95
	1.20	1.24	3.23

The inference time for the concrete crack quantification was 5940 milliseconds from the image uploading to the final reporting of the quantified results. The quantified width has been used in the classification of the crack in the checklist, which ultimately serves as the basis for damage in the member whether or not it should be discarded on the site. The system was assessed on static images, but the limitation of the system was that the distance between the camera and the concrete crack object had to be made constant by 12 inches from the crack for testing, and the training images and scale had to be set accordingly by equating the pixels to mm. Hence, a modification to the system to handle variable distances is considered as an area of further research. Moreover, the dataset should be updated to further enhance the accuracy of the system.

Author Contributions: Conceptualization, A.A. and S.L.; methodology, A.A.; software, A.A.; validation, S.L., J.M.N. and Y.C.; formal analysis, S.K.; investigation, A.A.; resources, A.A. and J.M.N.; data curation, A.A. and J.M.N.; writing—original draft preparation, A.A. and J.M.N.; writing—review and editing, A.A. and J.M.N.; visualization, A.A.; supervision, S.K.; project administration, S.L.; funding acquisition, Y.C. All authors have read and agreed to the published version of the manuscript.

Funding: This work is supported by the Korea Agency for Infrastructure Technology Advancement (KAIA) grant funded by the Ministry of Land, Infrastructure, and Transport (Grant 23ORPS-B158109-04). This work is financially supported by the Korean Ministry of Land, Infrastructure, and Transport (MOLIT) under the "Innovative Talent Education Program for Smart City".

Institutional Review Board Statement: Not applicable.

Informed Consent Statement: Not applicable.

Data Availability Statement: The data presented in this study are available on request from the corresponding author. The data are not publicly available due to confidentiality purposes.

Conflicts of Interest: The authors declare no conflict of interest.

References

1. Aytekin, B.; Mardani, A. Sustainable Materials: A Review of Recycled Concrete Aggregate Utilization as Pavement Material. *Transp. Res. Rec. J. Transp. Res. Board* **2022**, *2676*, 468–491. [CrossRef]
2. Guo, Y. Analysis on Concrete Construction Technology in Civil Engineering Construction. *J. Phys. Conf. Ser.* **2021**, *2011*, 12–23. [CrossRef]
3. Yuan, Y.; Ge, Z. Crack Length Measurement Using Convolutional Neural Networks and Image Processing. *Sensors* **2021**, *21*, 5894. [CrossRef] [PubMed]

4. Sinha, S.K.; Fieguth, P.W. Automated detection of cracks in buried concrete pipe images. *Autom. Constr.* **2006**, *15*, 58–72. [CrossRef]
5. Yang, X.; Li, H. Automatic Pixel-Level Crack Detection and Measurement using Fully Convolutional Network. *Comput.-Aided Civ. Infrastruct. Eng.* **2018**, *33*, 1090–1109. [CrossRef]
6. Lee, B.Y.; Kim, J.K.; Myung, H. Measurement, and Diagnosis of Concrete Cracks Using Image Processing and Fuzzy Theory. In *Safety, Reliability, and Risk of Structures, Infrastructure, and Engineering Systems*; Taylor & Francis Group: London, UK, 2010.
7. Kim, B.; Cho, S. Image-based concrete crack assessment using a mask and region-based convolutional neural network. *Struct. Control Health Monit.* **2019**, *26*, 23–81. [CrossRef]
8. Yu, L.; He, S. Intelligent Crack Detection and Quantification in the Concrete Bridge: A Deep Learning-Assisted Image Processing Approach. *Adv. Civ. Eng.* **2022**, *22*, 1813821. [CrossRef]
9. Sohn, H.-G.; Lim, Y.-M. Monitoring Crack Changes in Concrete Structures. *Comput.-Aided Civ. Infrastruct. Eng.* **2004**, *20*, 52–61. [CrossRef]
10. Chen, S.; Zheng, S. Concrete Crack Width Detecting System for Android Platform. *Open Civ. Eng. J.* **2015**, *9*, 846–851.
11. Lee, S.; Jeong, M. Deep Learning-Based PC Member Crack Detection and Quality Inspection Support Technology for the Precise Construction of OSC Projects. *Appl. Sci.* **2022**, *12*, 9810. [CrossRef]
12. Zhang, J.; Qian, S. Automated bridge crack detection method based on lightweight vision models. *Complex Intell. Syst.* **2023**, *9*, 1639–1652. [CrossRef]
13. Li, X.; Meng, Q. Identification of Underwater Structural Bridge Damage and BIM-Based Bridge Damage Management. *Appl. Sci.* **2023**, *13*, 1348. [CrossRef]

engineering proceedings

MDPI

Proceeding Paper

Prediction of Ultimate Bond Strength between Ultra-High Performance Concrete and Titanium Alloy Bars Using a Machine Learning Approach †

Mahesh Acharya [1,*], Luis Bedriñana [2], Jared Cantrell [1], Ankit Bhaukajee [3] and Mustafa Mashal [1]

1 Department of Civil & Environmental Engineering, Idaho State University, Pocatello, ID 83209, USA; cantjare@isu.edu (J.C.); mashmust@isu.edu (M.M.)
2 Department of Civil Engineering, Universidad de Ingenieria y Tecnologia—UTEC, Barranco 15063, Peru; lbedrinana@utec.edu.pe
3 Farm Bureau Mutual Insurance Company of Idaho, Pocatello, ID 83201, USA; ankitbhaukajee@gmail.com
* Correspondence: achamahe@isu.edu; Tel.: +1-(980)-339-0764

† Presented at the Second International Conference on Maintenance and Rehabilitation of Constructed Infrastructure Facilities, Honolulu, HI, USA, 16–19 August 2023.

Abstract: This research discusses the viability of the next-generation novel materials, e.g., titanium alloy bars (TiABs) and ultra-high-performance concrete (UHPC) that have potential to be utilized in civil infrastructures, e.g., bridges, in combination with machine learning (ML) techniques. Since UHPC and TiABs have been demonstrated to be a realistic alternative to traditional construction materials for civil infrastructures, it is important to characterize bond performance of reinforcing, i.e., TiABs embedded in UHPC. The research utilizes improvement of ML techniques, e.g., transfer learning (TL) to predict the bond strength of TiABs in UHPC.

Keywords: UHPC; titanium alloy bars; durability; bond strength; machine learning; transfer learning; concrete structures

Citation: Acharya, M.; Bedriñana, L.; Cantrell, J.; Bhaukajee, A.; Mashal, M. Prediction of Ultimate Bond Strength between Ultra-High Performance Concrete and Titanium Alloy Bars Using a Machine Learning Approach. *Eng. Proc.* **2023**, *36*, 16. https://doi.org/10.3390/engproc2023036016

Academic Editor: Hosin (David) Lee

Published: 3 July 2023

1. Background

TiABs are currently just limited to aerospace industry but becoming popular for applications in concrete structures because of their exceptionally outstanding advantages. Compared to conventional reinforcing, the advantages that TiABs offer include but are not limited to higher strength, superior fatigue performance, high strength-to-weight ratio, lighter weight, lower modulus of elasticity, reduction in rebar congestion, smaller inelastic residual deformation, and excellent corrosion resistance [1]. Similarly, UHPC offers numerous advantages critical for bridge application such as exceptionally higher compressive strength, superior mechanical properties, and excellent durability compared to conventional or high-strength concrete [2]. However, both UHPC and TiABs have not been widely used to date in structural elements due to the lack of knowledge of their structural behavior and failure mechanism. They can be considered for building a new structure and repairing of existing structures. However, it is important to characterize properly the bond performance of reinforcing, i.e., TiABs embedded in UHPC for the safety of UHPC structures. The cost, nevertheless, plays a huge role in the ability of researchers to carry out experimental studies, and identifying an alternative way to model the behavior of these materials which would be a key challenge.

Machine learning (ML) methods are currently used widely, including applications to concrete structures, e.g., compressive strength of concrete by Chopra et al. [3], compressive strength of FRP-sheet-confined cylindrical specimens by Naderpour et al. [4], shear capacity of concrete beams/slabs by Hoang and Ashour et al. [5,6]. Researchers (Dahou et al., Golafshani et al., Mousavi et al., Liang, Farouk et al.) have also utilized ML in predicting the bond behavior of reinforcing with concrete [7–11]. Despite these advancements, it is still challenging to develop data-driven models for materials such as UHPC and TiABs, where experimental data are scarce. This paper presents the description of collected dataset, utilization, and a framework to develop an explainable, data-driven model (based on ensemble learning) to accurately predict the ultimate bond strength between UHPC and TiABs.

2. Description of Dataset

An experimental dataset from the existing literature was collected to train ML models. A total of 350 experimental tests were collected, which include six input parameters (compressive strength, $f'c$; yield strength, f_y; tensile strength, f_u; bar diameter, d; embedment length, l; and concrete cover, c) and one output parameter (ultimate bond strength, τ). The input and output parameters are selected based on the availability of data from the existing literature.

3. Transfer Learning Methods

Transfer learning (TL) is the improvement of a learning ML model in a new task through the transfer of knowledge from a related task that has already been learned. Here, a model developed for a task is reused as the starting point for a model on a second task [12]. Figure 1 explains the transfer learning with the dataset used in the study. The entire dataset is first split into two different sets. The first set of data includes all the collected data for normal steel–normal concrete specimens and/or normal steel–UHPC specimens. The second set of data includes all the collected data for titanium-alloy-reinforced ultra-high-performance concrete (TARUHPC). Boundary condition in identifying the TARUHPC specimen is with the yield strength of TiABs (f_y: 130 ksi) and compressive strength of UHPC ($f'c$: 18 ksi). Both A and B datasets are trained separately and then the knowledge is transferred for a new task. Two major TL methods were selected for this study: (a) Domain-Weighted Support Vector Transfer Regression (DW-SVTR) and (b) Two-stage TrAdaBoost.R2.

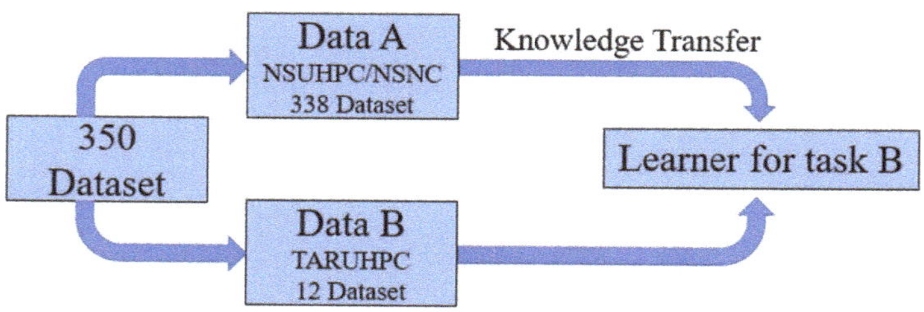

Figure 1. Transfer learning split of dataset.

3.1. Domain-Weighted Support Vector Transfer Regression (DW-SVTR)

Domain-weighted support vector transfer regression (DW-SVTR) is a novel regression-based transfer learning (TL) model which is a variant of least squares support vector machines for regression (LS-SVMR), coupling LS-SVMR with two weight functions [13]. Both weight functions have different effects. The first weight is obtained using kernel mean matching (KMM) to balance the source and target domains which offers more weight to source domain points that are relevant to target domain points. The second weight is a function of residuals to further reduce the negative interference of irrelevant source domain points corresponding to outliers for the target domain training sample. In short, the model first applies KMM to balance the source and target domains, then performs data augmentation to increase the number of training examples, and finally trains a weighted LS-SVMR model using the augmented training set. It then uses the trained model to predict the target values of the test set. The predicted target values are returned as the output.

3.2. *Two-Stage TrAdaBoost.R2*

Two-stage TrAdaBoost.R2 is a transfer learning (TL) model for regression problems modified by Paroe and Stone [14] from the existing boosting-based classification TL model TrAdaBoost [15]. For the prediction of bond strength, the model is obtained by implementing the Two-stage TrAdaBoost.R2 algorithm to train a decision tree regressor on a source domain and adapting it to a target domain. Best hyperparameters for transfer learning are then obtained. The model is then trained on the source domain data and a set of adapted models for each transfer learning step is produced. Finally, the adapted model is used to predict the target domain data to produce the predicted target value using the adapted model.

4. Results

The dataset was used to train and test the TL model in the target domain, i.e., TARUHPC specimens. The predicted bond strength from both DW-SVTR and TwoStage TrAdaBoost.R2 was then compared against the target bond strength that was obtained from the experimental results (Figure 2). The figure also presents the predicted output obtained from the empirical equations from ACI 318-19 (Equation (1)) [16] and Harajli (Equation (2)) [17]. It is evident from the plot that the empirical equations from ACI 318-19 and Harajli cannot accurately predict the bond strength of TARUHPC. However, the TL models can predict the bond strength more accurately for TARUHPC.

$$\tau = 5.91\frac{\sqrt{f'c}}{d}, \quad (1)$$

$$\tau = \sqrt{f'c}\left[1.2 + 3\frac{c}{d} + 50\frac{d}{l}\right]. \quad (2)$$

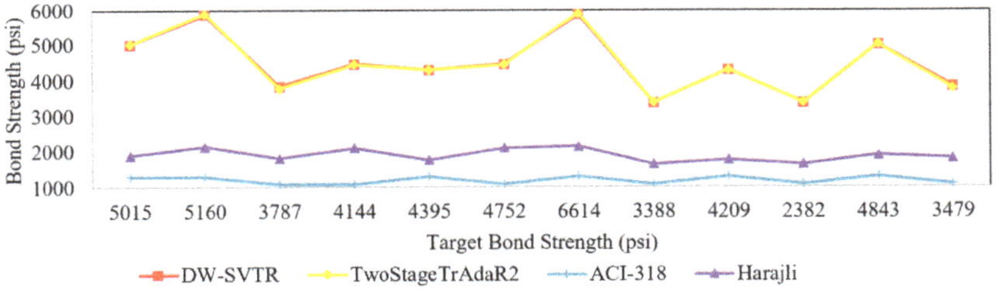

Figure 2. Comparison between target and predicted output for different models.

DW-SVTR, TwoStageTrAdaR2, ACI-318 and Harajli obtained an R^2 score (coefficient of determination) of 0.807, 0.809, −9.35 and −5.52, respectively. The R^2 score simply measures the amount of variance in the predictions explained by the dataset. Similarly, the MARD score (mean absolute relative deviation) was 0.0471, 0.0497, 0.722 and 0.570, respectively. MARD is interpreted as the average percentage error of the model in predicting the target variable.

5. Discussion

Titanium alloy bars (TiABs) and ultra-high-performance concrete (UHPC) are considered to be novel materials for civil infrastructures, e.g., bridges. The advantages associated with these materials are the main cause for them to be considered in sustainable infrastructure. TiABs is currently being studied by various researchers as a replacement of traditional steel reinforcing for concrete structures (beam and columns) and being utilized by various DOTs (Departments of Transportation) in the United States. UHPC as well are being studied and utilized as a replacement of traditional normal concrete for bridges, buildings and more. However, the combination of these two novel materials is yet to be studied. This research aimed to successfully predict the ultimate bond strength of TiABs with UHPC, which is an important factor in safety of concrete structures, e.g., structural load-bearing capacity, stiffness, or crack control. This research utilized two major TL techniques, DW-SVTR and TwoStageTrAdaR2, which very closely predicted the bond strength of TiABs with UHPC. On the other hand, empirical equations (ACI-318 and Harajli) need to be modified to successfully predict the bond strength of TiABs with UHPC. The research at Idaho State University is ongoing to accurately define the bond–slip relationship between TiABs and UHPC.

Author Contributions: Conceptualization, M.M. and J.C.; methodology, L.B., M.A. and M.M.; software, L.B., M.A. and A.B.; validation, M.A., L.B., A.B., M.M. and J.C.; formal analysis, M.A., L.B. and A.B.; investigation, M.A., L.B. and A.B.; resources, M.M. and J.C.; data curation, M.A., A.B. and L.B.; writing—original draft preparation, M.A. and A.B.; writing—review and editing, M.A., A.B., J.C., L.B. and M.M.; visualization, M.A., L.B. and A.B.; supervision, M.M., J.C. and L.B.; project administration, M.M. and J.C.; funding acquisition, M.M. and J.C. All authors have read and agreed to the published version of the manuscript.

Funding: This research was funded by Idaho State University Center for Advanced Energy Studies Seed Grant.

Institutional Review Board Statement: Not applicable.

Informed Consent Statement: Not applicable.

Data Availability Statement: The data are available upon request to the corresponding author.

Acknowledgments: This work was supported by State of Idaho, appropriated/funded for the Center for Advanced Energy Studies (CAES).

Conflicts of Interest: The authors declare no conflict of interest. The funders had no role in the design of the study; in the collection, analyses, or interpretation of data; in the writing of the manuscript; or in the decision to publish the results.

References

1. Acharya, M.; Khadka, R.; Mashal, M. Preliminary Bond Testing and Splicing of Titanium Alloy Bars. *Transp. Res. Rec. J. Transp. Res. Board* **2022**, *2676*, 410–427. [CrossRef]
2. Khadka, R.; Acharya, M.; LaBrier, D.; Mashal, M. Visualization of Macroscopic Structure of Concrete Based on X-ray Computed Tomography using Immersive Environments. In *HCII 2022*; Springer: Cham, Switzerland, 2022. [CrossRef]
3. Chopra, P.; Sharma, R.K.; Kumar, M. Prediction of Compressive Strength of Concrete Using Artificial Neural Network and Genetic Programming. *Adv. Mater. Sci. Eng.* **2016**, *2016*, 7648467. [CrossRef]
4. Naderpour, H.; Nagai, K.; Fakharian, P.; Haji, M. Innovative models for prediction of compressive strength of FRP-confined circular reinforced concrete columns using soft computing methods. *Compos. Struct.* **2019**, *215*, 69–84. [CrossRef]
5. Hoang, N.-D. Estimating punching shear capacity of steel fibre reinforced concrete slabs using sequential piecewise multiple linear regression and artificial neural network. *Measurement* **2019**, *137*, 58–70. [CrossRef]
6. Ashour, A.; Alvarez, L.; Toropov, V. Empirical modelling of shear strength of RC deep beams by genetic programming. *Comput. Struct.* **2003**, *81*, 331–338. [CrossRef]
7. Dahou, Z.; Sbartaï, Z.M.; Castel, A.; Ghomari, F. Artificial neural network model for steel–concrete bond prediction. *Eng. Struct.* **2009**, *31*, 1724–1733. [CrossRef]
8. Golafshani, E.M.; Rahai, A.; Sebt, M.H.; Akbarpour, H. Prediction of bond strength of spliced steel bars in concrete using artificial neural network and fuzzy logic. *Constr. Build. Mater.* **2012**, *36*, 411–418. [CrossRef]
9. Mousavi, S.M.; Peyma, A.B.; Moodi, Y. Predicting the Ultimate and Relative Bond Strength of Corroded Bars and Surrounding Concrete by Considering the Effect of Transverse Rebar Using Machine Learning. *Iran. J. Sci. Technol. Trans. Civ. Eng.* **2022**, *47*, 193–219. [CrossRef]
10. Liang, R.; Huang, Y.; Xu, Z. Experimental and Analytical Investigation of Bond Behavior of Deformed Steel Bar and Ultra-High Performance Concrete. *Buildings* **2022**, *12*, 460. [CrossRef]
11. Farouk, A.I.B.; Zhu, J.; Ding, J.; Haruna, S. Prediction and uncertainty quantification of ultimate bond strength between UHPC and reinforcing steel bar using a hybrid machine learning approach. *Constr. Build. Mater.* **2022**, *345*, 128360. [CrossRef]
12. Brownlee, J. A Gentle Introduction to Transfer Learning for Deep Learning. In Deep Learning for Computer Version; Machine Learning Mastery. Available online: https://machinelearningmastery.com/category/deep-learning-for-computer-vision/ (accessed on 15 November 2022).
13. Luo, H.; Paal, S.G. Reducing the effect of sample bias for small data sets with double-weighted support vector transfer regression. *Comput. Civ. Infrastruct. Eng.* **2020**, *36*, 248–263. [CrossRef]

14. Pardoe, D.; Stone, P. Boosting for regression transfer. In *Proceedings of the 27th International Conference on Machine Learning;* Fürnkranz, J., Joachims, T., Eds.; Omni Press: Madison, WI, USA, 2010.
15. Dai, W.; Yang, Q.; Xue, G.; Yu, Y. Boosting for transfer learning. In Proceedings of the 24th International Conference on Machine Learning. Ghahramani, A., Ed.; Omni Press: New York, NY, USA, 2007.
16. *ACI 318-19*; Building Code Requirements for Structural Concrete: (ACI 318-19); and Commentary (ACI 318R-19). American Concrete Institute: Farmington Hills, MI, USA, 2019.
17. Harajli, M.H. Development/splice strength of reinforcing bars embedded in plain and fiber reinforced concrete. *ACI Struct. J.* **1994**, *91*, 511–520.

engineering proceedings

MDPI

Proceeding Paper

Preliminary Mechanical Characterization of HMA Mixtures with a High Content of Recycled Materials †

Giulia Tarsi * and Cesare Sangiorgi

Department of Civil, Chemical, Environmental and Materials Engineering (DICAM), University of Bologna, 40131 Bologna, BO, Italy; cesare.sangiorgi4@unibo.it
* Correspondence: giulia.tarsi2@unibo.it

† Presented at the Second International Conference on Maintenance and Rehabilitation of Constructed Infrastructure Facilities, Honolulu, HI, USA, 16–19 August 2023.

Abstract: The use of recycled materials is necessary to realize the green transition towards carbon neutrality. Several waste products are highly valued materials that cannot be landfilled without exploiting their full potential. Promoting the circular economy concept, this study aims to produce more sustainable paving materials using selected recycled products in binders and asphalt mixes. Rubber (R) from End-of-Life Tyres (ELTs) and Re-refined Engine Oil Bottom (REOB), i.e., the by-product of waste lubricants refining, were employed to produce extended bitumens (25%wt. bitumen replacement) trying to solve the ELTs and REOBs large production, thus disposal, worldwide. In addition, recycled aggregates from various urban and industrial sources were used to halve the quantity of virgin mineral aggregates in the developed asphalt mixtures. Considering two different types of REOBs, two mass proportions of R and REOB and two production temperatures of extended bitumens, eight asphalt mixes containing about 50%wt. of recycled materials were manufactured and underwent to preliminary mechanical tests. The stiffness, tensile and moisture resistances of the greener asphalt concretes were evaluated and compared to two reference mixes: one mainly consisted of virgin materials, and another contained 50%wt. of recycled aggregates and neat bitumen. The eight greener mixes exhibited promising responses in terms of stiffness and tensile strength, showing better intermediate values than the reference ones, but more water susceptibility.

Keywords: rubber-REOB extenders; recycled aggregates; HMA; mechanical characterization

Citation: Tarsi, G.; Sangiorgi, C. Preliminary Mechanical Characterization of HMA Mixtures with a High Content of Recycled Materials. *Eng. Proc.* **2023**, *36*, 17. https://doi.org/10.3390/engproc2023036017

Academic Editor: Hosin (David) Lee

Published: 4 July 2023

1. Introduction

More than 90% of the European road network consists of asphalt mixtures [1], i.e., petroleum bitumen and virgin mineral aggregates. Hence, the asphalt industry needs to be restructured in order to substitute the conventional road products that come from non-renewable resources with suitable innovative materials. Pavement engineers have focused on eco-friendly road construction materials, encouraging the use of recycled raw materials to achieve the progressive, yet irreversible, green transition. In particular, recycled materials can be used as constituents of sustainable paving materials to partially or totally replace the bitumen and aggregates present in traditional asphalt concrete.

At the binder level, specific recycled products, such as waste polymers, resins and engine or bio-oils, can be used either as modifiers/additives for improving the performances of the final bituminous material, or as replacements for the virgin constituents themselves. The latter option represents the current trend of paving materials technology, which aims to produce extended bitumens or alternative binders, in which bitumen is replaced at a minimum percentage equal to 25% or 75%, respectively [2–4]. Among available recycled materials, this study employs rubber (R) from End-of-Life Tyres (ELTs) and Re-Refined Engine Oil Bottom (REOB), which is the by-product of the exhausted engine oils refinery process for producing second-hand lubricants, as constituents of extended bitumens. The recycling of these two wastes may help in reducing the disposal concerns related to their large-scale production worldwide. Only in Italy, the collection of ELTs exceeded 210,000 tons in 2019 [5]; while ≥180,000 tons of waste engine oils were collected in 2018 [6]. Recent studies have investigated the possibility of combining the use of waste polymers and oils to produce sustainable extended bitumen, increasing the recycling rate of bitumen. Their combination could exploit the positive effects of both recycled materials, while mitigating their drawbacks. In fact, the softening effect of REOB can represent a key factor to compensate for the increased stiffness of R-modified bitumens [7,8]. Moreover, R can improve the performance of the final binder at high temperatures and REOB at low temperatures [8–10]. At the asphalt mix level, the use of recycled aggregates can help the paving industry to achieve the sustainability goals due to the high energy demand of aggregates' production and greenhouse gas emissions [11]. Reclaimed asphalt pavement (RAP), construction and demolition (C&D) and various urban and industrial wastes are the main sources of recycled aggregates. These materials can substitute virgin materials, contributing to the circular economy [12]. In this research, the recycled aggregates from RAP, urban and industrial wastes are used to replace 50%wt. of the virgin ones.

The authors aim to evaluate and, eventually, assess the feasibility of using high contents of recycled materials at the binder and the asphalt mix levels for the production of greener paving materials, satisfying the specifications of traditional road products.

2. Materials

A 50/70 penetration grade bitumen (coded as B) was used for producing extended bitumens and greener asphalt mixes, and it is the reference material of the study. Powdered rubber (R) from ELTs and two Re-refined Engine Oil Bottoms (REOBs) from two distinct refining plants, namely O1 and O2, were used to replace 25%wt. of bitumen, producing extended bitumens. The size distribution of R varies between 0 and 0.42 mm. The two REOBs have similar density, about 1.0 g/cm^3, but O1 is less viscous than O2.

A total of eight R-REOB extenders were produced by varying the type of REOB (O1 and O2), the R-REOB mass proportion and the production temperature. In detail, R-REOB mass proportions equal to 1:1 and 1:2 (coded R and R2, respectively) were selected based on a preliminary study that can be found elsewhere [10]. The production temperatures were 160 °C and 180 °C. Pre-heated R and REOBs at each production temperature were mixed at a rate of 800 rpm for 1 h. REOBs were warmed up in oven for at least 1 h 30′, while R for 15′. While mixing, all blends were continuously heated by a heating plate. Once prepared, each R-REOB blend was incorporated into B by replacing its 25%wt. for producing extended bitumens. The B and extenders were preliminary heated in oven at either 160 °C or 180 °C for ≥1 h 30′ and 15′, respectively, keeping apart before combining them. Then, the blend was heated again for 15′ prior mixing at 800 rpm for 1 h.

The mix design of Hot Mix Asphalt (HMA) mixes for wearing course layers in compliant with Italian specifications was adopted. Eight greener HMA mixes and a control mixture containing 50%wt. of recycled aggregates (8% incinerated urban wastes, 22% industrial wastes and 20% RAP) and an optimum binder content of 5.7% on the weight of aggregates were produced using either the developed extended bitumens and the Pen 50/70. These mixes with 50% of recycled aggregates were coded as CAM (i.e., Minimum Environmental Criteria), followed by the name of the extended or neat bitumen used. Moreover, the CAM mixes were compared with a traditional HMA that mainly contains virgin materials (T 50/70). Its optimum binder content was 5.6%. Four samples per each mix were manufactured by using the gyratory compactor (EN 12697-31, 120 gyrations).

3. Mechanical Characterizations

Static and dynamic characterizations of all HMA mixes were performed. Two static mechanical tests were used to measure the Indirect Tensile Strength (ITS) and the Indirect Tensile Strength Ratio (ITSR) of mixes in compliance with EN 12697-23 and EN 12697-12 standards, respectively. The tensile strength of asphalt concretes was determined by applying a compression load at a constant speed rate of 51 mm/min. The ITS test was performed at 25 °C. The ITSR ratio measures the durability of the samples as it determines the effect of saturation and accelerated water conditioning. This investigation quantifies the ratio between the ITS values of an asphalt mix after water conditioning to that of a dry specimen. All mixes were conditioned in water bath according to Method A of the standard. Then, the samples were removed, dried and conditioned at 25 °C in a climate chamber before undergoing to ITS test. During these characterizations, the load was applied until a failure occurred; hence, the set of four specimens per each HMA mix were halved to perform the ITS test on dry and wet samples. Regarding dynamic characterization, all HMA mixes were subjected to the Indirect Tensile Stiffness Modulus (ITSM) test by using a servo-pneumatic testing machine. The stiffness modulus was established according to EN 12697-26 standard in the IT-CY configuration. Pulse loading was applied with a 124 ms rise time to generate a horizontal deformation of 5 ± 2 µm. As it is a non-destructive test, it was replicated at three different test temperatures, namely 10, 20 and 30 °C. Before being tested, all samples were conditioned at the test temperature for at least 4 h.

4. Results and Discussion

The preliminary mechanical responses of the innovative CAM mixes and the two control mixtures are reported in Figure 1. In general, the ITS test highlighted the intermediate response of the CAM mixes manufactured using the developed extended bitumens from the two control ones, i.e., T B and CAM B. Only the CAM BO1R 180 and CAM BO1R2 180 mixes showed less strength than T B. Nonetheless, all mixes met the Italian specification that requires an ITS value higher than 0.7 MPa. Regarding the water susceptibility, both control mixes exhibited a very low level of water susceptibility. Despite all HMA mixtures exceeded the minimum ITSR value requested by the Italian specification, that is usually equal to 90%, the CAM mixes produced using the eight extended bitumens were more susceptible to water damage than the control ones. The ITSM results showed that the CAM mixes with extended bitumens behaved similarly to T B, especially at 20 °C. At 10 °C, the CAM mixes with R-REOB-modified bitumens generally showed a similar or lower levels of stiffness than T B, which can be related to the softening effect caused by REOBs. The opposite results were obtained at 30 °C. The presence of R stiffened the final paving mixes. The T B and the CAM mixes with extended bitumens were less stiff than CAM B at all test temperatures. This trend can be related to the presence of recycled aggregates and, especially, RAP. A general increase in ITS, ITSR and ITSM values was observed when enhancing the content of R. On the other hand, the effects of the production temperature and the type of REOB were not clear.

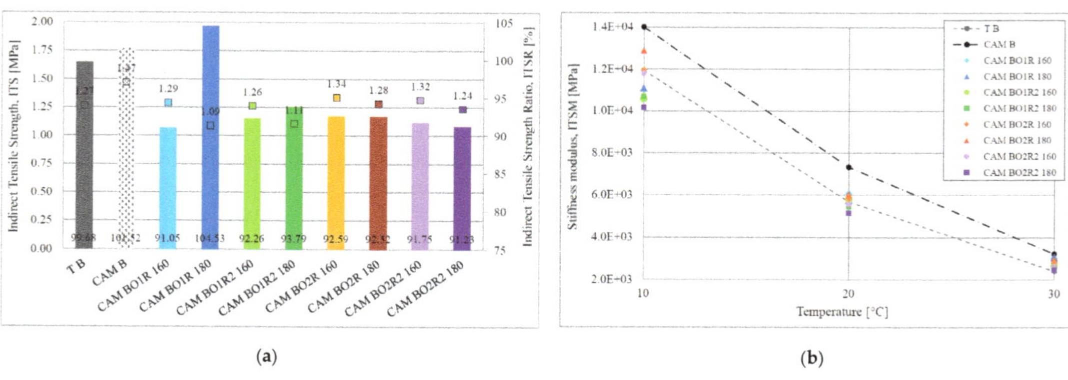

(**a**) (**b**)

Figure 1. Results of HMA mixes under mechanical analysis: (**a**) ITS and ITSR tests; (**b**) ITSM test.

5. Conclusions

The preliminary mechanical characterization of greener paving materials was conducted to evaluate and, eventually, assess the feasibility of the use up to about 50%wt. of recycled materials in HMA mixes for wearing course layers. Thanks to static and dynamic mechanical tests, eight CAM mixes containing the developed extended bitumens were analysed. The proposed greener HMA mixes achieved intermediate values between the control mixtures (T B and CAM B), highlighting similar behaviour of traditional road construction materials. This means that the use of R-REOB-modified bitumens can mitigate the increased stiffness caused by the use of recycled aggregates, especially RAP. The ITSM values confirmed the stiffening effect of R and its influence at high temperatures and the softening effect of REOB. Water conditioning has a negative effect on CAM mixtures. Albeit all mixes meet the Italian specification, the use of recycled materials seems to increase the water susceptibility of the final asphalt materials, which needs further investigations.

Author Contributions: Conceptualization, G.T. and C.S.; methodology, G.T. and C.S.; validation, G.T. and C.S.; formal analysis, G.T.; investigation, G.T.; data curation, G.T.; writing—original draft preparation, G.T.; writing—review and editing, G.T. and C.S.; supervision, C.S.; funding acquisition: C.S. All authors have read and agreed to the published version of the manuscript.

Funding: This research was founded by ECOPNEUS Scpa, Italy.

Institutional Review Board Statement: Not applicable.

Informed Consent Statement: Not applicable.

Data Availability Statement: The data presented in this research are available upon request.

Acknowledgments: The authors would thank ECOPNEUS Scpa for the financial support and for providing the rubber and ITELYUM Regeneration Srl for providing the REOBs.

Conflicts of Interest: The authors declare no conflict of interest. The founder had no role in the design of the study, in the analysis/interpretation of data, in writing/publication of the manuscript.

References

1. European Asphalt Pavement Association (EAPA). Asphalt—A Key Construction Product for the European Circular Economy. EAPA Position Paper. 2022. Available online: https://eapa.org/eapa-position-papers/ (accessed on 3 July 2023).
2. Raouf, M.A.; Williams, R.C. Temperature susceptibility of non-petroleum binders derived from bio-oils. In Proceedings of the 7th Asia Pacific Conference on Transportation and the Environment, Semarang, Indonesia, 3–5 June 2010.
3. Jimenez del Barco-Carrion, A.; Perez-Martinez, M.; Themeli, A.; Lo Presti, D.; Marsac, P.; Pouget, S.; Hammoum, F.; Chailleux, E.; Airey, G.D. Evaluation of bio-materials' rejuvenating effect on binders for high-reclaimed asphalt content mixtures. *Mater. Constr.* **2017**, *67*, 327. [CrossRef]
4. Aziz, M.d.M.A.; Rahman, M.T.; Hainin, M.R.; Bakar, W.A.W.A. Alternative binders for flexible pavement. *Constr. Build. Mater.* **2015**, *84*, 315–319. [CrossRef]
5. *Il Circolo Vizioso che fa Crescere il Paese—Report 2019*; Ecopneus Scpa: Milano, Italy, 2019.
6. *Rapporto di Sostenibilità 2018*; Consorzio Nazionale Oli Usati (CONOU): Rome, Italy, 2018.
7. Amini, A.; Imaninasab, R. Investigating the effectiveness of vacuum tower bottoms for asphalt rubber binder based on performance properties and statistical analysis. *J. Clean. Prod.* **2018**, *171*, 1101–1110. [CrossRef]
8. Fernandes, S.R.; Silva, H.M.; Oliveira, J.R. Developing enhanced modified bitumens with waste engine oil products combined with polymers. *Constr. Build. Mater.* **2018**, *160*, 714–724. [CrossRef]
9. Becker, Y.; Méndez, M.P.; Rodrìguez, Y. Polymer modified asphalt. *Vis. Tecnol.* **2001**, *9*, 39–50.
10. Tarsi, G.; Caputo, P.; Porto, M.; Sangiorgi, C. A Study of Rubber-REOB Extender to Produce Sustainable Modified Bitumens. *Appl. Sci.* **2020**, *10*, 1204. [CrossRef]
11. Ma, F.; Sha, A.; Lin, R.; Huang, Y.; Wang, C. Greenhouse Gas Emissions from Asphalt Pavement Construction: A Case Study in China. *Int. J. Environ. Res. Public Health* **2016**, *13*, 351. [CrossRef] [PubMed]
12. Martinho, F.C.G.; Picado-Santos, L.G.; Capitão, S.D. Feasibility Assessment of the Use of Recycled Aggregates for Asphalt Mixtures. *Sustainability* **2018**, *10*, 1737. [CrossRef]

engineering proceedings

MDPI

Proceeding Paper

On-Board Evaluation of Pavement Wetness from Water Spray †

Ebrahim Riahi [1], Wiyao Edjeou [2], Manuela Gennesseaux [2], Sebastien Buisson [2], Veronique Cerezo [2] and Minh-Tan Do [2,*]

1 TS2-LMA, Gustave Eiffel University, 13300 Salon de Provence, France; ebrahim.riahi@univ-eiffel.fr
2 AME-EASE, Gustave Eiffel University, 44344 Bouguenais, France; wiyao.edjeou@associated.ltu.se (W.E.); manuela.gennesseaux@univ-eiffel.fr (M.G.); sebastien.buisson@univ-eiffel.fr (S.B.); veronique.cerezo@univ-eiffel.fr (V.C.)
* Correspondence: minh-tan.do@univ-eiffel.fr

† Presented at the Second International Conference on Maintenance and Rehabilitation of Constructed Infrastructure Facilities, Honolulu, HI, USA, 16–19 August 2023.

Abstract: The paper presents an accelerometer-based system that can be implemented in a passenger car to estimate water depths on a road surface. Tests conducted on various road surfaces at different vehicle speeds allow the determination of an appropriate location of the accelerometers and define a relevant filtering of the recorded signals. Results show that the system can relate acceleration amplitudes to water depths lower than 0.5 mm. The potential use of the system to warn drivers under adverse weather conditions, especially when skid resistance drops during and after a precipitation, is presented and discussed.

Keywords: road safety; water depth; accelerometric signal; signal filtering; skid resistance

Citation: Riahi, E.; Edjeou, W.; Gennesseaux, M.; Buisson, S.; Cerezo, V.; Do, M.-T. On-Board Evaluation of Pavement Wetness from Water Spray. *Eng. Proc.* **2023**, *36*, 18. https://doi.org/10.3390/engproc2023036018

Academic Editor: Hosin (David) Lee

Published: 4 July 2023

1. Introduction

Skid resistance represents the contribution of road surfaces to the tire/road friction, which allows drivers to control their vehicles. A low skid resistance can induce accidents (lane departure, collisions) [1]. Among the factors that can reduce skid resistance, water is one of the most important. Previous works showed that the coefficient of friction decreases as the water depth on the road surface increases [2,3]. In [2], it was found that, beyond a critical water depth (around 0.2 mm), the reduction in skid resistance can be rapid.

Knowledge of the water depth on a road surface is then imperative to driver safety. Many attempts have been made on the development of a reduced-cost system that can be mounted on passenger cars to estimate water depths. The starting point of these works is the vibration of a car's wheel arch due to water spray. By disposing accelerometers [4,5] or capacitive transducers [6], on the backside of a wheel arch, it is possible to detect water depths from less than 0.5 mm [4,6] to 2 cm [5]. Riahi developed an accelerometer-based system and, from experiments conducted on a trailer equipped with a passenger car's tire [7], proved that it is possible not only to detect the presence of water, but also to estimate its thickness.

This paper constitutes the continuity of Riahi's work, by transposing the developed system to a real passenger car. In the following sections, experiments (sensors, test setup and test program) are first described. Results are then presented and discussed.

2. Methodology

As stated in [2], the rapid drop in skid resistance is observed for low water depths (around 0.2 mm). Therefore, the developed system is focused on the estimation of water depths lower than 1 mm. Four types of water flow result from a tire rolling on a wet road [8] (Figure 1): frontal splash, torrent spray, circumferential spray and side splash. As indicated in [9], the torrent and circumferential sprays reflect the presence of low water depths. The first step, then, is to measure vibrations caused by torrent and circumferential sprays. The second step is to develop a signal processing method to remove noise from accelerometric signals induced by car vibrations or road surface irregularities.

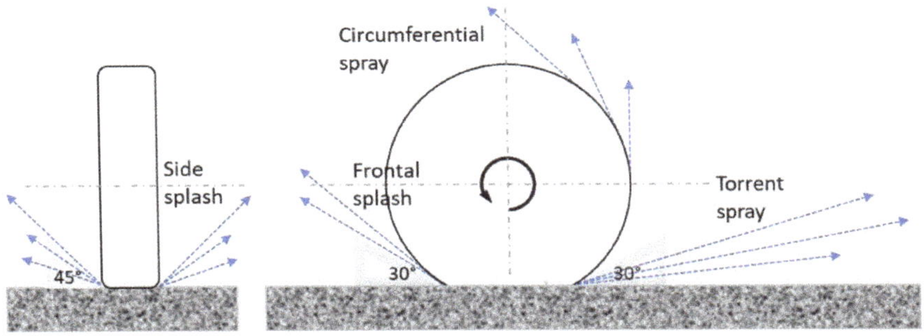

Figure 1. Flows generated by the ejection of water from the tire–road contact area ((**left**) front view of the tire; (**right**) side view of the tire; the direction is from the right to the left).

3. Experiments

3.1. Sensors

3.1.1. Accelerometers

Brüel & Kjær piezoelectric accelerometers (type 4507) are positioned on the back side of the right front wheel arch. To define an appropriate location for the accelerometers, four positions are tested (Figure 2a): at the wheel arch's bottom (defined as 0° position) and at 36°, 52° and 90°, respectively (it is not possible to position accelerometers A2 and A3 at 30° and 60°, respectively, due to the shape of the wheel arch). Measurements are performed with a sampling rate of 38 kHz, using a HBM data acquisition system.

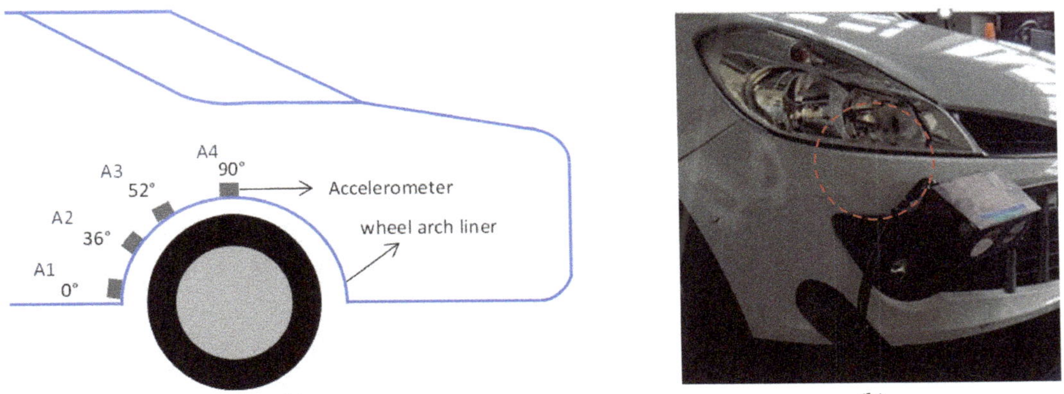

(**a**) (**b**)

Figure 2. Sensors used for the experiments: (**a**) position of the accelerometers on the front wheel arch (A1 to A4 represent the accelerometers); (**b**) Aquasens sensor mounted on the test car.

3.1.2. Optical Sensor to Measure Water Depths

The so-called Aquasens sensor [10] produces a white light source which illuminates the road surface. The reflected light, altered by the presence of water on the road surface, is analyzed to determine the water depth. The sensor is mounted on the front of the test car (Figure 2b, red circle) at 50 cm above the ground, with an inclination angle of 45°. The measurement area is located 1.5 m in front of the axis of the wheel.

3.2. Test Tracks

Two surfaces are tested (Figure 3a): "low texture" (C1, left) and "high texture" (E1, right) surfaces with a mean texture depth (MTD) of 0.38 mm and 0.95 mm, respectively. The test track contains a rotating irrigation system to simulate a wet road (Figure 3b). Tests were performed at dry and wet conditions. Six test speeds (20, 30, 40, 50, 60 and 70 km/h) were conducted. Each test configuration was repeated three times.

(**a**) (**b**)

Figure 3. Test tracks: (**a**) test surfaces; (**b**) wetting system.

4. Results

Signal Processing

Comparisons of dry and wet power spectral densities (PSDs) (Figure 4a) show that the system can detect the presence of water. The difference (wet/dry) is pronounced for frequencies between 2.6 and 6.5 kHz (grey zone in Figure 4b).

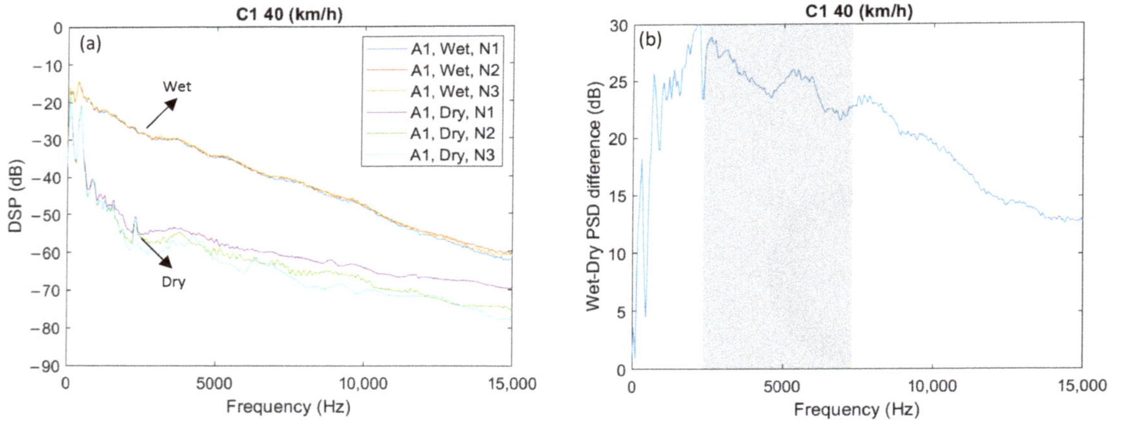

Figure 4. Power spectral density (PSD) of the recorded signals on C1 surface at 40 km/h: (**a**) comparison of dry vs. wet (N1 to 3 are the repeated runs); (**b**) difference between PSD (wet) and PSD (dry).

A Chebyshev filter is used with a passing band of 2–7 kHz to filter the raw signals. The best relationship between acceleration amplitudes and water depths is obtained with accelerometer A1 for speeds higher than 40 km/h (Figure 5). This result is promising, because wetness detection is most needed for high speeds. The water depth, and the corresponding acceleration amplitude, decrease when the macrotexture (MTD) increases.

This result would be expected, because there is less accumulated water on high-MTD surfaces and, hence, less water spray.

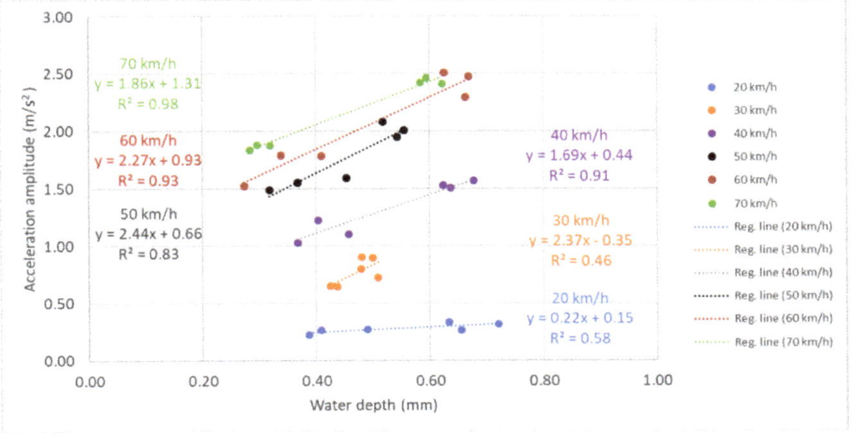

Figure 5. Relationship of acceleration amplitude to water depth (results from accelerometer A1).

5. Conclusions

Tests conducted on a passenger car (Clio 3) show that accelerometers can be positioned on the backside of the front wheel arch to record vibrations induced by water spray, and relate the acceleration amplitude to the water depth on the pavement. This finding opens the possibility of estimating wet friction and using this information for driver assistance (or, more generally, for autonomous driving) or road maintenance (detection of local slippery areas).

Author Contributions: Conceptualization, M.-T.D. and E.R.; methodology, E.R. and S.B.; formal analysis, E.R. and W.E.; writing—original draft preparation, M.-T.D. and E.R.; writing—review and editing, M.G. and V.C. All authors have read and agreed to the published version of the manuscript.

Funding: This study was performed in the framework of the ENA project (Autonomous Shuttle Experiments) financed by the French Government as part of the Future Investments Program now integrated into France 2030, and operated by the Environment and Energy Management Agency—ADEME. Project labeled by CARA European Cluster for Mobility Solutions.

Institutional Review Board Statement: Not applicable.

Informed Consent Statement: Not applicable.

Data Availability Statement: Data presented in this study are available on request from the first author. The data are not publicly available due to a confidentiality agreement within the framework of the ENA project.

Conflicts of Interest: The authors declare no conflict of interest.

References

1. McCarthy, R.; Flintsch, G.; de Leon Izeppi, E. Impact of skid resistance on dry and wet weather crashes. *J. Transp. Eng. Part B Pavements* **2021**, *147*, 04021029. [CrossRef]
2. Do, M.-T.; Cerezo, V.; Beautru, Y.; Kane, M. Modeling of the connection road surface microtexture/water depth/friction. *Wear* **2013**, *302*, 1426–1435. [CrossRef]
3. Yu, M.; You, Z.; Wu, G.; Kong, L.; Liu, C.; Gao, J. Measurement and modeling of skid resistance of asphalt pavement: A review. *Constr. Build. Mater.* **2020**, *260*, 119878. [CrossRef]
4. Prevost, D.; Cerezo, V.; Do, M.T.; Chabanon, C. On-board estimation of water depth using low-cost sensors. In Proceedings of the SURF 2012 (7th Symposium on Pavement Surface Characteristics), Norfolk, VA, USA, 19–22 September 2012.

5. Schmiedel, B.; Gauterin, F.; Unrau, H.-J. Road wetness quantification via tyre spray. *Proc. Inst. Mech. Eng. Part D J. Automob. Eng.* **2018**, *233*, 28–37. [CrossRef]
6. Döring, J.; Beering, A.; Scholtyssek, J.; Krieger, K.-L. Road Surface Wetness Quantification Using a Capacitive Sensor System. *IEEE Access* **2021**, *9*, 145498–145512. [CrossRef]
7. Riahi, E.; Edjeou, W.; Buisson, S.; Gennesseaux, M.; Do, M.-T. Estimation of water depth on road surfaces using accelerometric signals. *Sensors* **2022**, *22*, 8940. [CrossRef] [PubMed]
8. Weir, D.H.; Strange, J.F.; Heffley, R.K. *Reduction of Adverse Aerodynamic Effects of Large Trucks*; Report FHWA-RD-79-84; Federal Highway Administration: Washington, DC, USA, 1978.
9. Schmiedel, B.; Gauterin, F. Tire Splash and Spray Directly before and during Hydroplaning. *Tire Sci. Technol.* **2019**, *47*, 141–159. [CrossRef]
10. Holzwarth, F.; Eichhorn, U. Non-contact sensors for road conditions. *Sens. Actuators A Phys.* **1993**, *37–38*, 121–127. [CrossRef]

engineering proceedings

Proceeding Paper

How Drivers Feel When Traversing Speed Humps under a Variety of Driving Conditions †

Hong-jun Cho, Ho-hyuk Na and Do-Gyeong Kim *

Department of Traffic Engineering, University of Seoul, 163 Seoulsiripdae-ro, Dongdaemun-gu, Seoul 02504, Republic of Korea; peterchol@uos.ac.kr (H.-j.C.); skghgur@uos.ac.kr (H.-h.N.)
* Correspondence: dokkang@uos.ac.kr

† Presented at the Second International Conference on Maintenance and Rehabilitation of Constructed Infrastructure Facilities, Honolulu, HI, USA, 16–19 August 2023.

Abstract: Speeding is known to be one of the main causes of traffic crashes. Therefore, various speed management techniques are applied to prevent accidents caused by speeding in many countries. Among them, speed humps are regarded as a cost-effective facility compared to other speed management techniques. Although the development of speed bumps continues through various studies, most studies focus on reducing the amount of impact, so there is no study that can comprehensively determine the driver's feeling. Therefore, in this study, a generalized ordered regression model was used to demonstrate the driver's feelings when passing a speed hump under various driving conditions through field experiments.

Keywords: speed humps; user preference; generalized ordered logit model; sinusoidal; speed management

Citation: Cho, H.-j.; Na, H.-h.; Kim, D.-G. How Drivers Feel When Traversing Speed Humps under a Variety of Driving Conditions. *Eng. Proc.* **2023**, *36*, 19. https://doi.org/10.3390/engproc2023036019

Academic Editor: Hosin (David) Lee

Published: 4 July 2023

1. Introduction

Speeding is known to be one of the main causes of traffic accidents. According to Injury Facts, speeding was a factor accounting for 29% of all road fatalities in 2020, killing 11,258 people, representing an average of more than 30 deaths per day [1]. Therefore, many states in the United States are applying various speed-reducing traffic calming techniques to prevent accidents caused by speeding [2]. There are various techniques that can be applied to traffic calming techniques, such as speed bumps, cameras, and signs [3]. Among them, speed bumps are considered the most excellent facility for speed reduction compared to other speed management technologies [4]. They cause problems such as discomfort due to vehicle passing impact [5], vehicle damage [6], occupant injury [7], and noise pollution [8]. Among these problems, several studies have been conducted to develop various types of speed bumps in order to reduce the amount of impact applied to the driver [8–10].

However, most studies only deal with how the impact amount changes depending on the shape, so there is no research to measure the degree of comfort felt by the driver even if the impact amount is reduced. Therefore, the purpose of this study is not only to reduce the amount of impact by changing the shape, but also to determine whether the comfort or discomfort increases or decreases under what conditions because the comfort felt by the driver can vary depending on the shape and the passing condition. In this study, a generalized ordered logit model that can be applied by relaxing the assumption of the parallel line that the influence of the dependent variable is the same is used to reflect the size of the influence of the driving conditions according to the driver's emotional state [11].

2. Methodology

The ordered logit model is a model suitable for verifying the effect of an independent variable on a dependent variable when the dependent variable has a sequence relationship [12]. It is used when there is a hierarchy of dependent variables, such as accident severity (injury, minor injury, serious injury, death) or satisfaction level (very dissatisfied, dissatisfied, neutral, satisfied, very satisfied) [13]. The biggest feature of the ordered logit model is that it assumes parallelism. In the model, the inclination coefficients of the independent variables have the same value, and only the intercept terms of the regression lines are different. That is, they have different parallel regression lines. For this reason, the ordered logit model is also known as the proportional odds model.

However, the parallel regression line assumption is not always satisfied. This assumption is frequently violated, in which case the use of sequence logit models is inappropriate. Therefore, in this study, the "autofit" command (STATA) was used to determine whether the assumption of parallelism of various variables was violated. In addition, the omodel command was used to determine the significance of the ordered logit model results. If it determines that the sequence logit model results are not significant, it can provide a basis for applying a generalized ordinal logit model with relaxed parallelism assumptions.

The generalized ordered logit model is a model that considers that the inclination may not be parallel by alleviating the parallelism assumption of the ordered logit model [14]. In this study, we ran a generalized ordered logit model using the STATA module gologit2 and checked the *p*-values of each variable to determine whether the variables were parallel. After this process, we interpreted the results of the generalized ordered logit model.

3. Study Procedure

3.1. Variables

The dependent variable of the data is Satisfaction (SAT), which is categorized into Dissatisfied, Neutral, and Satisfied. The independent variables were career, car, speed, shock, and bump type. Car type and bump type were set as categorical variables and used as dummy variables; car_1 means a van, car_2 means a truck, and type_A, type_B, and type_C mean s-type bump types A, B, and C. (Table 1)

Table 1. Variable description.

Variables	Description	Measurement
Satisfaction	Driver's satisfaction through poll	Dissatisfy = 1, Neutral = 2, Satisfy = 3
speed humps type_A type_B type_C	4 type of speed humps passed by drivers (dummy variable)	Passing type A = 1, else = 0 Passing type B = 1, else = 0 Passing type C = 1, else = 0 (Passing type O: type_A, type_B, type_C = 0)
Car type car_1 car_2	3 type of cars with drivers (dummy variable)	Ride in a van = 1, else = 0 Ride in a truck = 1, else = 0 (Ride in a passenger car: car_1, car_1 = 0)
Career	Driver's driving experience	years
speed	Car's speed passing speed hump	km/h
shock	impact the driver receives	$m/s^{1.75}$

3.2. Data

We conducted an experiment at Automobile Safety Research Institute of Korea Transportation Safety Authority. For the field experiment, four types of speed bumps were installed, including parabolic ones. In addition, a field experiment was conducted by setting a human body impulse measurer and a satisfaction survey questionnaire. The experimental vehicles consisted of passenger cars, SUVs, and trucks, and a total of 10 drivers drove through four types of speed bumps at a speed of 10 km/h to 50 km/h and had them fill out a satisfaction questionnaire. And research was conducted. (Figure 1)

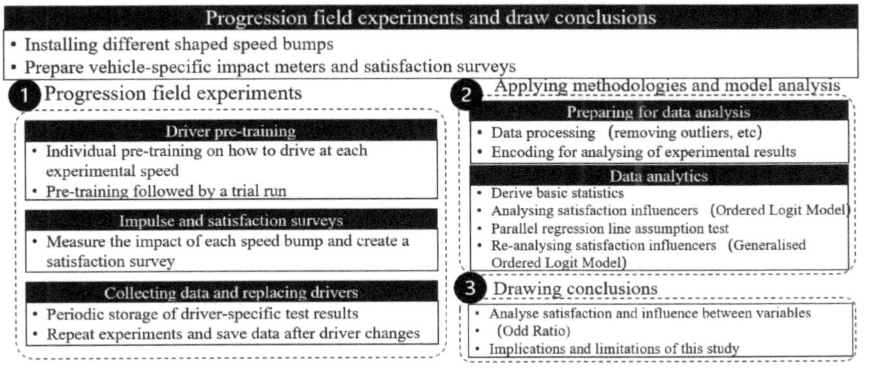

Figure 1. Research Progression.

4. Result

4.1. Result of Ordered Logit Model

In this study, the results of the ordered logit model were first drawn. We utilized STATA's module ologit and showed the odds ratio of each variable (Table 2).

Table 2. Result of ordered logit model.

Variables	type_A	type_B	type_C	car_1	car_2	Career	Speed	Shock	/cut1	/cut2
Coefficient	0.3297	0.2665	0.6329	−1.2217	−2.4531	0.1763	−0.1207	−0.7192	−6.3696	−4.8559
Odds ratio	1.3905	1.3054	1.8831	0.2947	0.0860	1.1927	0.8863	0.4872	−6.3696	−4.8559
$p > \lvert z \rvert$	0.273	0.370	0.036	0	0	0	0	0		

4.2. Assessing Parallel-Line Assumption

However, since the ordered logit model assumes that the regression lines are all parallel, it is necessary to test whether the results of this model are significant. Therefore, in this study, the parallel regression line assumption was tested using the autofit module as shown in Table 3. Among all the variables, the car_2, speed, and shock variables had p-values below 0.05, so we rejected the assumption of parallelism. For the remaining variables, the p-value was greater than or equal to 0.05, so the assumption of parallelism was accepted, confirming that the slopes of the regression lines are equal.

Table 3. Autofit result of Parallel-Line Assumption.

Variables	type_A	type_B	type_C	car_1	car_2	Career	Speed	Shock
p Value	0.0873	0.1906	0.8832	0.8147	0.0089	0.3633	0.0001	0.0109

Prob > chi2 = 0.3104.

4.3. Result of Generalized Ordered Logit Model

As shown in Table 4, the variables for types A and B are not statistically significant ($p > 0.05$), while all other variables are statistically significant ($p < 0.05$). First, looking at the relationship between speed bump type and satisfaction, when passing type C rather than type O, the possibility that the driver chooses a higher level of satisfaction increases by about 96% ((1.9655 − 1) × 100 = 96). When selecting a truck, the probability that the driver chooses satisfaction over dissatisfaction decreases by about 96%, and the probability of choosing satisfaction over dissatisfaction or average decreases by about 86%. That is, passenger cars, vans, and trucks showed high satisfaction in that order [14].

Table 4. Result of generalized ordered logit model.

	Variables	type_A	type_B	type_C	car_1	car_2	Career	Speed	Shock	Cons
1	Coefficient	0.3199	0.2793	0.6757	0.179	−1.1583	−3.202	−0.1592	−0.4809	7.5232
	Odds ratio	1.3769	1.3222	1.9655	1.1961	0.314	0.0407	0.8528	0.6182	1850.54
	$p > \|z\|$	0.29	0.35	0.026	0	0	0	0	0.003	0
2	Coefficient	0.3199	0.2793	0.6757	0.179	−1.1583	−1.9609	−0.0985	−0.9804	4.4392
	Odds ratio	1.3769	1.3222	1.9655	1.1961	0.314	0.1407	0.9062	0.3752	84.7063
	$p > \|z\|$	0.29	0.35	0.026	0	0	0	0	0	0

5. Conclusions

First of all, the level of satisfaction with type C of the s-shaped bumps was higher than that of the original arc type. Therefore, replacing the existing arc-shaped bumps with type C bumps can be one of the countermeasures to reduce the driver's discomfort due to the existing speed bumps. In addition, the satisfaction level of vans and trucks was lower than that of passenger cars. Therefore, when crossing a speed bump, the driver of a van or truck must reduce the speed more than a passenger car. Through this, if you make them aware of this fact during driving education, it will be possible to reduce the inconvenience caused by speed bumps to drivers. Finally, satisfaction decreased as passing speed increased. Through this, it was confirmed numerically that the speed bumps, whose purpose is to reduce the speed of passing vehicles, are effective.

Author Contributions: Conceptualization, D.-G.K.; methodology, H.-j.C.; Supervision, D.-G.K.; Validation, H.-j.C.; Investigation, H.-h.N.; Writing-original draft preparation, H.-j.C.; Writing-review and editing, H.-h.N. and D.-G.K. All authors have read and agreed to the published version of the manuscript.

Funding: This research was supported by a grant (RS-2021-KA160861) from Technology Business Innovation Program(TBIP) funded by Ministry of Land, Infrastructure and Transport of Korea government.

Institutional Review Board Statement: Not applicable.

Informed Consent Statement: Informed consent was obtained from all subjects involved in the study.

Data Availability Statement: Not applicable.

Conflicts of Interest: The authors declare no conflict of interest.

References

1. NHTSA. Available online: https://www.nhtsa.gov/risky-driving/speeding (accessed on 28 February 2023).
2. U.S. Department of Transportation Federal Highway Administration. Available online: https://highways.dot.gov/safety/speed-management/uslimits2/traffic-calming-eprimer/module-1-purpose-and-organization-eprimer (accessed on 28 February 2023).
3. Ewing, R. *Traffic Calming: State of the Practice ITE/FHWA*; Semantic Scholar: Seattle, WA, USA, 1999; pp. 17–65.
4. Amirarsalan, M.M.; Ali, A.K. Optimization of Speed Hump Profiles Based on Vehicle Dynamic Performance Modeling. *J. Transp. Eng.* **2014**, *140*, 04014035. [CrossRef]
5. Philip, A.W.; John, P.B. Toward a North America Geometric Design Standard for Speed Humps. *ITE J.* **2000**, *70*, 30–34.
6. Ufuk, K. Vibration Levels Exposed at Speed Bump and Speed Hump Transitions. *AKU J. Sci. Eng.* **2022**, *22*, 332–341.
7. Milan, A. Speed Hump Spine Fractures: Injury Mechanism and Case Series. *J. Spinal Disord. Tech.* **2011**, *24*, 332–341.
8. Emad, K.; Mohammed, A. Performance analysis of speed control humps and dips based on health and comfort criteria. *J. Veh. Noise Vib.* **2017**, *13*, 295.
9. Abdulgazi, G.; Ertugrul, B.; Abdullah, H.L. An investigation into the effect of parabolic speed hump profiles on ride comfort and driving safety under variable vehicle speeds: A campus experience. *Sustain. Cites Soc.* **2019**, *45*, 413–421.
10. Ali, A.K.; Amirarsalan, M.M. Simulation Modeling of Dynamic Response of Vehicles to Different Types of Speed Control Humps. In *T&DI Congress 2014*; Amiy, V., Geoffrey, D.G., Eds.; ASCE Library: Reston, VI, USA, 2014; pp. 533–542.

11. Getahun, E.A.; Morgan, H.S. Use of generalized ordered logistic regression for the analysis of multidrug resistance data. *Prev. Vet. Med.* **2015**, *121*, 374–379.
12. Kang, E.N.; Maeng, J.Y. A Study on Factors Influencing Job-Seeking Behavior of Unemployed People with Disabilities: Application of ICF Analysis Framework. *Disabil. Employ.* **2011**, *21*, 103–129.
13. Ha, S.S.; Park, B.H. Comparative Analysis of Accident Severity Using Ordinal Logit Model, 「National Territorial Planning」. *J. Korean Natl. Land Urban Plan. Assoc.* **2011**, *46*, 183–192.
14. Cho, J.Y. Factors Affecting Food Insecurity in the Elderly: Application of Ordinal Logit Regression Analysis. *Health Soc. Res.* **2015**, *35*, 375–406.

engineering proceedings

Proceeding Paper

Development of Plant-Mix-Type Modified Mixture with Excellent Flexibility and Stress Relaxation Properties for Ensuring High Resistance to Cracking †

Nhat Thanh Tran * and Masashige Aoki

Institute of Research and Development, Taisei Rotec Corporation, Kamiya Kounosu, Saitama 1456, Japan; masashige_aoki@taiseirotec.co.jp
* Correspondence: tran_thanh_nhat@taiseirotec.co.jp

† Presented at the Second International Conference on Maintenance and Rehabilitation of Constructed Infrastructure Facilities, Honolulu, HI, USA, 16–19 August 2023.

Abstract: Over the past few years, in order to extend the service life of pavements, the authors developed a special asphalt mixture that adequately prevents the ingress of water from the foundation layer of the asphalt into the top layers. The highly flexible mixture with a premix-type modified asphalt has been applied on a road, and the mixture showed excellent cracking performance in serviceability. Since the highly flexible mixture uses a premix-type modified asphalt that is transported by an asphalt tanker truck, it is difficult to manufacture in small quantities for responding to small-scale and emergency constructions. In order to deal with this difficulty, the present study developed a plant-mix-type modified asphalt mixture using a special additive at the plant. The laboratory tests and field evaluation results indicated that both the premix-type asphalt mixture and the plant-mix-type modified asphalt mixture exhibited excellent flexibility and stress relaxation properties.

Keywords: asphalt mixture; premix; plant-mix; reflection cracks; low-temperature cracking; flexibility; stress relaxation

Citation: Tran, N.T.; Aoki, M. Development of Plant-Mix-Type Modified Mixture with Excellent Flexibility and Stress Relaxation Properties for Ensuring High Resistance to Cracking. *Eng. Proc.* **2023**, *36*, 20. https://doi.org/10.3390/engproc2023036020

Academic Editor: Hosin (David) Lee

Published: 4 July 2023

1. Introduction

An increase in the traffic volume of heavy vehicles may increase cracks in the asphalt layer, resulting in allowing water to ingress from the foundation layer to the asphalt layer. This also weakens the foundation layer and subgrade, thereby reducing the rutting resistance, cracking resistance, and interlayer adhesion of the pavement [1,2]. In Japan, there are large differences in temperatures and rainfall among the four seasons, potentially leading to rainwater infiltration, rutting in summer, and cracking in winter. Therefore, the authors have developed a special asphalt mixture using a premix-type asphalt binder, which effectively prevents reflection cracks and has high resistance to low-temperature cracking and rutting. The highly flexible mixture has been applied in the field, and the mixture shows excellent cracking resistance on construction sites [3]. However, because the highly flexible mixture uses a premix-type modified asphalt binder that is transported by an asphalt tanker truck, it is challenging to produce small quantities to supply for small-scale and emergency constructions.

The main objective of this study was to develop a plant-mix-type modified asphalt mixture by mixing special additives when the mixture is manufactured at the plant. This study also presented the laboratory test results and field evaluation results of the premix-type asphalt mixture and the plant-mix-type modified asphalt mixture that exhibited excellent flexibility and stress relaxation properties.

2. Concept Development of the Special Asphalt Binder

In order to design the properties of the asphalt binder, the various types of binder tests and asphalt mixture tests were investigated based on the common causes of cracks in asphalt pavement [3]. Styrene-butadiene-styrene (SBS) and special petroleum resin were used together to modify the premixed asphalt [3]. The special process oils were also applied to improve the ductile performance of the butadiene in SBS [3]. By using these materials, it is possible to achieve a mixture that has both high cracking resistance and rutting resistance [3].

The plant-mix-type additive, which is added to polymer-modified asphalt type II (PMA II), is a resin that has high flexibility and stress relaxation. Because the gel in the additive shows extremely high viscosity, the plant-mix-type additive is wrapped in hot-melt resin, making this additive easier to convey, weigh, and store in a plant. The shape of the plant-mix-type additive is shown in Figure 1.

Figure 1. The shape of plant-mix-type additive.

3. Experimental Work, Results and Discussion

3.1. Characterization Tests of Asphalt Binder

The plant-mix-type modified asphalt binder (plant-mix binder) was prepared by adding the plant-mix-type additive into the PMA II. Table 1 shows the results of the asphalt binder test for each type of binder. The penetration index (PI) values of plant-mix binder and premix binder, which indicates the temperature sensitivity of the asphalt binder, were higher than that of the PMA II. This means there is little change in performance between low and high temperatures for the highly flexible mixture. In addition, the viscosity of the plant-mix binder and premix binder is higher than that of the PMA II at 60 °C, thereby ensuring a high rutting resistance of the highly flexible mixture. Furthermore, the results of the fragility point, bending test, and $|G^*|\sin\delta$ indicated the highly flexible mixture may have high resistance to fatigue cracking and low-temperature cracking.

Table 1. The binder test results of plant-mix binder, premix binder, and PMA II.

Properties		Plant-Mix Binder	Premix Binder	PMA II
Penetration	1/10 mm	146	135	54
Penetration index PI		8.9	9.3	1.6
Viscosity at 60 °C	Pa·s	7320	9670	2574
Fragility point	°C	−30	−34	−14
Fracture energy (−20 °C)	$\times 10^{-3}$ MPa	1583	1028	157
Fracture toughness (−20 °C)	MPa	40	32	270
Dynamic shear $|G^*|\sin\delta$ (25 °C) *	10^5 Pa	0.9	0.6	9.5

* The DSR test was conducted using 8 mm plates, 10 rad/s, 1% strain, and 1 mm samples.

3.2. Characterization Tests of Asphalt Mixture

Table 2 presents the characteristics of aggregate gradation investigated in this study. The aggregate gradation was typically designed to investigate a dense-graded mixture (13F) with a 12.5 mm nominal maximum particle size. In the design, the asphalt content (5.4%) of the highly flexible mixture (plant-mix binder and premix binder) and the PAM II mixture was determined in accordance with the Japanese pavement construction handbook [4], which was the same as the common protocol for asphalt mixtures. As shown in Table 3, the resistance to cracking at low temperature, fatigue, and rutting of these mixtures were evaluated in the laboratory by three-point bending beam (TPBB) test [5,6], four-point bending fatigue (FPBF) test [5], and wheel tracking test (WTT) [5,7].

Table 2. The aggregate gradation for all asphalt mixtures.

Sieve size (mm)	19.0	13.2	4.75	2.36	0.600	0.300	0.075
Criteria (%)	100	95~100	52~72	40~60	25~45	16~33	8~13
Percent passing (%)	100	99.8	71.3	54.3	33.3	20.6	8.0

Table 3. Tests and evaluations in the laboratory.

Evaluation	Test	Standard
Cracking resistance	TPBB test (−15 °C~15 °C, 6.25 × 10^{-3} 1/s strain rate)	JRA-B005
Fatigue resistance	FPBF test (5 °C, 5 Hz, 400 μm)	JRA-B018T
Rutting resistance	WTT (60 °C, 686 N, 42 cycles/min)	JRA-B003

Figure 2 shows the relationships of temperature with tension stress and tension strain of the TPBB test. The results shown in Figure 2a indicated that the brittleness temperature of the plant-mix flexible mixture, the premix flexible mixture, and the PMA II mixture were −10 °C, −15 °C or less, and 5 °C, respectively. This means that under the same loading, the highly flexible mixtures have lower failure temperatures than those of the PMA II mixture. Therefore, the highly flexible mixtures showed ductile performance in a wide temperature range on the low-temperature side. In addition, the results shown in Figure 2b illustrated that the tension strain values of the highly flexible mixtures tended to be more than twice higher than those of the PMA II mixture at the brittle region. Therefore, the highly flexible mixtures potentially showed excellent flexibility at low temperatures.

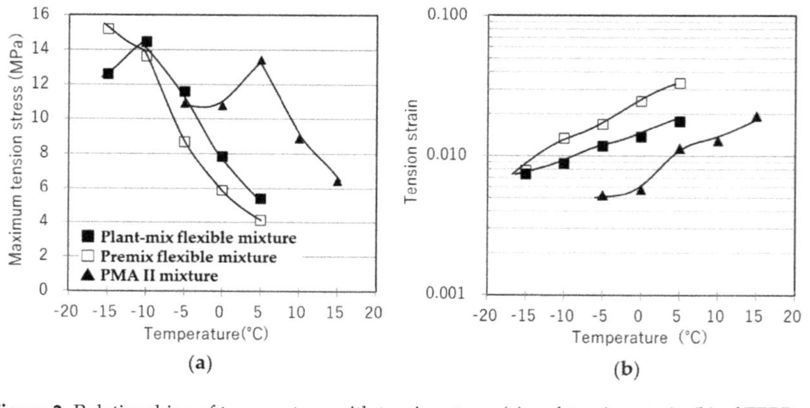

Figure 2. Relationships of temperature with tension stress (**a**) and tension strain (**b**) of TPBB test.

The results of FPBF test demonstrated that the number of cycles to failure of the plant-mix flexible mixture, the premix flexible mixture, and the PMA II mixture were 770,000 cycles, 830,000 cycles, and 14,000 cycles, respectively. There was no significant difference in the results between the plant-mix flexible mixture and the premix flexible mixture. The results also indicated that the fatigue resistance of the highly flexible mixtures was more than 50 times higher than that of the PMA II mixture.

The results of WTT showed that the dynamic stability (DS) values (which indicates the passing wheel load per 1.0 mm of rut depth [7]) of the plant-mix flexible mixture and the premix flexible mixture were greater than or equal to 6000 cycles/mm, which were the same as the results obtained from the PMA II mixture.

4. Plant Results and Discussion

Figure 3 shows the mixing steps of the plant-mix-type modified asphalt binder for the flexible mixture at the plant. The FPBF test results demonstrated that the number of cycles to failure of the plant-mix flexible mixture produced at the plant was more than 1 million cycles (the test was set up to stop at 1 million cycles), which was higher than the result obtained from the laboratory mentioned in Section 3.2. In addition, the WTT results indicated that the DS values of the plant-mix flexible mixture were greater than or equal to 6000 cycles/mm, which were the same as the results obtained from the laboratory.

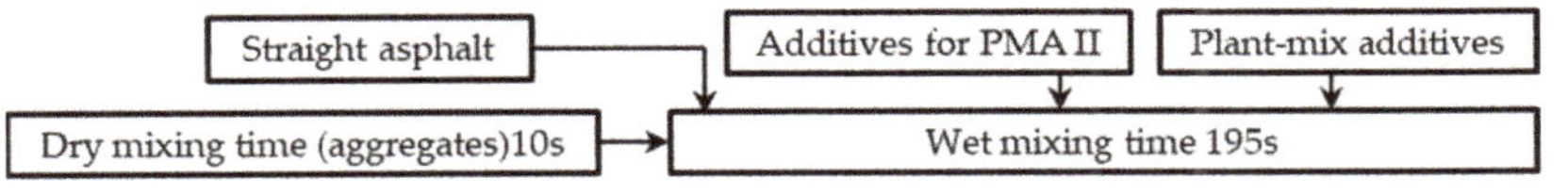

Figure 3. Mixing process of the plant-mix flexible mixture in the plant.

5. Field Test Section for Validation of the Plant-Mix Flexible Mixture

In 2003, the present study applied the premix flexible mixture to construct a national highway in cold regions. After 18 years, the crack growth rate of the premix flexible mixture on construction sites was ten times lower than that of the PMA II mixture, indicating the excellent cracking resistance of the premix flexible mixture [8].

Since 2019, the plant-mix flexible asphalt mixture has been also manufactured in small quantities (about 20 tons) for responding to small-scale and emergency constructions. As shown in Figure 4, after 2 years, the cracking and rutting performance of the test sections in Ishikawa Prefecture Japan were monitored. The results demonstrated that rutting and cracking did not occur in the test sections.

(a) (b)

Figure 4. Views of the test section in Ishikawa Prefecture ((**a**) 2019 and (**b**) 2021).

6. Conclusions

The findings of the present study can be summarized as follows:

- The present study developed a premix flexible mixture and a plant-mix flexible mixture that have both high cracking resistance and high rutting resistance.
- Using the plant-mix flexible asphalt mixture is an effective solution for manufacturing a highly flexible asphalt mixture in small quantities that is able to respond to small-scale and emergency constructions.

Author Contributions: Conceptualization, N.T.T. and M.A.; methodology, N.T.T. and M.A.; formal analysis, N.T.T.; investigation, N.T.T. and M.A.; data curation, N.T.T. and M.A.; writing—original draft preparation, N.T.T.; writing—review and editing, N.T.T. and M.A.; visualization, M.A.; supervision, M.A. All authors have read and agreed to the published version of the manuscript.

Funding: This research received no external funding.

Institutional Review Board Statement: Not applicable.

Informed Consent Statement: Not applicable.

Data Availability Statement: No new data were created or analyzed in this study. Data sharing is not applicable to this article.

Conflicts of Interest: The authors declare no conflict of interest.

References

1. Japanese Ministry of Land, Infrastructure, Transport and Tourism. *Pavement Inspection Guidelines*; MILT: Tokyo, Japan, 2016.
2. Takahashi, S.; Ono, Y.; Sato, M. Project for detailed instructions on asphalt pavement. *Pavement* **2016**, *51*, 13–19.
3. Konno, M.; Otomo, N.; Isayama, H.; Kokufuda, Y. Development of asphalt binder with high rutting resistance and high cracking resistance. *Road Constr.* **2017**, *765*, 63–69.
4. Japan Road Association. *Pavement Construction Handbook*; Japan Road Association: Tokyo, Japan, 2006.
5. Japan Road Association. *Pavement Handbook and Test Method Handbook*; Japan Road Association: Tokyo, Japan, 2019; Volume 3.
6. Tran, N.T.; Takahashi, O. A comparative analysis of the fracture performance of wearing course mixtures under different geometries, compactions, and aggregate gradations. *Int. J. Pavement Eng.* **2020**, *21*, 1703–1715. [CrossRef]
7. Tran, N.T.; Takahashi, O. A comprehensive evaluation of the effects of aggregate gradation on the shear strength properties of wearing course mixtures. *Int. J. Pavement Eng.* **2021**, *22*, 550–559. [CrossRef]
8. Karaki, K.; Kizawa, S.; Okajima, H. Long-term performance of specially modified asphalt mixture with excellent cracking resistance. In Proceedings of the 34th Japan Road Conference, Tokyo, Japan, 4–5 November 2021. Paper 3018.

engineering proceedings

Proceeding Paper

Large-Scale Test Setup of Concrete Pavement Slabs Jointed by Carbon Fiber-Reinforced Polymer Dowel Bars as Load Transfer Devices †

Taha Ahmed [1,*], Ahmad Saad [1], Abdulhadi Kazem [1], Ali Radwan [2], Ali AlMutairi [1,*] and Sarah Ashkanani [1]

1 Civil Engineering Department, College of Engineering, Australian University, Safat 13015, Kuwait; a.said@au.edu.kw (A.S.); a.kazem@au.edu.kw (A.K.); 1900863@au.edu.kw (S.A.)
2 Civil & Architectural Engineering Department, International University—Kuwait, Mohamad Bin Qasim Street, Ardiya 92400, Kuwait; aliradwan96@outlook.com
* Correspondence: t.ahmed@au.edu.kw (T.A.); 1821215@au.edu.kw (A.A.)
† Presented at the Second International Conference on Maintenance and Rehabilitation of Constructed Infrastructure Facilities, Honolulu, HI, USA, 16–19 August 2023.

Abstract: Conventional steel bars are mostly used as the main load transfer mechanism between jointed slabs in rigid pavements; however, they are generally prone to corrosion which reduces the load transfer efficiency at the joints. This study evaluates the performance of steel bars wrapped with Carbon Fiber Reinforced Polymer (CFRP) sheets, introducing a corrosion-free alternative to conventional steel bars while maintaining the required strength. This paper explains the test setup of large-scale shear strength and load transfer efficiency tests that are currently conducted on the slab samples to evaluate the structural performance of the proposed dowel bars and concrete mix designs.

Keywords: concrete pavement; fiber-reinforced polymer; load transfer efficiency; shear strength

Citation: Ahmed, T.; Saad, A.; Kazem, A.; Radwan, A.; AlMutairi, A.; Ashkanani, S. Large-Scale Test Setup of Concrete Pavement Slabs Jointed by Carbon Fiber-Reinforced Polymer Dowel Bars as Load Transfer Devices. *Eng. Proc.* **2023**, *36*, 21. https://doi.org/10.3390/engproc2023036021

Academic Editor: Hosin (David) Lee

Published: 4 July 2023

1. Introduction

The performance of jointed concrete pavements is often closely related to the load transfer capacity of dowel bars at the pavement joints. Faulting is frequently seen in pavement joints without dowel bars because the provided load transfer method by aggregate interlock alone is insufficient. Steel dowels are commonly used to enhance load transfer in concrete pavements. However, the issue of high bearing stress and corrosion of steel dowels has a significant impact on their long-term performance. There have been efforts to improve dowel durability using alternative shapes (other than round) to further reduce dowel–concrete bearing stresses and to use alternative materials for improved corrosion resistance. Fiber Reinforced Polymer (FRP) dowel bars are among those alternative materials. They have been given considerable attention because of their high strength and excellent corrosion resistance. There are several benefits of using Carbon Fiber Reinforced Polymer (CFRP) materials in construction, including their lightweight nature, having high strength and greater stiffness per unit weight, improving the strength of concrete, and increasing the life span of facilities [1–3].

2. Research Methodology

This study evaluates the performance of steel bars wrapped with CFRP sheets, introducing a corrosion-free alternative to conventional steel bars while maintaining the required strength. This paper explains the test setup of large-scale shear strength and load transfer efficiency tests that are currently conducted on rigid pavement slab samples to evaluate their structural performance of the proposed dowel bars and concrete mix designs.

2.1. Load Transfer Efficiency (LTE) Test

The main function of dowel bars in rigid pavement slabs is to efficiently transfer the applied loads from the loaded slab to the neighboring unloaded slab. As a result, both loaded and unloaded slabs bend. Dowel bars significantly reduce the stresses and deformations that are induced at the loaded slabs compared to concrete pavements without dowel bars. Load transfer efficiency influences the magnitude of reduced stresses and deformations. The LTE can be calculated as per Equation (1) [4]:

$$\text{LTE}, \% = \frac{d_u}{d_l} * 100 \tag{1}$$

where d_u and d_l are the joint's vertical displacement for unloaded and loaded concrete slabs, respectively measured on top of the joint's edge.

The objective of this test is to evaluate the LTE of the steel dowel bars wrapped with CFRP sheets compared to the conventional steel dowel bars.

2.2. Shear Strength Test

The shear test is intended to apply uniform pressure on the dowel bars jointing the concrete slabs. This shear pressure will make the dowel bars reach a sliding failure along the parallel plane to the applied shear pressure. The objective of this test is to evaluate the shear strength of steel dowel bars wrapped with CFRP sheets compared to conventional steel dowel bars. The shear strength test method was adopted from the AASHTO T-253-76 test setup, as shown in Figure 1 [5,6]. To be able to conduct the shear test on the slab specimens, a few modifications were introduced to the AASHTO T253-76 test setup as explained in Section 3.2 below.

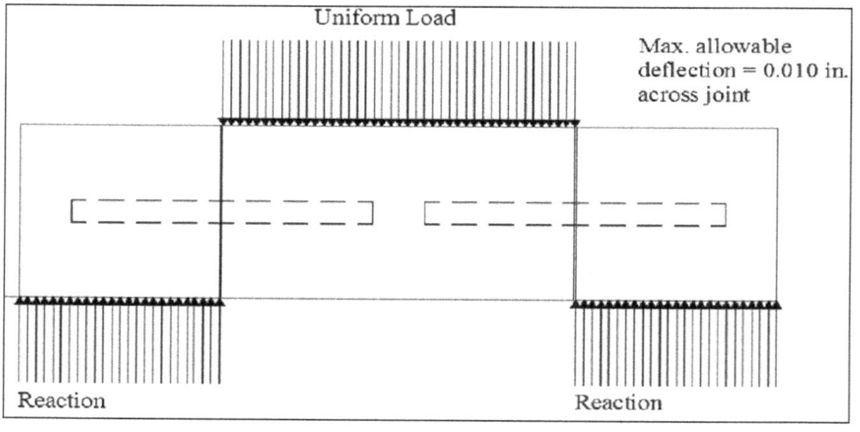

Figure 1. AASHTO T253-76 Test Setup [5].

3. Results and Discussion

3.1. Load Transfer Efficiency (LTE) Test Setup

Figures 2 and 3 show the LTE test setup details. The concrete pavement slabs were designed as per the general specifications of FHWA [5]. Each slab has dimensions of 2400 mm length, 1000 mm width, and a thickness of 200 mm with four dowel bars of 25 mm diameter embedded in the middle of the slab thickness.

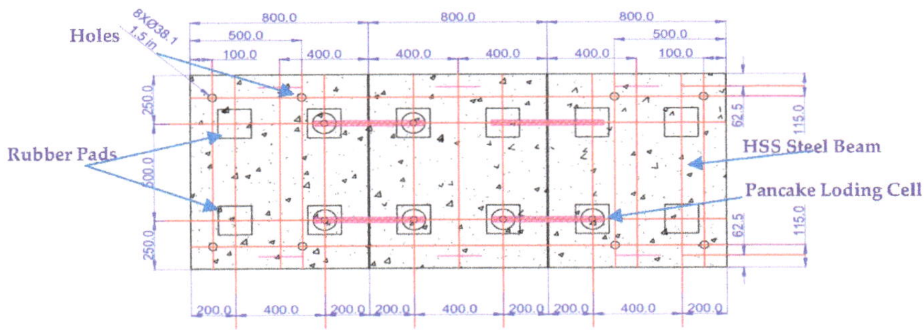

Figure 2. LTE Test Set-up Details of the Concrete Slab Specimen with Dimensions in mm.

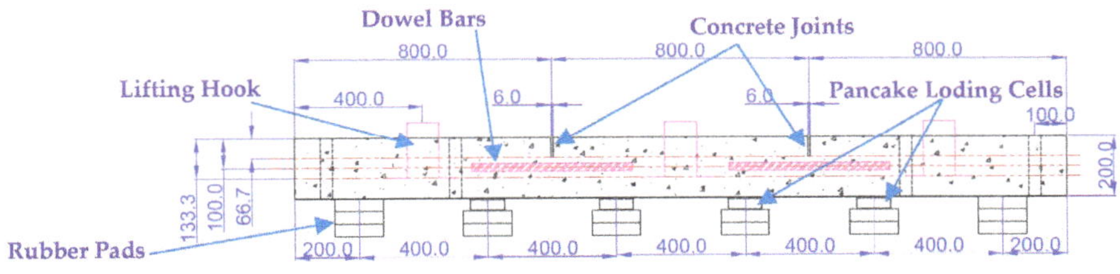

Figure 3. Slab Side View of the Concrete Slab Specimen with Dimensions in mm.

A set of 180 mm thickness rubber pads were placed underneath the slabs at specific locations to simulate the actual soil in a real pavement section (see Figure 3). The physical and mechanical properties of these pads were determined to ensure that they can represent the required soil modulus of subgrade reaction. Additionally, six pancake-loading cells were placed underneath the concrete slabs where the dowel bars are located as shown in Figure 2. The pancake loading cells are used to measure the applied load on both loaded and unloaded slabs during the test. Also, the joint's vertical displacement for the unloaded and loaded concrete slabs is measured using Linear Variable Differential Transformers (LVDT) attached to the surface of the concrete slabs. Moreover, each slab has eight 38.1 mm (1.5-inch) diameter holes located along the slab edges and matching the holes in the large-scale loading steel frame (see Figures 4 and 5). These holes are used to connect eight different threaded bars to both the loading frame and the slab test sample during the shear test.

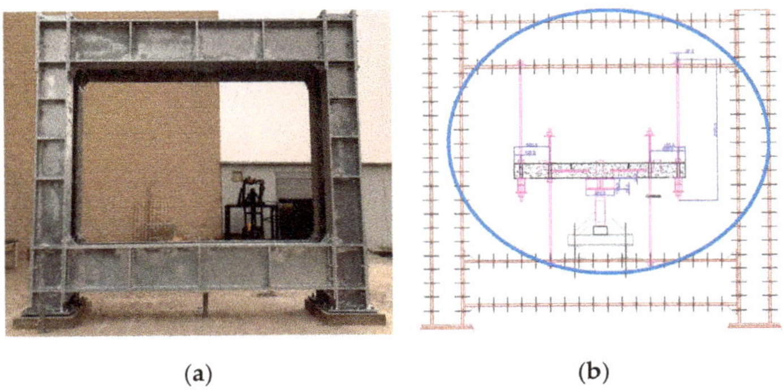

(**a**) (**b**)

Figure 4. (**a**) Large-Scale Loading Steel Frame, (**b**) Shear Strength Test Set-up.

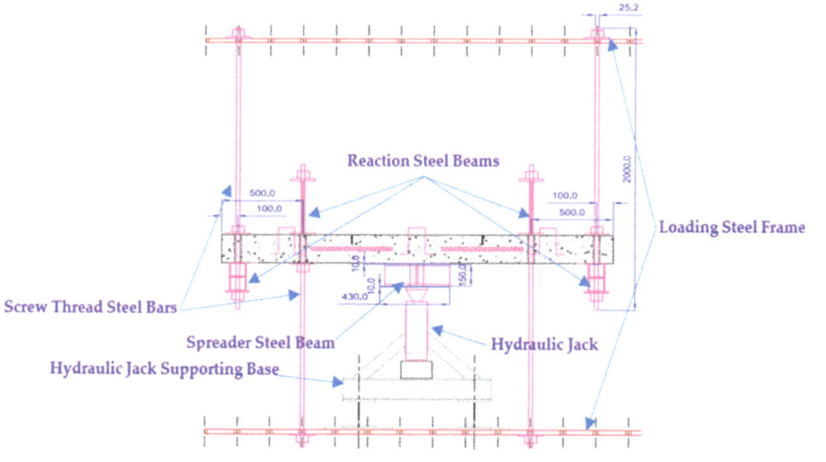

Figure 5. Close up of Shear Strength Test Set-up of the Concrete Slab Specimen.

3.2. Shear Strength Test Set-Up

The shear test set-up was installed to a large-scale loading steel frame, Figures 4 and 5 show the shear test set-up details.

The shear load is applied on the concrete slab specimen using a 50-ton hydraulic jack fixed underneath the concrete slab. A spreader steel beam is placed between the hydraulic jack and the concrete slab to transfer the point load from the hydraulic jack into a uniform pressure on the slab middle segment. Additionally, the concrete slab is connected to four reaction beams at specific locations on top and below the slab specimen (as shown in Figure 5) to prevent any deflection in the concrete slab and forcing it to fail in shear only.

4. Conclusions and Ongoing Work

This paper presented the testing setup for load transfer efficiency (LTE) and shear strength tests of dowel bars in rigid pavement slabs. In addition, different configurations of CFRP-wrapped steel bars are included to assess their performance. Wrapping of the bars was chosen as an efficient method to protect dowel bars from corrosion and other environmental exposures. The testing setup was successfully completed and confirmed for both LTE and shear strength tests. Currently, the research team is conducting the actual LTE and shear strength tests on the prepared slab specimens. Different concrete mix designs incorporating crumb rubber will be evaluated in the study as well.

Author Contributions: T.A., A.S. and A.K.: Conceptualization, methodology, data curation, Supervision, writing—review and editing. A.R., A.A. and S.A.: Investigation, methodology, conceptualization, data curation, and writing—original draft. All authors have read and agreed to the published version of the manuscript.

Funding: This research received no external funding.

Institutional Review Board Statement: Not applicable.

Informed Consent Statement: Not applicable.

Data Availability Statement: Data are available by the corresponding authors upon request.

Acknowledgments: The authors would like to acknowledge the great help and support provided by the Australian University–Kuwait.

Conflicts of Interest: The authors declare no conflict of interest.

References

1. Larson, R.M.; Smith, K.D. *Evaluation of Alternative Dowel bar Materials and Coatings;* Ohio Department of Transportation: Columbus, OH, USA, 2011.
2. Hu, C.; Ma, J.; Zhao, J.; Leng, Z.; Jelagin, D. Experimental Study of Dowel Bar Alternatives Based on Similarity Model Test. *Hindawi Adv. Mater. Sci. Eng.* **2017**, *2017*, 3981908. [CrossRef]
3. Larson, R.M.; Smith, K.D. *Evaluating the Use of Fiber-Reinforced Polymer Bars in Continuously Reinforced Concrete Pavement;* CPTP Tech Brief (FHWA-HRT-05-081); Federal Highway Administration (FHWA): Washington, DC, USA, 2017.
4. Sadeghi, V.; Hesami, S. Investigation of load transfer efficiency in jointed plain concrete pavements (JPCP) using FEM. *Int. J. Pavement Res. Technol.* **2018**, *11*, 245–252. [CrossRef]
5. American Association of State Highway and Transportation Officials (AASHTO). *Guide for Design of Pavement Structures;* AASHTO: Washington, DC, USA, 1993.
6. Harrington, J.F. Comparison of Alternative Laboratory Dowel Bar Testing Procedures. Master's Thesis, Iowa State University, Ames, IA, USA, 2006.

engineering proceedings

MDPI

Proceeding Paper

Study on Traffic Incident Management Boundary Based on Gis and Its Historical Travel Time Data †

Dong Hyeop Kim and Jin-Tae Kim *

Department of Transportation Systems Engineering, Korea National University of Transportation, 157 Cheoldobangmulkwan-ro, Uiwang 16106, Republic of Korea; kimdh1110@hanmail.net

* Correspondence: jtkim@ut.ac.kr; Tel.: +82-31-461-8736

† Presented at the Second International Conference on Maintenance and Rehabilitation of Constructed Infrastructure Facilities, Honolulu, HI, USA, 16–19 August 2023.

Abstract: This study proposes a method to determine a spatial boundary of traffic operation and management techniques in strategic schemes against sudden traffic incidents based on historical data in the Seoul metropolitan area. Through the combination of data analysis and a geographical information system, it was found that there were general tendencies after the occurrence of an incident pertaining to its significance and how long the effects of incidents last. We classified the properties of accidents based on their duration and the space left available within the relevant road lane. This study found that the longer the incident's duration, the greater the effect of the traffic incident. When the number of available lanes was one, the impact of the traffic accident was greater. In the case of two or more available lanes, the spatial boundary tended to be identical, while changes in travel speed were affected by incident type.

Keywords: incident; accident; traffic operation; traffic management

1. Introduction and Objectives

Traffic incidents can be classified into two types: predictable and unpredictable. To reduce the likelihood of such incidents, construction sites' specific traffic management strategies and tactics can be prepared in advance. In the case of an unpredictable accident, the development of relevant traffic management measures would be difficult and delayed by the amount of time needed to detect the accident. To deal with such unexpected incidents, traffic management schemes should be established based on the time–space boundary of their effects. A potential approach would be (1) dispersion of traffic influx in that space boundary to reduce the inbound traffic and (2) additional capacity to increase the throughput of the outbound traffic. Previous studies have suggested utilizing traffic information provision techniques, such as variable message signs (VMS), car navigation systems (CNSs), and lane control systems (LCSs). Although it has always been necessary in order to prepare a strategy to deal with traffic congestion expanding over time, no guideline has been made available to determine the time–space boundary of unexpected traffic accidents.

This study intends to analyze the historical changing trends of the time–space boundary of urban highway networks affected by such traffic incidents that have occurred in Seoul's metropolitan areas. When specific patterns of accidents' time–space boundaries exist, and such patterns can be explained by certain factors of traffic accidents, understanding these patterns would be fundamental to the development of future traffic management strategies.

Citation: Kim, D.H.; Kim, J.-T. Study on Traffic Incident Management Boundary Based on Gis and Its Historical Travel Time Data. *Eng. Proc.* **2023**, *36*, 22. https://doi.org/10.3390/engproc2023036022

Academic Editor: Hosin (David) Lee

Published: 4 July 2023

2. Data Collection and Analysis

2.1. Materials and Methods

This study employed the historical travel speed data, which were obtained through 'OpenAPI' from the Seoul transportation operation and information service center (TOPIS). The travel speed data were collected every 5 min for each of 5358 urban highway links for a duration of 2 months (1 March 2020–30 April 2022). The total number of collected data items was 94,129,344. The collected TOPIS data were refined based on the Korean national standard node-link scheme.

During that time period, it was officially reported that a total of 2281 traffic accidents occurred. To analyze the influence of traffic accidents based on the data, we attempted to set a boundary for the traffic accidents. The size of the boundary was defined to count from one to up to five upstream links, where the traffic accident occurred (see Figure 1).

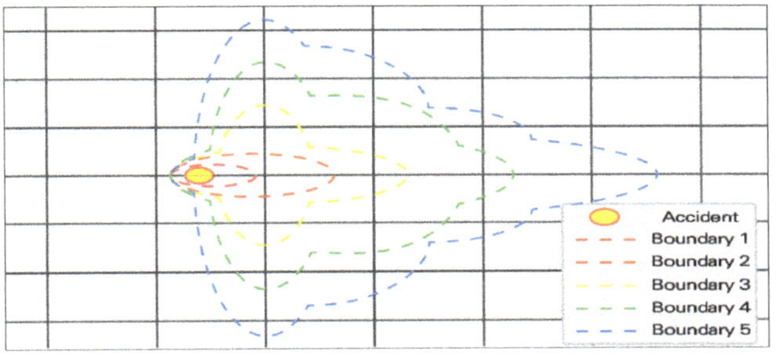

Figure 1. Expected boundary of network area affected by incident.

The traffic accidents were classified by the number of available lanes and the duration of the accident in order to analyze the impact of a traffic accident. The number of available lanes was classified into one lane, two lanes, and more than three lanes by calculating the number of lanes of the accident link and the number of lanes controlled due to the occurrence of a traffic accident (see Figure 2). In the case of the duration of the accident, the analysis was performed by classifying the time taken from the occurrence of a traffic accident to the end time, into 30 min or less, 30 min to 1 h, and over 1 h.

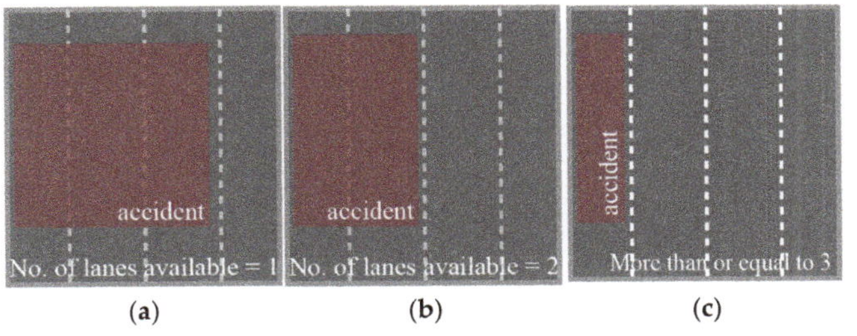

Figure 2. (**a**) Available number of lanes is one; (**b**) available number of lanes is two; (**c**) available number of lanes is three.

To analyze the change in cruising speed according to the occurrence of a traffic accident, the speed up to 40 min after the occurrence of a traffic accident was analyzed every 5 min. The boundary of a traffic accident's impact was extracted from the 5-min-interval-based analysis of percent speed changes. Tables 1–3 show average speed decrease every 5 min in each case. The minus sign represents the percentage of speed decrease, and the plus represents the percentage of speed increase.

Table 1. Percentage of speed decrease after accident occurrence in Case 1.

Incident Duration (min)	Boundary	Time Period (min) after Accident Occurrence							
		0~5	5~10	10~15	15~20	20~25	25~30	30~35	35~40
>30	1	−9.4%	1.6%	0.7%	12.0%	2.2%	8.8%	0.8%	1.8%
	2	−3.6%	−1.6%	3.9%	2.2%	5.1%	5.4%	5.4%	1.6%
	3	−2.1%	3.1%	1.0%	1.7%	2.8%	6.2%	2.2%	7.5%
	4	−1.3%	1.1%	3.8%	2.4%	2.5%	4.7%	2.7%	3.9%
	5	−3.7%	3.1%	3.4%	3.6%	2.8%	6.1%	6.2%	4.7%
30~60	1	−10.2%	0.5%	1.4%	−1.1%	1.6%	5.9%	1.5%	7.1%
	2	−14.9%	4.1%	2.2%	5.8%	14.8%	1.5%	1.8%	7.4%
	3	−6.4%	7.1%	0.1%	3.2%	0.3%	8.3%	3.9%	3.1%
	4	−2.7%	0.9%	3.3%	6.4%	2.2%	2.7%	3.3%	4.6%
	5	−2.7%	3.7%	3.2%	2.6%	3.8%	3.1%	2.4%	3.8%
60 ≤	1	−18.6%	−0.2%	−4.1%	−3.8%	14.8%	46.8%	2.0%	25.2%
	2	−7.5%	6.3%	−5.9%	0.9%	9.5%	3.0%	2.0%	3.6%
	3	−6.5%	1.0%	7.2%	−1.5%	1.7%	5.2%	8.0%	3.0%
	4	−5.7%	3.6%	4.7%	5.0%	3.0%	0.2%	5.7%	−0.1%
	5	−4.9%	2.9%	1.9%	7.4%	1.3%	2.4%	5.3%	4.5%

Table 2. Percent speed decrease after accident occurrence in Case 2.

Incident Duration (min)	Boundary	Time Period (min) after Accident Occurrence							
		0~5	5~10	10~15	15~20	20~25	25~30	30~35	35~40
>30	1	−9.6%	3.0%	3.5%	4.7%	1.8%	5.9%	−0.3%	3.1%
	2	−8.6%	3.9%	2.9%	7.2%	3.3%	6.8%	8.1%	2.7%
	3	−1.7%	2.4%	3.4%	4.5%	2.6%	1.3%	5.3%	5.5%
	4	−2.5%	2.4%	3.0%	4.5%	3.7%	2.9%	6.1%	2.3%
	5	−3.9%	3.1%	4.4%	3.5%	2.4%	3.4%	3.1%	2.8%
30~60	1	−13.0%	2.3%	12.1%	2.1%	1.1%	4.9%	0.1%	7.5%
	2	−6.7%	4.1%	−1.0%	5.1%	1.7%	8.9%	4.9%	4.8%
	3	−3.2%	1.9%	3.4%	2.0%	2.3%	2.1%	2.3%	3.9%
	4	−3.5%	3.1%	4.8%	1.9%	3.0%	3.3%	3.9%	2.9%
	5	−2.2%	3.2%	2.2%	3.5%	3.0%	3.3%	2.8%	2.7%
60 ≤	1	−17.3%	−1.1%	−5.4%	4.8%	1.5%	2.6%	1.2%	2.0%
	2	−8.0%	−2.9%	0.6%	3.0%	7.3%	0.2%	1.4%	−0.3%
	3	−2.1%	1.5%	1.7%	2.1%	3.8%	8.1%	2.1%	1.0%
	4	−3.8%	2.3%	4.4%	3.1%	3.7%	2.2%	2.8%	2.2%
	5	−2.6%	1.7%	2.9%	2.5%	2.6%	3.2%	2.9%	3.6%

Table 3. Percent speed decrease after accident occurrence in Case 3.

Incident Duration (min)	Boundary	Time Period (min) after Accident Occurrence							
		0~5	5~10	10~15	15~20	20~25	25~30	30~35	35~40
>30	1	−6.4%	2.4%	3.3%	5.5%	3.4%	8.0%	4.8%	4.2%
	2	−6.4%	2.5%	0.7%	5.6%	4.5%	6.0%	5.1%	2.1%
	3	−3.8%	1.7%	2.8%	2.8%	3.9%	3.6%	4.2%	4.9%
	4	−2.7%	2.8%	1.2%	3.3%	2.7%	3.2%	3.4%	4.2%
	5	−1.9%	2.8%	2.1%	2.4%	2.6%	3.1%	2.2%	3.7%
30~60	1	−9.5%	−0.6%	1.3%	1.4%	2.5%	4.9%	2.0%	5.7%
	2	−5.9%	−0.5%	8.6%	−0.4%	5.1%	5.6%	1.5%	2.7%
	3	−1.3%	−0.6%	1.5%	2.8%	3.8%	3.2%	3.4%	4.3%
	4	−3.3%	2.1%	2.9%	1.4%	3.3%	1.7%	2.7%	2.4%
	5	−2.2%	2.0%	2.2%	3.0%	2.2%	3.8%	2.9%	1.5%
60 ≤	1	−12.9%	3.5%	−2.6%	5.4%	1.4%	6.8%	2.8%	5.6%
	2	−8.0%	4.2%	−0.3%	12.5%	3.8%	4.6%	2.4%	3.7%
	3	−6.8%	1.0%	3.6%	2.6%	0.8%	2.4%	−0.1%	2.3%
	4	−1.8%	2.6%	7.3%	1.8%	1.5%	1.8%	2.8%	2.3%
	5	−2.9%	2.9%	1.8%	5.3%	1.8%	3.4%	3.4%	3.0%

2.2. Results

2.2.1. CASE 1 (Number of Available Lanes Is One)

The results show that the incident management boundary differed for each incident case. When incident duration was over 60 min, the time–space boundary lasted until 15–20 min after the accident. It was found that there was an effect on up to three upstream intersections in the spatial range (See Table 1).

2.2.2. CASE 2 (Number of Available Lanes Is Two)

When the number of available lanes was two, it was found that the change in traffic flow was less affected than when only one lane was available. The time–space boundary lasted until 10–15 min after the accident. It was found that there was an effect on up to two upstream intersections in the spatial range (See Table 2).

2.2.3. CASE 3 (Number of Available Lanes Is More than Three)

When the number of available lanes was more than three, it was found that the traffic flow was affected up to 5 min and then recovered. Traffic flow was affected at the intersection where the accident occurred or one upstream intersection in the spatial range (See Table 3).

3. Conclusions

This study has identified changes in traffic flow conditions due to accidents, based on actual accident and traffic flow data, to determine a boundary of traffic management measures for accidents on urban roads. To analyze traffic flow condition changes caused by accidents, traffic accident information and traffic flow information were collected and analyzed, and the boundary of traffic management measures was presented according to the type of accident. The types of accidents were classified based on their duration (less than 30 min, 30–60 min, and over 60 min) and the space left available (by the number of lanes available: one, two, and more than three).

It was determined that the longer the duration of the accident, the greater its impact. The impact was also greater when the number of available lanes was one. In the case of two and three available lanes, it was found that the speed reduction rate due to an accident was higher if the number of available lanes was two, but there was no significant difference in the management boundary. When developing a traffic control plan in accordance with these suggestions, in the event of a traffic accident in the future, the boundary of management measures should vary by accident type, but in line with the results of this analysis, it is suggested to set the boundary of management (1) up to 15 min in time, and (2) up to three upstream intersections in space.

Author Contributions: Conceptualization, J.-T.K. and D.H.K.; methodology, J.-T.K.; software, D.H.K.; validation, J.-T.K. and D.H.K.; formal analysis, J.-T.K.; investigation, D.H.K.; resources, D.H.K.; supervision, J.-T.K.; project administration; J.-T.K. All authors have read and agreed to the published version of the manuscript.

Funding: This work was supported by a Korea Institute of Police Technology (KIPoT) grant funded by the Korea government (KNPA) (No.092021C29S02000, Development of on-site control technology for road traffic network control in the event of an incident or disaster).

Institutional Review Board Statement: Not applicable.

Informed Consent Statement: Not applicable.

Data Availability Statement: Not applicable.

Conflicts of Interest: The authors declare no conflict of interest.

engineering proceedings

Proceeding Paper

Toward the Determination of the Appropriate Capturing Resolution of Surface Textures in Relation to Pavement Friction †

Malal Kane * and Minh-Tan Do

Campus de Nantes, Université Gustave Eiffel, 44340 Bouguenais, France; minh-tan.do@univ-eiffel.fr
* Correspondence: malal.kane@univ-eiffel.fr; Tel.: +33-240-84-58-39

† Presented at the Second International Conference on Maintenance and Rehabilitation of Constructed Infrastructure Facilities, Honolulu, HI, USA, 16–19 August 2023.

Abstract: The objective of this work is to determine the appropriate resolution of pavement textures representative of their surface skid resistance. The friction and the texture of different pavement surfaces were first measured. Then, the friction of these pavements was computed using the dynamic friction model (DFM) and the resampled textures at different resolutions. Finally, a comparison of the experimental and model results is made possible to determine the optimal resolution, bringing them as close as possible. After analyzing the results, it was found that the optimal resolution is 500 μm in these study conditions.

Keywords: texture; friction; skid resistance; resolution; pavement; surface

1. Introduction

Pavement friction is one of the most important parameters for the safety of road users. Indeed, this pavement property allows vehicles to minimize their stopping distance and thus, minimize collision risks. The most important pavement parameter with respect to friction is its surface texture. Indeed, this texture, composed of several wavelengths generally classified in two groups (microtexture and macrotexture), and which allow, via its penetration of tire tread, to oppose the movement of the tire. However, to our knowledge and despite the definition of these two texture ranges, no study or expert has stated the resolution at which these surface textures should be captured [1–6].

And yet, defining this resolution would make it possible to completely dispense with actual measurements of tire–pavement friction. Which, instead of reporting the pavement contribution only, will also strongly depend on the tire, as well as the operating conditions of the contact [7]. The appropriate captured texture would allow the definition of the texture parameters correlating with pavement friction. It is obvious that the same texture parameter of the same surface will depend on the resolution with which the surface is captured. Hence, the need to define the appropriate resolution of pavement texture representative of its friction is obvious.

2. Approach

To explore the texture resolution effect, the friction and the texture of a set of pavement surfaces are first measured. Then, in a second step, from these measured textures, different new resolutions are created by resampling. The friction of these resampled textures is calculated using the dynamic friction model (DFM [8–13]), a model already validated. In a third step, a comparison between calculated friction (with the DFM) and the experimentally measured friction on the set of pavement surfaces are conducted in order to determine which of the resolutions (obtained after resampling) allows the most accurate prediction.

Citation: Kane, M.; Do, M.-T. Toward the Determination of the Appropriate Capturing Resolution of Surface Textures in Relation to Pavement Friction. *Eng. Proc.* **2023**, *36*, 23. https://doi.org/10.3390/engproc2023036023

Academic Editor: Hosin (David) Lee

Published: 4 July 2023

3. Dynamic Friction Model

The DFM or dynamic friction model is a friction model of a rubber pad/pavement based on the modeling of the contact of these two bodies [8–13]. This model considers the type of rubber, the roughness of the road, the level of wetting and the operating conditions (load, sliding speed, etc.). It has been validated several times and has thus shown its ability to predict the friction between a rubber pad and a pavement surface. This model has also not evolved since its development, as it is a fairly comprehensive representation of tire/pavement friction, considering the characteristics of the tire as its geometry, tread, slip rate, etc. For all calculations performed with the DFM, the contribution of wavelength scales is considered below the resolution as being μ_{adh} = 0.145. (This adhesion friction coefficient corresponds to the optimum value which, inputted to the DMF, outputs the best prediction of the experimental friction.) [8–13].

4. Raw Experimental Data

Thirty different surfaces were tested. These 30 surfaces were obtained from 5 different types of aggregates; for each of these aggregates, 1 asphalt and 1 mosaic sample were made [14]. Each of these samples underwent three levels of polishing using the polishing head of the Wehner–Schultz machine [14,15], which makes three states of roughness for each sample. So, in summary: 3 aggregates times 2 types of samples times 3 levels of polishing. Each of these 30 surfaces underwent friction measurement using the friction head of the WS machine.

Additionally, each texture of these 30 surfaces was captured at 10 μm resolution using a high resolute profilometer. Each original texture is composed of 15 parallel profiles, 76 mm long, spaced at 0.5 mm and sampled every 10 μm. The other 4 resolutions were generated from the original texture via resampling as follows. To achieve 100 μm resolution (and, respectively for 500 μm, 1000 μm, and 5000 μm), 1 in 10 points (50 points, 100 points, and 500 points, respectively) was retained from the original profile. So, in addition to the original resolution, 4 other resolutions of 100 μm, 500 μm, 1000 μm, and 5000 μm of each surface were obtained from the one at 10 μm (Figure 1).

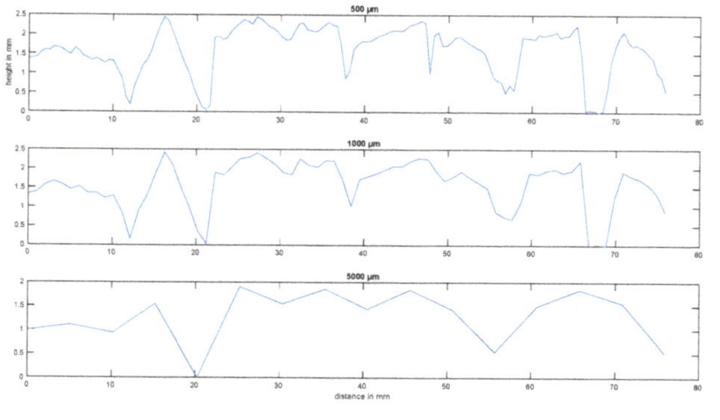

Figure 1. Example of an original profile and the 4 others resolution of that profile.

5. Results and Discussion

The selected surfaces and the polishing carried out on them allowed us to cover a wide range of friction ranging from 0.2 to 0.6. The figures below compare the experimental results with the computational results from the DFM. The curves in Figure 2 show the correlation lines between measured and calculated friction for the five resolutions. For each resolution, the R^2 was evaluated. Figure 3 shows the variation of R^2 as a function of resolution.

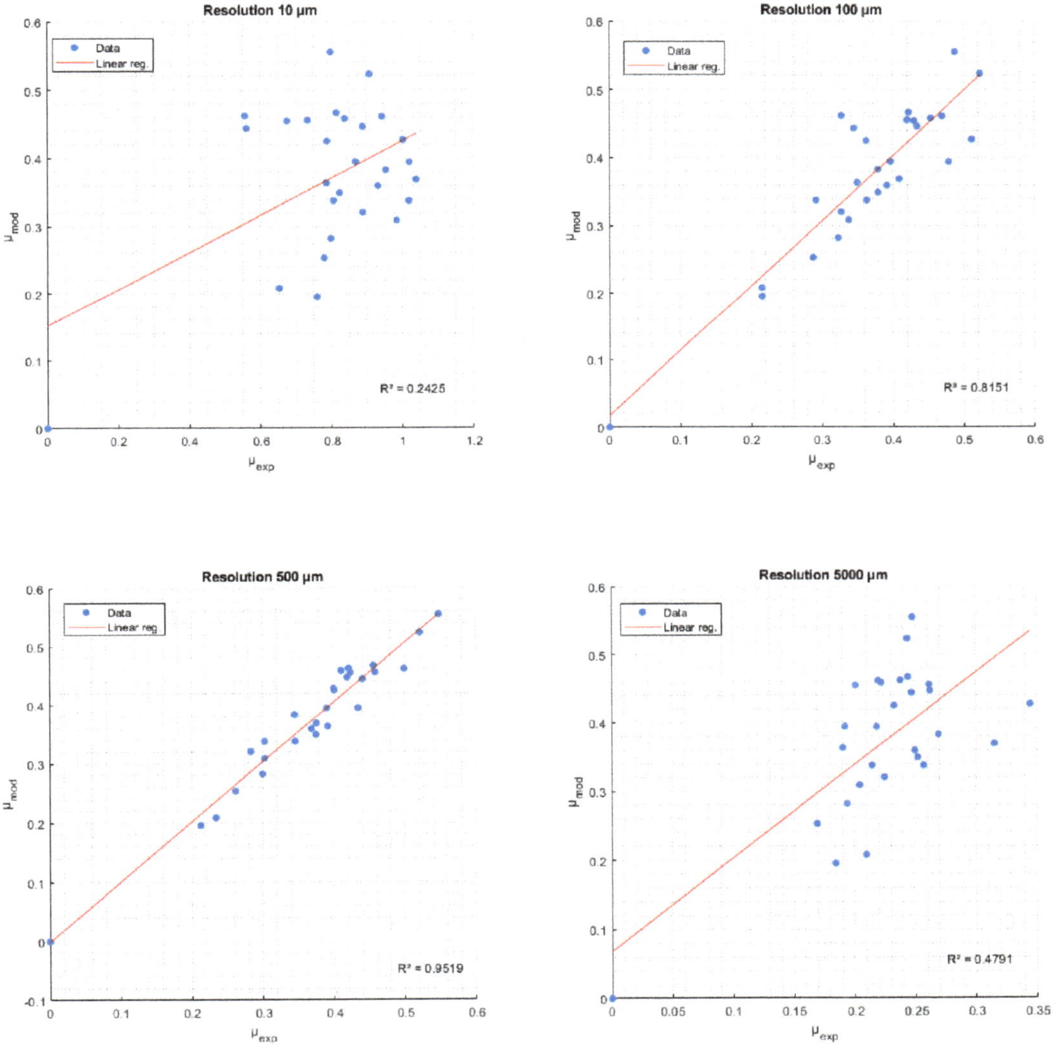

Figure 2. Comparison between the experimental results and the computational results from the DFM at the five resolutions (only four are displayed here).

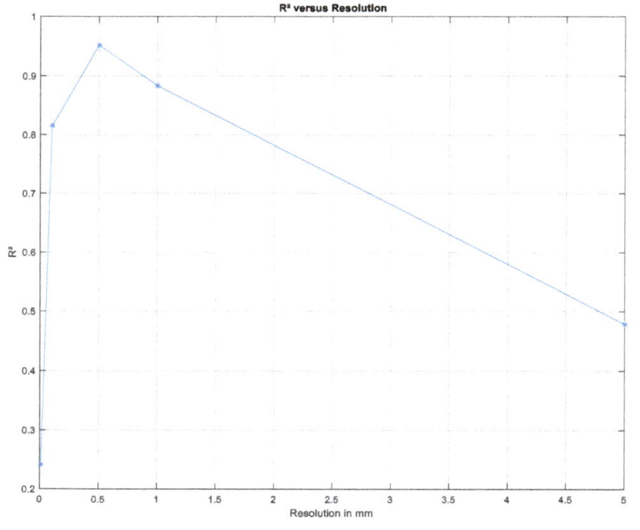

Figure 3. Variation of R^2 as a function of resolution.

When analyzing these curves, we can conclude that the representative resolution of the texture, in relation to the pavement friction, is 500 µm. This is a bit surprising, considering all the efforts made to implement technologies capable of capturing road surfaces in higher resolutions. Any texture parameter to be computed in the future to report on this pavement characteristic must then be computed from a texture captured at this resolution.

6. Conclusions

The objective of this work was to determine the appropriate resolution of pavement textures representative of their frictions. A set of pavement surfaces have been selected and their frictions and textures have been measured. Then, from these measured textures at different resolutions, the friction of each surface was computed using the DFM. The comparison between the experimental results and model allowed us to conclude that the appropriate resolution is 500 µm. However, it has been noted that the sliding speed in question in this study was 60 km/h and that optimal resolution could not be the same at other speeds.

Author Contributions: Conceptualization, M.K.; methodology, M.K.; software, M.K.; validation, M.K. and M.-T.D.; formal analysis, M.K.; investigation, M.K.; writing—review and editing, M.K. and M.-T.D.; visualization, M.K.; supervision, M.K. All authors have read and agreed to the published version of the manuscript.

Funding: This research received no external funding.

Institutional Review Board Statement: Not applicable.

Informed Consent Statement: Not applicable.

Data Availability Statement: The data presented are openly available at: Mendeley Data, V1, doi: 10.17632/kkcztnxzph.1 (https://data.mendeley.com/datasets/kkcztnxzph/1).

Conflicts of Interest: The authors declare no conflict of interest.

References

1. Beautru, Y.; Kane, M.; Do, M.-T.; Cerezo, V. Influence of road surface microtexture on thin water film traction. In Proceedings of the 7th International Conference on Maintenance and Rehabilitation of Pavements and Technological Control, MAIREPAV 2012, Auckland, New Zealand, 28–30 August 2012.
2. Kane, M. A contribution of the analysis of the road macrotexture and microtexture roles vis-à-vis skid resistance. *J. Test. Eval.* **2022**, *50*, 20210047. [CrossRef]
3. Kane, M.; Edmondson, V. Skid resistance: Understanding the role of road texture scales using a signal decomposition technique and a friction model. *Int. J. Pavement Eng.* **2022**, *23*, 499–513. [CrossRef]
4. Kane, M.; Rado, Z.; Timmons, A. Exploring the texture-friction relationship: From texture empirical decomposition to pavement friction. *Int. J. Pavement Eng.* **2015**, *16*, 919–928. [CrossRef]
5. Do, M.-T.; Cerezo, V.; Beautru, Y.; Kane, M. Modeling of the connection road surface microtexture/water depth/friction. *Wear* **2013**, *302*, 1426–1435. [CrossRef]
6. Edmondson, V.; Woodward, J.; Lim, M.; Kane, M.; Martin, J.; Shyha, I. Improved non-contact 3D field and processing techniques to achieve macrotexture characterisation of pavements. *Constr. Build. Mater.* **2019**, *227*, 116693. [CrossRef]
7. Rasol, M.; Schmidt, F.; Ientile, S.; Adelaide, L.; Nedjar, B.; Kane, M.; Chevalier, C. Progress and monitoring opportunities of skid resistance in road transport: A critical review and road sensors. *Remote Sens.* **2021**, *13*, 3729. [CrossRef]
8. Kane, M.; Artamendi, I.; Scarpas, T. Long-term skid resistance of asphalt surfacings: Correlation between Wehner-Schulze friction values and the mineralogical composition of the aggregates. *Wear* **2013**, *303*, 235–243. [CrossRef]
9. Kane, M.; Cerezo, V. A contribution to tyre/road friction modeling: From a simplified dynamic frictional contact model to a "Dynamic Friction Tester" model. *Wear* **2015**, *342–343*, 163–171. [CrossRef]
10. Kane, M.; Do, M.-T.; Cerezo, V.; Rado, Z.; Khelifi, C. Contribution to pavement friction modelling: An introduction of the wetting effect. *Int. J. Pavement Eng.* **2019**, *20*, 965–976. [CrossRef]
11. Kane, M.; Edmondson, V. Tyre/road friction prediction: Introduction a simplified numerical tool based on contact modelling. *Veh. Syst. Dyn.* **2022**, *60*, 770–789. [CrossRef]
12. Edmondson, V.; Ardill, O.; Martin, J.; Lim, M.; Kane, M.; Woodward, J. Quantifying long-term rates of texture change on road networks. *Int. J. Pavement Eng.* **2022**, *23*, 1957–1969. [CrossRef]
13. Kane, M.; Lim, M.; Tan Do, M.; Edmonsond, V. A new predictive skid resistance model (PSRM) for pavement evolution due to texture polishing by traffic. *Constr. Build. Mater.* **2022**, *342*, 128052. [CrossRef]
14. Kane, M. Road polishing: High-resolution textures dataset of three mosaics made of different aggregates at different polishing stages. *Data Brief* **2022**, *45*, 108594. [CrossRef] [PubMed]
15. Do, M.-T.; Tang, Z.; Kane, M.; de Larrard, F. Pavement polishing-Development of a dedicated laboratory test and its correlation with road results. *Wear* **2007**, *263*, 36–42. [CrossRef]

engineering proceedings

MDPI

Proceeding Paper

Study of Long-Term Field Performance of Chip Seal in Washington State †

Angelique Umutoniwase [1,*], Haifang Wen [2] and Kevin Littleton [2]

1 Washington State Department of Transportation, Spokane, WA 99207, USA
2 Department of Civil and Environmental Engineering, Washington State University, Pullman, WA 99164, USA; haifang_wen@wsu.edu (H.W.); kevin.littleton@wsu.edu (K.L.)
* Correspondence: umutona@wsdot.wa.gov; Tel.: +1-(509)-595-9498

† Presented at the Second International Conference on Maintenance and Rehabilitation of Constructed Infrastructure Facilities, Honolulu, HI, USA, 16–19 August 2023.

Abstract: Chip seal is a pavement preservation treatment that is usually used on lower-volume routes by States and local agencies. Chip seal provides a low-cost surface treatment, improves skid resistance, and reduces the deterioration of the existing underlying pavement. Chip seals consist of a layer of aggregate (chip) spread evenly on top of an emulsion that is sprayed onto the deteriorated surface. The performance of chip seals is significantly affected by the existing conditions and pretreatment carried out before the chip seal is applied. A study was conducted by retrieving long-term performance of chip seals from the Washington pavement management system (WSPMS). In collaboration with Washington State Department of Transportation (WSDOT), twelve (12) road sections previously paved with hot mix asphalt (HMA) or chip seal and later maintained with chip seal were selected, and their performance was analyzed in terms of cracking, rutting, and roughness index (IRI). It was found that chip seal greatly reduced the cracking and slowed down the crack growth when compared to HMA overlay. However, no improvement in rutting or international roughness index (IRI) caused by the chip seal was found.

Keywords: chip seal; WSPMS; crack growth; rutting; international roughness index (IRI)

Citation: Umutoniwase, A.; Wen, H.; Littleton, K. Study of Long-Term Field Performance of Chip Seal in Washington State. *Eng. Proc.* **2023**, *36*, 24. https://doi.org/10.3390/engproc2023036024

Academic Editor: Hosin (David) Lee

Published: 6 July 2023

1. Introduction

A chip seal, often referred to as a bituminous surface treatment (BST), is a pavement preservation treatment technique that is usually used on lower-volume routes by States and many local agencies. In addition to providing a low-cost surface treatment, other benefits of chip seals include improving skid resistance and reducing the deterioration of the existing underlying pavement. Roads treated with chip seal, however, do deteriorate over time due to construction quality, traffic, and environmental effects. The performance of chip seals is significantly affected by the existing conditions and the pretreatment carried out before the chip seal is applied. Therefore, to achieve the desired performance, it is recommended to apply the right treatment on the right pavement at the right time [1].

Several researchers have studied the field performance of chip seal over time. Jalali and Vargas-Nordcbeck (2021) assessed whether chip seal application extends the pavement lifespan based on the initial conditions and the type of chip seal applied. In their study, six different chip seal treatments placed on LR-159 in Auburn, Alabama, were evaluated and compared to a control section which was not treated by chip seal. The treatments included: (1) single chip seal, (2) single chip seal with crack sealing, (3) double chip seal, (4) triple chip seal, (5) fiber mat chip seal, and (6) scrub seal. Cracking performance data collected since 2012 were analyzed. Findings showed that, generally, treated sections performed better than untreated sections. It was also found that the pavements that were in good condition and those whose cracks were filled before chip seal was applied can last at least

10 years, depending on traffic volume. Moreover, multilayered chip seal performed better than a single layer, with a median time to failure (MTTF) of more than 10 years [2].

In Louisiana, Mousa et al. (2020) also studied the pavement life extension when chip seal is applied. The researchers analyzed cracking, roughness, and overall pavement condition of 47 flexible and composite pavement sections. The study concluded that chip seal treatments extended pavement life by 4 to 17 years depending on the pretreatment condition, as well as the type of pavement treated (flexible or composite) [3].

Mamlouk and Dosa (2014) evaluated the life-extending benefits of single chip seal using data from the Long-Term Pavement Performance (LTTP) [4]. Pavement initial conditions were classified as rough, medium, or smooth, based on the international roughness index (IRI). As in the previous study, treated sections were compared with untreated sections, and the results showed a better performance for treated sections than untreated sections. This study also found a high life extension for the pavements that were smooth at the time chip seal was applied. The results showed a life extension of 4–7 years, 2–3 years, and 0–1 year for smooth, medium, and rough pavements, respectively. A similar study was conducted by the Federal Highway Administration (FHWA) in 2010 with the aim of highlighting the degree to which pavement preservation treatments extend the service life of pavements [5]. Performance data was collected from six target states (California, Kansas, Michigan, Minnesota, Texas, and Washington) selected based on their developed pavement management system or other information gathering systems. Results showed that chip seal treatments extended pavement service life for 3 to 8 years.

2. Methodology

The pavement management system is a system for recording all data on pavement condition to monitor deterioration, thus planning for maintenance and repair as well as mobilizing funds on time. Data are mostly managed through a computer program in which all information about roads and highways in a certain state is recorded. Like other States, WSDOT has a pavement management system known as the Washington pavement management system (WSPMS). In this program, various data components such as cracking, rutting, roughness, lane configuration, cross section, location, and construction history, among others, can be retrieved. In this study, twelve (12) road sections in Eastern Washington which were previously paved with hot mix asphalt (HMA) or chip seal and later maintained with chip seal were selected for analysis. An assessment of how chip seals impacted their performance was carried out in terms of cracking, rutting, and roughness index (IRI) over a period of 23 years (1996–2019), which included performance of pavement prior to chip seals and the performance of chip seals.

3. Results

Based on the construction history, all sections were retrieved and the results of rutting, IRI, and cracking were plotted. An example of the data plots is shown in Figure 1 below:

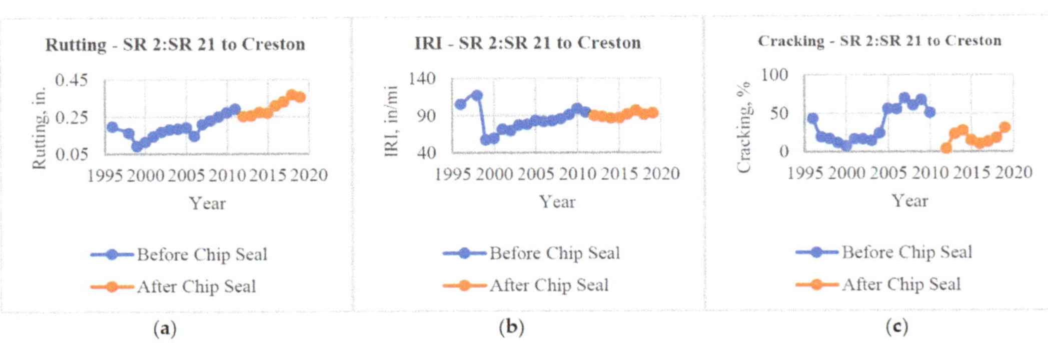

Figure 1. Distresses for SR 2: SR 21 to Creston: (**a**) Rutting; (**b**) IRI (**c**) Cracking.

For most of the sections, rutting results indicated a slight decrease after the chip seal was applied, then rose again. This soaring can be explained by the presence of rutting in the old pavement. While IRI was found to be slightly improved after chip seal was applied, a substantial decrease in cracking was observed. The growth rate for each section was calculated before and after chip seal by computing the slope between two data points then averaging them. The immediate reduction after the application of chip seal was included in the calculation of the growth rate. Results are shown in Table 1 below. The downward trend (negative data) obtained for the cracking and IRI can be justified by maintenance activities that were performed after the application of chip seal.

Table 1. Growth rates of Rutting, IRI, and Cracking.

Road Section and Location	Growth Rate before and after Chip Seal					
	Rutting, in/Year		IRI, in/mi/Year		Cracking, %/Year	
	Before	After	Before	After	Before	After
SR 2/SR 21 to Creston	1.64	1.49	3.04	0.52	4.61	3.85
SR 2/Creston to Rocklyn Rd	2.78	2.44	3.88	2.15	2.81	0.79
SR 2/Rocklyn Rd to Davenport	2.25	1.99	3.53	2.39	6.88	2.26
SR 2/Pend Oreille Co Line to SR 211	1.06	1.72	1.72	−0.40	0.07	−0.17
SR 2/SR 211 to Newport	2.29	1.26	1.82	−0.91	3.64	0.63
SR 20/Walker Hill Rd to Graves Mt Rd	1.75	0.03	5.04	4.10	1.28	1.09
SR 26/Dusty to Colfax	0.37	1.65	1.14	0.14	1.51	1.49
SR 27/Freeman to 32nd Ave	1.66	1.26	2.04	3.27	3.49	1.71
SR 127/Church Hill Rd to Dusty	2.47	1.10	1.13	3.85	0.79	0.49
SR 211/US2 to SR 20	0.13	0.95	1.09	0.25	0.61	−0.88
US 195/Idaho to Colton	3.18	2.85	1.47	4.06	−0.09	2.30
US 395/Columbia River Bridge to Boyd's	1.55	0.59	2.11	0.92	−0.27	−0.29

Paired t-test was performed to determine the effectiveness of chip seal on the growth rates of rutting, IRI, and cracking, for the projects converted from ACP to chip seal. Project 4 (Pend Oreille Co Line to SR 211) was not included in the analysis because the conversion was from chip seal to chip seal. Results showed that chip seal significantly affected crack growth rate. However, it did not affect the roughness and rutting growth rate. p-values were 0.15 for rutting, 0.37 for IRI, and 0.02 for cracking.

4. Conclusions

Generally, results showed considerable improvement in terms of cracking for all projects. This improvement can be attributed to one of functions of chip seal, which is to seal minor cracks.

Rutting slightly decreased for a few sections and for others an upward trend was noticed. This change in rutting may have resulted from the type of rutting in the existing pavement. The studded tire wear may be the major cause of the rutting. The placement of chip seal does not mitigate the studded tire wear.

A slight decrease in the roughness was also observed. This minor drop can be justified by the texture of chip seal. Cover aggregate in chip seal are macrotexture and result in a rough surface which improve the skid resistance.

The pavements analyzed had different service life length. Some were less than 10 years old, while others were more than 10 years. It was hard to compare how chip seal improved the pavement life span. Therefore, the growth rate was used instead of pavement life. However, slowed growth rates of cracking, rutting, and roughness can result in increased service life.

Author Contributions: Conceptualization, A.U. and H.W.; methodology, A.U. and H.W.; software, A.U., H.W. and K.L.; validation, A.U., H.W. and K.L.; formal analysis, A.U. and H.W.; investigation, A.U., H.W. and K.L.; resources, A.U., H.W. and K.L.; data curation, A.U.; writing—original draft preparation, A.U.; writing—review and editing, A.U., H.W. and K.L.; visualization, A.U.; supervision, H.W.; project administration, H.W. All authors have read and agreed to the published version of the manuscript.

Funding: This research received no external funding.

Institutional Review Board Statement: Not applicable.

Informed Consent Statement: Not applicable.

Data Availability Statement: Data is contained within the article.

Conflicts of Interest: The authors declare no conflict of interest.

References

1. Peshkin, D.G.; Hoerner, T.E.; Zimmernan, K.A. *Optimal Timing of Pavement Preventive Maintenance Treatment Applications*; NCHRP Report 523; Transportation Research Board: Washington, DC, USA, 2004.
2. Jalali, F.; Vargas-Nordcbeck, A. Life-extending benefit of chip sealing for pavement preservation. *Transp. Res. Rec.* **2021**, *2675*, 104–116. [CrossRef]
3. Mousa, M.R.; Elseifi, M.A.; Bashar, M.Z.; Zhang, Z.; Gaspard, K. Short and Long-Term Field Performances and Optimal Timing of Chip Seal in Hot and Humid Climates. *Transp. Res. Rec.* **2020**, *2674*, 33–43. [CrossRef]
4. Mamlouk, M.S.; Dosa, M. Verification of effectiveness of chip seal as a pavement preventive maintenance treatment using LTPP data. *Int. J. Pavement Eng.* **2014**, *15*, 879–888. [CrossRef]
5. Wu, Z.; Groeger, J.L.; Simpson, A.L.; Hicks, R.G. *Performance Evaluation of Various Rehabilitation and Preservation Treatments*; No. FHWA-HIF-10-020; U.S. Department of Transportation, Federal Highway Administration, Office of Asset Management: Washington, DC, USA, 2010.

engineering proceedings

MDPI

Proceeding Paper

Comparison between Two Different Deployment Types of Road-Side Devices Reducing Incident-Related Potential Conflicts †

Jaehyeong Lee [1] and Jin-Tae Kim [2,*]

[1] Department of Transportation Policy and System Engineering, Korea National University of Transportation, 157 Cheoldobangmulkwan-ro, Uiwang 16106, Republic of Korea; imleejh0421@naver.com
[2] Department of Transportation System Engineering, Korea National University of Transportation, 157 Cheoldobangmulkwan-ro, Uiwang 16106, Republic of Korea
* Correspondence: jtkim@ut.ac.kr; Tel.: +82-31-461-8736

† Presented at the Second International Conference on Maintenance and Rehabilitation of Constructed Infrastructure Facilities, Honolulu, HI, USA, 16–19 August 2023.

Abstract: The carpet-Type materials painted yellow in pedestrian waiting areas are safety devices used to prevent collisions with vehicles and children. They are primarily recommended to be placed on both the pavement and the wall sides, but local governments in the Republic of Korea place them only on the pavement side, excluding the wall part to reduce the deployment costs and increase the number of deployment sites. The difference in effectiveness between these Types may be different, but this has not yet been proven. This study examines the difference between these Types found in the field test. The test employed two different measures: (1) the level of driver visibility and (2) the vehicle deceleration rate. As a result of the analysis, the yellow carpet installed on both the pavement and the wall sides was comparatively more effective. The test results suggest that this yellow carpet Type should be installed on both pavement and wall sides, particularly for school zone areas, to maximize its safety effects.

Keywords: safety device; visibility; vehicle deceleration

Citation: Lee, J.; Kim, J.-T. Comparison between Two Different Deployment Types of Road-Side Devices Reducing Incident-Related Potential Conflicts. *Eng. Proc.* **2023**, *36*, 25. https://doi.org/10.3390/engproc2023036025

Academic Editor: Hosin (David) Lee

Published: 4 July 2023

1. Introduction

In the Republic of Korea, various efforts have been made to reduce the number of traffic accident victims of a young age. In order to prevent such encounters between children and vehicles in advance, local highways near (a 300 m radius) elementary schools and kindergartens are designated as school zones in the Republic of Korea. Various traffic safety devices, such as speed humps and traffic safety signs, are installed in these areas. Among them, the yellow carpet ensures the drivers' visibility to children by installing a waiting space in yellow before the crossing.

According to the yellow carpet installation standards prepared by the Ministry of Public Administration and Security, yellow materials should be painted on the walls and pavements, or yellow temporary surfaces should be installed if there are no walls [1]. However, due to local governments, they are sometimes installed only on the pavements for various reasons, such as reducing installation.

The yellow carpet's effect of promoting 'traffic safety for children' may be different when the 'yellow blocks' that draw attention to child pedestrians are (1) installed only on the pavement and (2) installed on walls built together with the pavement. If 'yellow blocks' are installed only on the pavement without the walls, it may be difficult to easily spot children standing on the yellow carpet from the driver's point of view. However, the effect of promoting children's traffic safety as a result of these two situations has not been examined. Hence, this study compares the difference in the effectiveness of children's traffic safety when yellow carpets are installed in two different ways.

2. Research Approach

In this study, two Types of yellow carpets were installed as shown in Table 1, with Type A installed on the pavement and rear walls at the same time and Type B installed only on the pavement.

Table 1. Comparison of yellow carpet Types deployed in the field.

Type	Area Where Yellow Carpet Is Deployed		Remarks
	Horizontal Bottom	Vertical Background	
A	○	○	Standard installation
B	○	X	Modified in practice

The two Types of indicators in Table 2 were applied to analyze the difference in effects observed in the above two morphological facilities.

Table 2. Analysis measures and data collection methods.

Measures	Data Collection Methods
Drivers' visibility	Measured via VAS with a set of field images
Driving speed	Measured via image processing with drone video

2.1. Drivers' Visibility Analysis Method

In this study, the VAS (Visual Attention Software) analysis method performed in previous studies was applied to quantify the 'level of drivers' visibility'. As mentioned above, VAS is a technology that quantifies the degree to which the eyes are dispersed or concentrated on products arranged in a specific place as ordinary consumers enter the store. This study estimates the degree to which drivers focus their attention in the direction of the 'yellow carpet' as they enter the school zone by applying this technology.

VAS numerically expresses the probability that a person will focus on a specific object. As shown in Figure 1, the high concentration point is expressed as 'red', the middle concentration point is expressed as 'yellow', and the low concentration point is expressed as 'dark' [2]. Through this color scheme, the concentration of the gaze is expressed with a heating map.

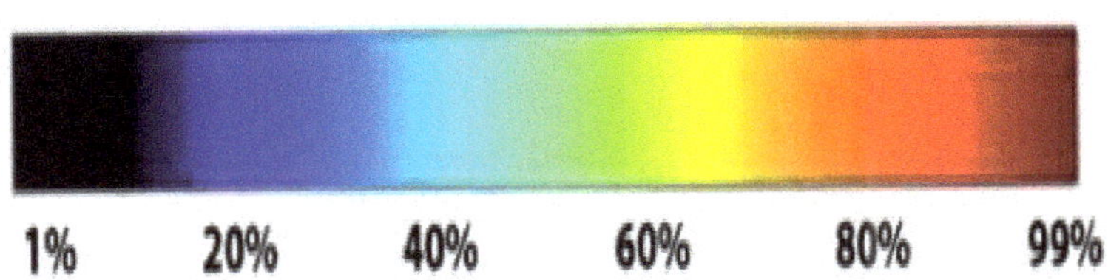

Figure 1. VAS heat map visibility range.

Figure 2 is an example of an analysis after setting a yellow carpet for the 'area of interest'. Figure 2a is a 'drivers' visibility concentration' heat map derived through VAS analysis, and Figure 2b presents the results of the 'drivers' visibility concentration' analysis related to a specific point in connection.

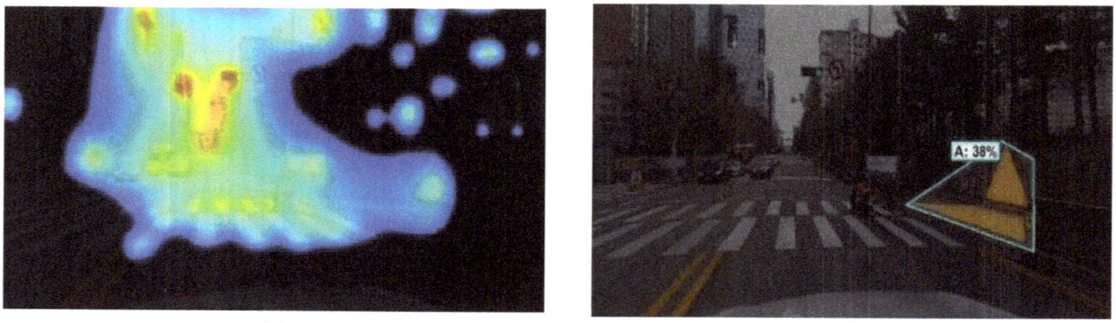

(**a**) (**b**)

Figure 2. (**a**) Intermediate heat map analysis results. (**b**) Drivers' visibility.

2.2. Cruising Speed Analysis Method

The point speed collection method was not adopted to collect data on the change in cruising speed, but the interval speed collection method was selected in this study to collect data on the change in cruising speed.

The field video data collection method using a drone was applied as a method of collecting driving speed indicator data in a school zone. After collecting on-site video data through drone aerial shooting, driving speed data were extracted by referring to location and time information obtained per image frame using the Vegas Pro 17 image-processing program. Based on the collected data, the driving trajectory data of individual vehicles were analyzed, and through this, data on changes in driving speed were collected. Night shots of drones are not institutionally allowed. Therefore, this analysis excluded a night driving speed analysis from the scope of the study.

3. Results

A paired *t*-test was conducted to confirm whether the analysis results were statistically significant. The null hypothesis and the opposing hypothesis are outlined below. The hypothesis test was conducted at the 95% confidence level ($\alpha = 0.05$):

H_0: No difference in visibility and speed after installation of the yellow carpet.

H_α: Visibility and deceleration effects after installation of the yellow carpet.

3.1. Visibility

3.1.1. Daytime

Table 3 shows the drivers' visibility concentration before and after installation according to the yellow carpet Type in the daytime. Both yellow carpet forms A and B rejected the null hypothesis and confirmed that visibility was improved.

Table 3. Drivers' visibility *t*-test according to yellow carpet installation Types A and B—daytime.

Type	Average		*p*-Value	Result
	Before	After		
A	0.33	0.45	0.00	Reject
B	0.31	0.39	0.00	Reject

3.1.2. Night-Time

Table 4 shows the drivers' visibility concentration before and after installation according to the yellow carpet Type at night-time. As a result of the analysis, only the yellow carpet installed in the form of Type A was rejected. Therefore, it was found that only Type A has an effect in enhancing the drivers' visibility.

Table 4. Drivers' visibility t-test according to yellow carpet installation Types A and B—night-time.

Type	Average		p-Value	Result
	Before	After		
A	0.57	0.70	0.01	Reject
B	0.34	0.32	0.41	Not reject

3.2. Cruising Speed

Table 5 shows the cruising speed before and after installation according to the Type of yellow carpet. Both yellow carpet Types A and B rejected the null hypothesis, confirming that the cruising speed decreased.

Table 5. Cruising speed t-test according to yellow carpet installation Types A and B.

Type	Average		p-Value	Result
	Before	After		
A	28.3	24.9	0.00	Reject
B	40.9	37.2	0.00	Reject

3.3. Overall Result

Table 6 summarizes the results derived from this study.

Table 6. Yellow carpet installation effect (95% confidence level).

Indicator	Experimental Conditions		Analysis Results
Drivers' visibility	Day	Type A	Statistically enhanced (+42%)
		Type B	Statistically enhanced (+26%)
	Night	Type A	Statistically enhanced (+23%)
		Type B	Statistically no difference (meaningless)
Cruising speed	Day	Type A	Statistically reduced (−12%)
		Type B	Statistically reduced (−9%)

It was determined that the on-site yellow carpet was effective compared to the situation where there was no yellow carpet (95% confidence level), without distinguishing between Type A and Type B. However, it was determined that the installation of 'Form A' had a positive effect in night situations, while 'Form B' had no safety promotion effect because it was statistically the same as the situation without the 'Yellow Carpet' (95% confidence level).

Referring only to the figures derived from this study, it was found that the installation effect of 'Yellow Carpet Type B' was half that of 'Yellow Carpet Type A', and the effect could not be evaluated in the night-time.

When safety management in a child protection zone is not limited to daytime but also includes night-time, it would be desirable to install a yellow background color in a form that can be effective for night-time. It is a "Type A" that applies to both the pedestrian waiting space pavement and the wall.

4. Conclusions

In preparation for the absence of a yellow carpet, situations in which the yellow carpet is installed in the form of two different Types were classified, compared, and analyzed. In the case of yellow carpets, it was determined that full installation, including the wall, rather than installation on the pavement alone, met the goal of promoting traffic safety in the school zone. Therefore, when local governments install yellow carpets, additional efforts are needed to create a safe transportation environment for children by installing the materials on pavements and walls.

Author Contributions: Conceptualization, J.-T.K. and J.L.; methodology, J.-T.K.; software, J.L.; validation, J.-T.K. and J.L.; formal analysis, J.-T.K.; investigation, J.L.; resources, J.L.; supervision, J.-T.K.; project administration; J.-T.K. All authors have read and agreed to the published version of the manuscript.

Funding: This work was supported by a Korea Institute of Police Technology (KIPoT) grant funded by the Korea government (KNPA) (No.092021C29S02000, Development of on-site control technology for road traffic network control in the event of an incident or disaster).

Institutional Review Board Statement: Not applicable.

Informed Consent Statement: Not applicable.

Data Availability Statement: Not applicable.

Conflicts of Interest: The authors declare no conflict of interest.

References

1. Ministry of Public Safety. Yellow-Carpet Fabrication and Installation Guidelines, Republic of Korea. 2018. Available online: http://incrc.org/resources/?pageid=7&mod=document&uid=349 (accessed on 28 June 2023).
2. Ahn, H.S.; Kim, J.T. Study on Visual Recognition Enhancement of Yellow-carpet Placed at Near Pedestrian Crossing Areas: Visual Attention Software Implementation. *J. Inf. Technol. Serv.* **2016**, *15*, 73–83.

engineering proceedings

MDPI

Proceeding Paper

Evaluating Remediation Techniques for Fouled Ballast on Army Installations †

Charles E. Williams, Jr. * and Thomas J. Beasley

U.S. Army Engineering Research and Development Center, 3909 Halls Ferry Road, Vicksburg, MI 39180, USA; thomas.j.beasley@erdc.dren.mil

* Correspondence: charles.e.williams@usace.army.mil

† Presented at the Second International Conference on Maintenance and Rehabilitation of Constructed Infrastructure Facilities, Honolulu, HI, USA, 16–19 August 2023.

Abstract: Rail transport is a vital asset for U.S. Army distribution networks for movements of oversized and overweight vehicles and munitions. As the rail infrastructure ages on military installations, the reliability of these rail systems is a critical concern to support military power projection requirements. The Engineering Research and Development Center (ERDC) evaluates over 1500 miles of Army track approximately every four years. Many Army installations have significant problems with fouled ballast. These installations have few resources to remediate fouled ballast, and/or may not be aware of remediation techniques. The primary objective of this paper is to provide a list of remediation techniques for installations to implement with efforts to reduce fouled ballast and improve track maintenance.

Keywords: rail maintenance; fouled ballast; remediation techniques; track inspection

Citation: Williams, C.E., Jr.; Beasley, T.J. Evaluating Remediation Techniques for Fouled Ballast on Army Installations. *Eng. Proc.* **2023**, *36*, 26. https://doi.org/10.3390/engproc2023036026

Academic Editor: Hosin (David) Lee

Published: 7 July 2023

1. Introduction

Fully inspecting, evaluating, and understanding rail substructure is vital to safe transport of military goods and sustaining critical missions. Ballast is a critical component for track stability and performance. Rail ballast has four primary functions: (1) provide adequate track drainage; (2) restrain the track laterally, longitudinally, and vertically; and (3) support the loads of the track and trains while distributing to the subgrade; and (4) allow adjustment of track line and surface/profile [1]. Fouled ballast is a significant maintenance issue that jeopardizes the performance of the rail system. Ballast fouling is often a result of poor drainage flow along the track. When fouling material accumulates over time and loose soil particles start filling voids, the ballast starts to degrade [2,3].

Figure 1 presents a photograph of fouled ballast at an Army installation. The ballast is covered with accumulated fines. Unfortunately, many military installations have limited resources to remediate fouled ballast and may not be aware of remediation techniques to implement within their budgets.

EDRC's rail inspection team inspects over 1500 miles of Army rail track every four years. Each year ten to twelve Army installations are inspected, and a report is provided to the installation identifying the track defects. This paper focuses on one primary defect, fouled ballast, which is identified within these reports. We also elucidate recommended fouled ballast remediation techniques that we would begin presenting to installations at final inspection report briefings. The overall objective is to better inform installations of available remediation techniques to improve track maintenance and performance.

Figure 1. Fouled Ballast located at an Army Installation.

2. Evaluation Methods

Figure 2 displays the average ballast defect density percentage identified by ERDC's rail inspection team from inspecting 42 Army installations. The percentage of fouled ballast identified by the track inspector is subjective but based on experience and training. The track inspector provides each section a fouled ballast density percentage ranging from 0% to 100%. The average fouled ballast density percentage is calculated by dividing the percentage provided to each section by the total number of sections inspected.

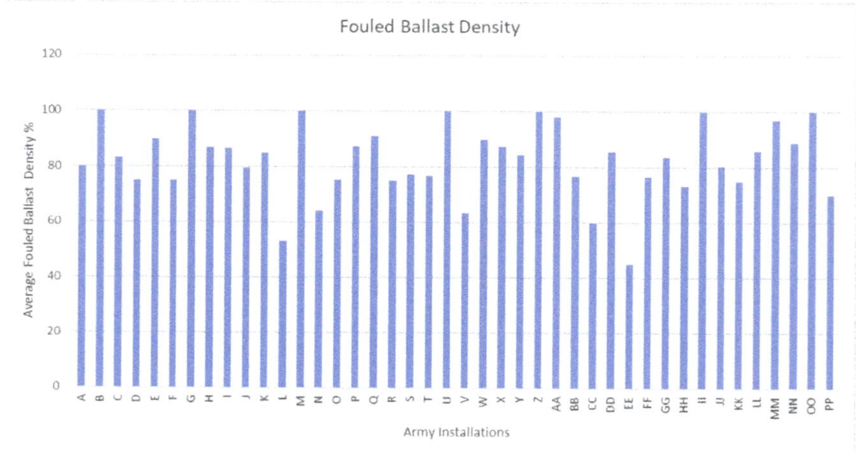

Figure 2. Fouled Ballast Density inspected on Army Installations.

Table 1 displays results by assessing the fouled ballast identified and determining the serviceability and reliability for continued track use. When the fouled ballast density falls in the range of (100–85%), it is classified as failed ballast. About 50% of the installations inspected met this criterion. The second highest fouled ballast density between (85–70%), is classified as serious, and 35.7% of the installations inspected met this criterion. None of the installations met the criteria for very fair, satisfactory, or good. These results show that fouled ballast is a significant maintenance issue for Army installations.

Table 1. Fouled Ballast Condition Results.

Fouled Ballast Condition Density Index (%)	Ballast Condition Descriptor	% Of Installations Meeting Criteria
100–85 Failed	Overall degradation is total.	50
85–70 Serious	Extreme serviceability or reliability reduction.	35.7
70–55 Very Poor	Unsatisfactory serviceability or reliability reduction.	9.5
55–40 Poor	Significant serviceability or reliability loss.	4.8
40–25 Very Fair	Serviceability or reliability is noticeably degraded.	0
25–10 Satisfactory	Serviceability or reliability is degraded but adequate.	0
10–0 Good	Slight or no serviceability or reliability reduction.	0

3. Techniques for Remediating Fouled Ballast

Ballast performance is greatly affected by the accumulation of fouling materials and water retained by fouling particles [2,3]. When ballast degradation occurs under constant rail traffic over time, it is called breakdown fouling [2,3]. The result of the degradation causes the settlement rate of the ballast to increase [3,4]. Ballast degradation can also increase maintenance cost, a significant factor for installations when making budget decisions. However, implementing timely remediation techniques can lower maintenance costs over the lifetime of the ballast [3].

Techniques for remediating fouled ballast are not a defined standard of practice for the Army or industry. However, our discussions with rail industry representatives, academia, and certified track inspectors reveal that more emphasis on fouled ballast remediation is needed. Although there are techniques available for implementation, funding plays a major role in addressing fouled ballast in the short and long term.

A non-destructive evaluation method that can be used to identify fouled ballast layers under the rail is Ground Penetrating Radar (GPR) [5]. While the technology has some limitations, GPR has become a cost-efficient method for determining fouled ballast layer depth. The capability to quantify the amount to fouled ballast with greater certainty allows for integrating remediation techniques to improve maintenance planning practices.

The following is a list of proven remediation techniques that will be provided to the army for review:

1. Track Undercutting—Track undercutting is performed by removing existing ballast beneath the rail ties. Once the ballast is removed from the track. The fines are removed from the ballast and can be either reused or discarded. Undercutting is a valuable method when the track is not raisable due to possible overhead utilities or bridge structures [1].
2. Repair/Improve Track Side Drainage—poor drainage is a problem that increases ballast fouling. The ability to divert rainwater away from the track lessens the accumulation of fines. The ability to maintain proper drainage structures, side ditches, and other drainage appurtenances reduces subgrade problems, track deterioration, and ballast fouling.
3. Shoulder Cleaning—Shoulder cleaning is a process of removing the existing ballast outside the rail ties, mechanically cleaning the fines from the ballast, and replacing good ballast back onto the shoulder area. This is a cost-effective processes that is less disturbing to the existing ballast sections. It can be performed on longer sections, and typically several miles can be cleaned in one day.

4. Geotextile Fabrics—geotextiles are typically used for reinforcement and separation between two materials [6]. The use of geotextile fabric between the sub-ballast and sub-grade reduces the accumulation of fines entering the sub-ballast and fouling.

4. Recommendations/Conclusions

Many factors can influence ballast fouling, such as the gross weight of the load, type of ballast materials, transportation type (freight or passenger), regional climate, and the environmental factors [5]. All of these factors make predicting ballast fouling and ballast layer degradation difficult [5]. The Army has recognized that ballast fouling is a serious issue for its rail system. This paper evaluated past inspections reports from 42 Army installations, focusing on fouled ballast density, as identified by the track inspectors. Nearly 50% of the installations inspected had a condition index ranging from (100–85%), which is classified as ballast failure.

This paper provided recommendation techniques for the Army to review, and possibly implement to extend the life of their rail systems. Track undercutting and shoulder cleaning are proven remediation techniques, but each require labor and funding resources. As rail infrastructure continues to age, implementing maintenance strategies is critical to improve the performance of the rail system over time. Track rehabilitation is costly and identifying funding resources is difficult. The ability to maintain track systems over time will allow the Army to sustain their critical missions with minimal interruptions to rail service.

Author Contributions: Each author contributed to the conceptualization, methodology, investigation, data analysis, and manuscript writing. All authors have read and agreed to the published version of the manuscript.

Funding: This research received no external funding.

Institutional Review Board Statement: Not applicable.

Informed Consent Statement: Not applicable.

Data Availability Statement: Data results are provided within the manuscript from an ERDC vendor. The datasets are not publicly provided to protect the privacy of the vendor.

Acknowledgments: The authors would like to acknowledge Barret Baldwin and ERDC's rail inspection team for providing the ballast data.

Conflicts of Interest: The authors declare no conflict of interest.

References

1. UFC 4-860-03 Railroad Track Maintenance and Safety Standards. 2008. Available online: https://wbdg.org/ (accessed on 9 February 2023).
2. Hyslip, J.; Kashani, H.F. Ballast Life and Effective Parameters. In Proceedings of the 2018 Joint Rail Conference, Pittsburg, PA, USA, 18–20 April 2018.
3. Selig, E.T.; Waters, J.M. *Track Geotechnology and Substructure Management*; Thomas Telford Publications: New York, NY, USA, 1994.
4. Hyslip, J.; Kashani, H.F.; Trosino, M. Ballast State of Good Repair. In Proceeding of the 2017 AREMA Annual Conference and Exposition, Indianapolis, IN, USA, 17–20 September 2017.
5. Guo, Y.; Liu, G.; Jing, G.; Qu, J.; Wang, S.; Qiang, W. Ballast fouling inspection and quantification with ground penetrating radar (GPR). *Int. J. Rail Transp.* **2022**, *11*, 151–168. [CrossRef]
6. Rollings, M.P.; Rollings, R.S. *Geotechnical Materials in Construction*; McGraw-Hill: New York, NY, USA, 1996.

engineering proceedings

MDPI

Proceeding Paper

Bio-Oils as Asphalt Bitumen Rejuvenators †

Amir Tabaković [1,2,*], Dave van Vliet [1], Kirsten Roetert-Steenbruggen [1] and Greet Leegwater [1]

1 Netherlands Organization for Applied Research (TNO), Department of Building Materials and Structure, Unit Mobility and Built Environment, NEXT Delft, Molengraaffsingel 8, 2629 Delft, The Netherlands
2 Faculty of Civil Engineering and Geosciences, Materials and Environment, Delft University of Technology, Stevinweg 1, 2628 Delft, The Netherlands
* Correspondence: amir.tabakovic@tno.nl; Tel.: +353-85-128-4310

† Presented at the Second International Conference on Maintenance and Rehabilitation of Constructed Infrastructure Facilities, Honolulu, HI, USA, 16–19 August 2023.

Abstract: Bitumen rejuvenators are used to improve or restore the physical and mechanical performance of aged bitumen. Traditional bitumen rejuvenators are a product of crude oil. As crude oil production declines and the environmental and financial costs of crude oil increase, there is an urgent need to identify more environmentally sustainable bitumen rejuvenator alternatives. Bitumen rejuvenators generated from biological sources offer an environmentally friendly and economically viable alternative to the crude oil-based bitumen rejuvenators. This paper describes a study wherein microalgae oil was used as an aged bitumen rejuvenator. The chemical, physical, and mechanical effects of microalgae oil on 70/100pen aged bitumen were investigated. The results indicate that microalgae oil has the potential to be used as an aged bitumen rejuvenator.

Keywords: asphalt materials; bitumen rejuvenation; bio-bitumen rejuvenators; microalgae oil

Citation: Tabaković, A.; van Vliet, D.; Roetert-Steenbruggen, K.; Leegwater, G. Bio-Oils as Asphalt Bitumen Rejuvenators. *Eng. Proc.* **2023**, *36*, 27. https://doi.org/10.3390/engproc2023036027

Academic Editor: Hosin (David) Lee

Published: 7 July 2023

1. Introduction

The current drive to improve the sustainability of the asphalt industry is driving research in bio-based materials, specifically about whether they are suitable for use in road construction. Traditionally, crude oil products have been used as bitumen modifiers (softeners and strengtheners) in asphalt production. Recent studies [1,2] have shown that oils extracted from microalgae biomass can also be used to modify bitumen. Audo et al. [1] demonstrated that microalgae biomass can be converted into a water-insoluble material made of a mixture of an oily component and solid residues, which is ideal for the production of bio-binder [3]. Maximum bio-oil yields in the range of 40–50% have been reported [1], though they are in the initial stages of development. The BioRePavation project [4] demonstrated that microalgae oil can be used as an asphalt binder in the construction of new asphalt mixtures and in the asphalt recycling process.

This paper presents the results of a study conducted to assess the performance of microalgae oil as a rejuvenator for aged asphalt bitumen. Microalgae oil and biomass were tested for their chemical, physical, and mechanical compatibility and effects on the 70/100pen aged asphalt binder. Two plant-based oils (soy bean and palm oil) were used as comparative bio-oils, and a commercially available bio-based rejuvenator was used as a control material. The blends were prepared by adding rejuvenating oil (3% by weight) to aged bitumen. Fourier-Transform Infrared Spectroscopy (FTIR) was used to investigate the prevalence of chemical functional groups and possible differences between the pure and blended bitumen samples. Dynamic Shear Rheometer (DSR) was used to investigate the mechanical performance. Optical microscopy was used to investigate the physical appearance of the bitumen blends due to ageing and also as rejuvenating oils were added.

2. Materials and Methods

2.1. Materials

Microalgae biomass (Nannochloropsis Oceanica paste) and oil extracted from Nannochloropsis Oceanica paste biomass were supplied by Wageningen University. Two additional bio-based oils were selected for the study: soy bean oil (purchased from Ademia Cosmetiques) and palm oil (purchased from Cosmetica Natural Oils Ltd., Ipswich, UK). A standard, commercially available bio-based rejuvenator, Neomex® Hr, was supplied by Latexfalt B.V., and bitumen 70/100pen was supplied by Breedongroup PLC.

2.2. Methods

2.2.1. Bitumen Ageing Process

Fresh 70/100pen bitumen was aged in accordance with the protocol outlined by Kliewer [5]. The goal was to simulate 15 years of bitumen ageing. One kilogram of bitumen was poured into a stainless steel tray lined with greaseproof paper. The thickness of the bitumen layer in the tray was 10 mm. The tray was then placed in a forced air draft oven and subjected to a two-step ageing process as follows: step 1—4 h at 135 °C; step 2—4 days at 85 °C. Because the bitumen ageing was insufficient following steps 1 and 2, we extended step 2 by 16 days.

2.2.2. Blending Process

Five different blends containing 70/100pen of aged bitumen, four different rejuvenating oils, and a blend containing microalgae biomass were made. The blends were prepared by adding rejuvenating oil (3% by weight) to aged bitumen. The microalgae biomass blend was designed according to the lipid content in the microalgae biomass. The microalgae biomass lipid analysis was tested according to the Folch method [6]. Our analysis demonstrated that the dry weight biomass lipid content was 20%, and the moisture content was 70%. Table 1 summarizes the blend constituent weights.

Table 1. Bitumen blends, weight 200g.

Blends	Bitumen (g)	Rejuvenator/Oil (g)	Biomass (Paste) (g)
Bitumen + Rejuvenating Oils	194	6	—
Bitumen + Microalgae Biomass	194	—	100

Low shear blending was adopted for the blends preparations. The blend sample preparation process began by heating blend constituents, aged 70/100pen bitumen, and rejuvenating oil separately at 160 °C for 30 min before blending. Once the bitumen and oil reached 160 °C, the bitumen was placed on a temperature-controlled hot plate, and the oil was added to the bitumen. The blending temperature was held constant at 160 °C and mixed for 30 min at 600 RPM. In order to prevent foaming of the bitumen blend, the moisture was removed from the microalgae biomass by placing it in the oven at 100 °C for 60 min prior the blending.

2.2.3. Fourier-Transform Infrared Spectroscopy (FTIR)

The FTIR test method was employed to study the prevalence of chemical functional groups and possible differences between the pure and blended bitumen samples. The investigation was undertaken according to the protocol described by Hofko et al. [7].

2.2.4. Dynamic Shear Rheometer (DSR)

The DSR approach was employed to investigate bitumen stiffness change. The Dynamic Share Rheometer (DSR) test was conducted in accordance with EN 14770:2012.

2.2.5. Microscopic Analysis

To investigate the microstructure characteristics of the blends, a Zeiss Axioplan digital microscope with a Zeiss Plan-Neofluar lens was used at 200 times magnification. The goal was to map the dispersion of the rejuvenating oils throughout the blends and link the possible changes in physical properties to changes in the microstructure (occurring due to chemical interactions within the blends).

3. Results

3.1. Fourier-Transform Infrared Spectroscopy (FTIR)

Figure 1 shows the chemical composition with respect to functional groups in comparison to neat bitumen, aged bitumen, and rejuvenated aged bitumen blends (detected via FTIR). All seven blends demonstrate a very similar spectrum. At the lower end of the wavenumber, the blend with microalgae biomass shows a closer response to the neat bitumen, whereas all of the blends containing rejuvenating oils show the same response. This indicates that microalgae oil blends have a very similar chemical composition to blends containing Neomex Hr (a standard bio-based rejuvenator) and plant-based oils.

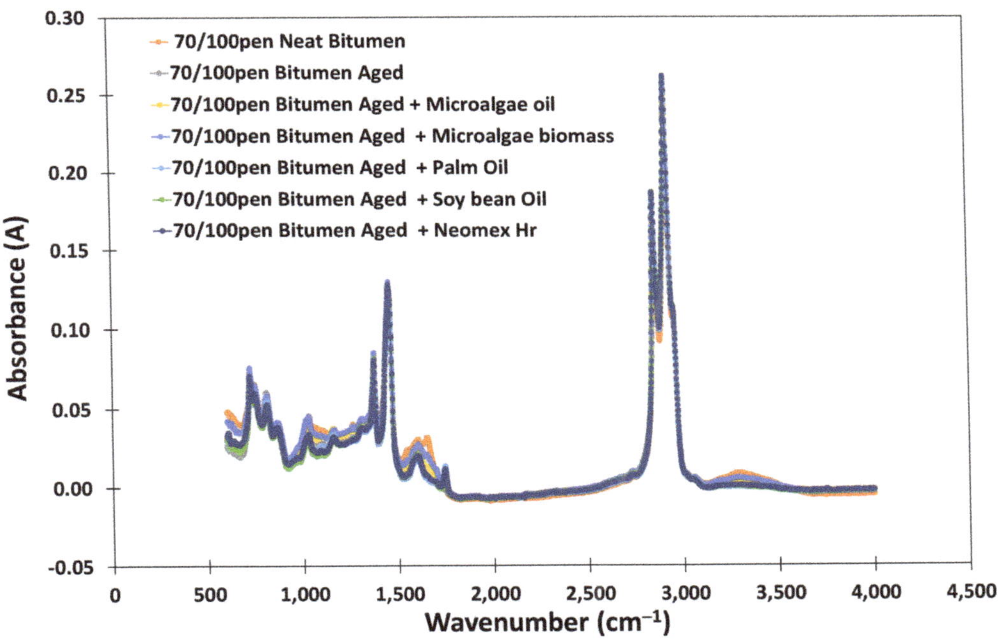

Figure 1. IR-spectra of 70/100pen aged bitumen and neat, aged, and rejuvenated aged bitumen.

3.2. Dynamic Shear Rheometer (DSR)

Figure 2 shows the results from the DSR test. The results show that the rejuvenating oils in aged bitumen blends reduce the complex shear modulus and stiffness of aged bitumen. The blends containing the microalgae oil and soy bean oil have very similar responses, whereas palm oil showed the strongest softening effect on the aged binder, exhibiting an even better softening effect than Neomex Hr (a standard bio-based rejuvenator). The blend containing microalgae biomass resulted in a higher stiffness than the aged bitumen.

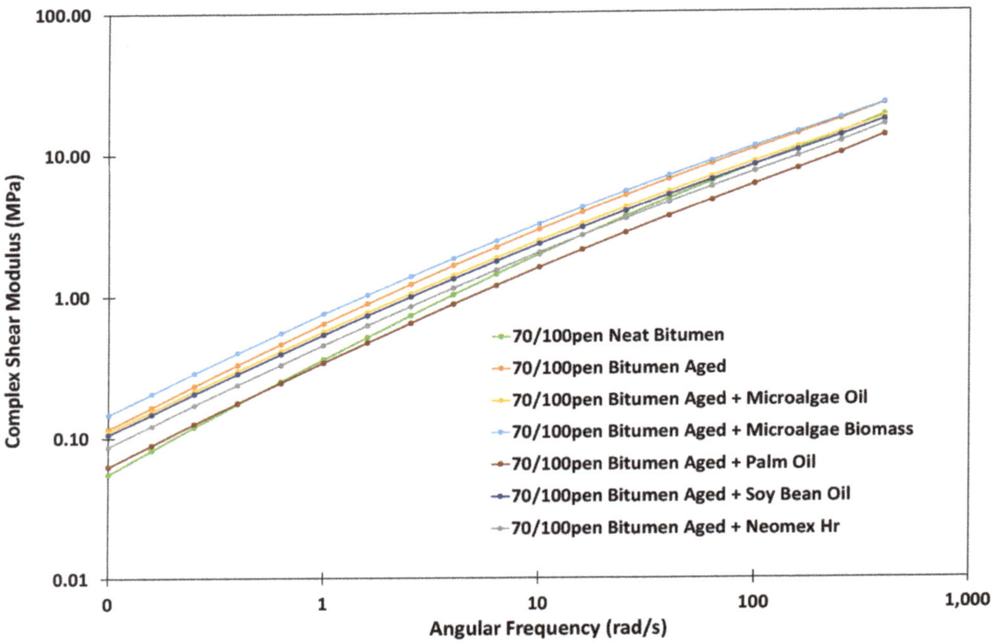

Figure 2. DSR data for 70/100pen aged bitumen and rejuvenated aged bitumen.

3.3. Microscopic Analysis

Figure 3 illustrates the optical microscope images of 70/100pen bitumen (neat, aged, and blends containing rejuvenating oils and microalgae biomass). The images show that the ageing and rejuvenating oils cause the bitumen to become pale and much lighter in appearance (Figure 3c–f). However, the blend containing microalgae biomass shows a darker appearance with light specs throughout the test sample. These specs are the organic material particles in the blend. Nevertheless, the images show the full homogenous structure of the material, indicating a good blending process and a lack of separation of rejuvenating oils and aged bitumen.

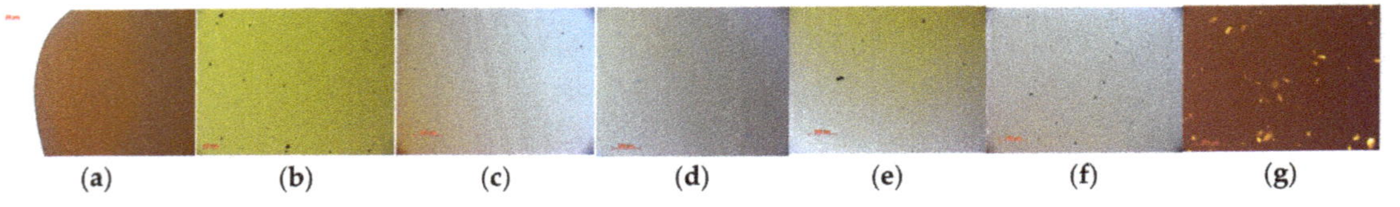

Figure 3. Microscopic images of 70/100pen bitumen: (**a**) neat, (**b**) aged, (**c**) aged with Neomex Hr, (**d**) aged with microalgae oil, (**e**) aged with palm oil, (**f**) soy bean oil and (**g**) aged bitumen with microalgae biomass.

4. Conclusions

The results reported in this study demonstrate that microalgae oil performs well as an aged bitumen rejuvenator in comparison to Neomex Hr (a standard bio-based rejuvenator) and plant-based (soy and palm) oils. The results illustrate that microalgae oil matches the chemical and mechanical performance of plant-based oils and Neomex Hr (a standard bio-based rejuvenator), indicating that microalgae oil can be used as an aged bitumen rejuvenator.

Upon request from the microalgae biomass and oil suppliers (Wageningen University), the microalgae biomass was also investigated, and the results were surprising. The biomass showed good chemical compatibility with the bitumen. Regarding mechanical performance, the complex shear modulus of the blend did increase but not to a significant level. This finding is promising because if microalgae biomass could be inserted directly into the bitumen mix, it would reduce the financial cost of implementing bio-based rejuvenators into asphalt mix production by eliminating the need for biomass processing, i.e., biomass oil extraction.

A future research direction is to investigate the performance of these oils and microalgae biomass as rejuvenators within a full asphalt mix.

Author Contributions: Conceptualization, A.T.; methodology, A.T.; formal analysis, A.T., D.v.V. and K.R.-S.; investigation, A.T.; resources, D.v.V., K.R.-S. and G.L.; data curation, A.T.; writing—original draft preparation, A.T.; writing—review and editing, A.T., D.v.V., K.R.-S. and G.L.; visualization, A. T.; supervision, G.L.; project administration, G.L.; funding acquisition, A.T. and G.L. All authors have read and agreed to the published version of the manuscript.

Funding: This research was funded by European Commission H2020 Marie Skłodowska Curie Individual Fellowship (IF) research funding programme, project: Advanced Solutions for Asphalt Pavements (ASAP), grant number 886828.

Institutional Review Board Statement: Not applicable.

Informed Consent Statement: Not applicable.

Data Availability Statement: The data generated in the study are fully presented within this article.

Acknowledgments: The authors would like to thank Declan O'Pray from Breedongrouop PLC for supplying the bitumen material, Hans Hendrikse from Latexfalt BV for supplying Neomex Hr rejuvenator, and Rene Wiffels, Iago Dominguez Teles and Rafael Cubero, from Wageningen University for supplying microalgae biomass and microalgae oil.

Conflicts of Interest: The authors declare no conflict of interest. The funders had no role in the design of the study; in the collection, analyses, or interpretation of data; in the writing of the manuscript; or in the decision to publish the results.

References

1. Audo, M.; Paraschiv, M.; Queffélec, C.; Louvet, I.; Hémez, J.; Fayon, F.; Lépine, O.; Legrand, J.; Tazerout, M.; Chailleux, E.; et al. Subcritical Hydrothermal Liquefaction of Microalgae Residues as a Green Route to Alternative Road Binders. *ACS Sustain. Chem. Eng.* **2015**, *3*, 8. [CrossRef]
2. Penki, R.; Rout, S.K. Next-generation bitumen: A review on challenges and recent developments in bio-bitumen preparation and usage. *Biomass Conv. Bioref.* **2021**. [CrossRef]
3. Tabaković, A. Bio-Binder—Innovative Asphalt Technology. *Appl. Sci.* **2020**, *10*, 8655. [CrossRef]
4. French Institute of Science and Technology Devoted to Transport Planning and Networks (IFSTTAR). *BioRePavation*. 2020. Available online: http://biorepavation.ifsttar.fr/ (accessed on 17 February 2022).
5. Kliewer, J.E.; Bell, C.A.; Sosnovske, D.A. Investigation of the relationship between field performance and laboratory ageing properties of asphalt mixtures. In *Engineering Properties of Asphalt Mixtures and the Relationship to Their Performance*; Huber, G.A., Decker, D.S., Eds.; ASTM International: West Conshohocken, PA, USA, 1995.
6. Axelsson, M.; Gentili, F. A Single-Step Method for Rapid Extraction of Total Lipids from Green Microalgae. *PLoS ONE* **2014**, *9*, e89643. [CrossRef] [PubMed]
7. Hofko, B.; Alavi, M.Z.; Grothe, H.; Jones, D.; Harvey, J. Repeatability and sensitivity of FTIR ATR spectral analysis methods for bituminous binders. *Mater. Struct.* **2017**, *50*, 187. [CrossRef]

engineering proceedings

MDPI

Proceeding Paper

Driving Speed Analysis Using Real-Time Traffic Light Status Information at Signalized Intersections †

Eunjin Choi *, Hyangmi Han, Ockhee Jeon, Seungcheol Lee and Kwangyoung Ko

Korea Road Traffic Authority, Department of Advanced Traffic, Hyeocksin-ro, Wonju-si 26466, Gangwon-do, Republic of Korea; hyan9m1@naver.com (H.H.); okki@koroad.or.kr (O.J.); leesc@koroad.or.kr (S.L.); gwangyoung@gmail.com (K.K.)

* Correspondence: toryjin1532@naver.com; Tel.: +82-33-74-5454

† Presented at the Second International Conference on Maintenance and Rehabilitation of Constructed Infrastructure Facilities, Honolulu, HI, USA, 16–19 August 2023.

Abstract: This study aims to analyze driver behavior when traffic light status information is provided to the in-vehicle systems of individual vehicles. In the case where signal information was provided when the vehicle was approaching an intersection in a red-light state, a statistically significant difference in both the driving speed and standard deviation of the speed was observed. The driving speed was 2.770 km/h, and the standard deviation of the speed increased by 0.153 km/h. In addition, an average speed increase of 2.751 km/h was observed when the remaining time information was provided, then when it was not. When only light was provided, the speed increased by 1.549 km/h; this was statistically insignificant.

Keywords: traffic light information; remaining time of the current traffic light; traffic signal with countdown indicator; driving behavior; SPaT message

Citation: Choi, E.; Han, H.; Jeon, O.; Lee, S.; Ko, K. Driving Speed Analysis Using Real-Time Traffic Light Status Information at Signalized Intersections. *Eng. Proc.* **2023**, *36*, 28. https://doi.org/10.3390/engproc2023036028

Academic Editor: Hosin (David) Lee

Published: 10 July 2023

1. Introduction

Recently, South Korea has enhanced its transport information services by implementing signal phase and timing (SPaT) messages through the C-ITS project. SPaT messages convey the state of the signal information, including the remaining time for the direction in which each vehicle intends to follow. Based on this message, individual drivers can receive warnings about the danger of traffic violations or suggestions for safer intersection-approaching speeds [1]. This study aims to analyze driver behavior when traffic light status information is provided to the in-vehicle systems of individual vehicles.

2. Experimental Design

To set up a testing environment, a vehicle was equipped with an automotive black box and a device that could provide signal information to the driver. Subsequently, a driving test was conducted at an intersection of a public road where vehicles were controlled. The speed, gear state, and brake state of the vehicle were recorded in 100 ms increments, and the signal information device was used to simultaneously record the lighting state (red, yellow, or green) and the remaining time of the traffic lights (Figure 1).

In total, 12 scenarios were designed for the driving test, considering three external factors: the presence or absence of a preceding vehicle, initial lighting conditions (red/green), and provision of information on the status (Table 1). The test set comprised 60 participants, of which 20 were in their 20 s and 30 s, 20 were in their 40 s and 50 s, and 20 were aged 60 years and older.

The main objective of the test is to determine the driver's behavior based on the provision of information on traffic light status. For this purpose, the case in which information was provided was further divided into one case in which only the state of lighting was provided and another case in which information on both the state of lighting and the remaining time was provided.

(**a**) (**b**)

Figure 1. Experimental vehicle (**a**); in-vehicle system (**b**).

Table 1. External factor of the driving test scenario.

Preceding Vehicle (2)	Initial Lightning Condition (2)	Type of Information (3) [1]
Presence	Red	Not provided
Absence	Green	Only lightning conditions
		Lighting conditions with the remaining time

[1] Traffic light status information.

Additionally, the test was further divided into two cases to investigate a scenario in which the vehicle approaches an intersection: when the state of the signal light was red and when it was green. This is because different actions are required depending on the color of the signal when a vehicle approaches an intersection.

3. Data Analysis

3.1. Red Light State

For the case in which signal information was provided when the vehicle was approaching an intersection **in a red-light state**, a statistically significant difference in both the driving speed and standard deviation of the speed was observed, where the driving speed was 2.770 km/h, and the standard deviation of the speed was increased by 0.153 km/h (Table 2). Hence, providing information on the status of the traffic light and the remaining time was believed to affect the reaction time required by drivers to freely select the speed of approaching and passing the intersection.

Table 2. Results of statistical analysis for the average difference in the vehicle speed according to whether traffic light status information is provided (t-test; $p < 0.01$).

	Information	Mean	Std.D	Std.E	t	df	Sig.	Mean Diff.	95%CI		N
									Lower	Upper	
V (kph)	Not provided	24.707	11.765	0.346	−6.658	2244	0.000	−2.770	−3.585	−1.954	1156
	Provision *	27.477	12.764	0.231							3057
Std.V	Not provided	1.287	0.843	0.025	−5.116	2270	0.000	−0.153	−0.212	−0.094	1155
	Provision *	1.441	0.926	0.017							3054

* In experiments where drivers received real-time traffic light state information through an in-vehicle system.

To analyze this effect based on the form in which information was provided, the effect was divided into three categories: non-providing, providing only the light condition, and simultaneously providing the light condition and remaining time. Consequently, an average speed increase of 2.751 km/h was observed when the remaining time information was provided, then when it was not (Table 3).

Table 3. ANOVA table for vehicle driving characteristics for the type of information at red light state.

	Information	N	Mean	Std.D	Std.E	Sum of Square	df	Mean Square	F	Sig.
V (kph)	None	686	24.530	12.978	0.495					
	Light condition	672	25.733	13.633	0.526					
	With remaining time	606	27.281	13.202	0.536					
	Between-group					2439.066	2	1219.533	6.921	0.001
	Within group					345,522.821	1961	176.197		
	Total					347,961.887	1963			
Std.V	None	685	1.356	0.806	0.031					
	Light condition	672	1.491	0.876	0.034					
	With remaining time	605	1.553	0.889	0.036					
	Between-group					13.263	2	6.632	9.041	0.000
	Within group					1436.916	1959	0.733		
	Total					1450.179	1961			

Notably, providing and utilizing the red remaining time information of vehicles approaching an intersection during a red light increased the frequency of vehicles passing the intersection without stopping.

When only the light condition was provided, the speed increased by 1.549 km/h; therefore, the statistical significance was minimal (Figure 2).

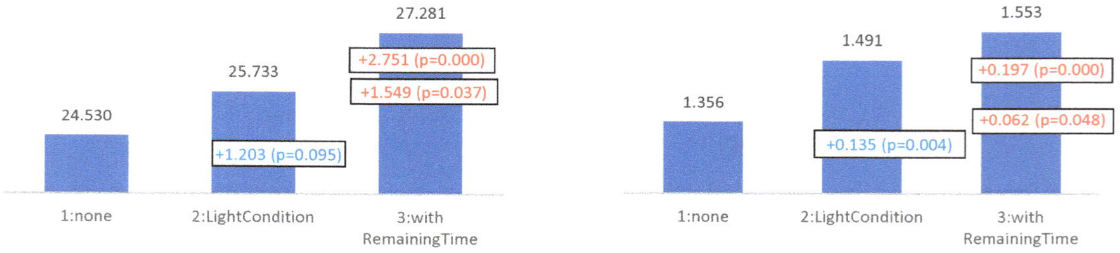

(**a**) Average vehicle speed (**b**) Standard deviation of vehicle speed

Figure 2. Vehicle driving characteristics by type of information provided at red light state.

However, during driving without any visibility restrictions, displaying the traffic lights inside the vehicle in a manner similar to that at the intersection did not significantly affect the driver's behavior. Thus, the case where only the light condition information was provided in the car was useful when visually recognizing the actual traffic light was difficult owing to distance limitations such as the preceding vehicle, fog, or heavy rain.

3.2. Green Light State

When approaching an intersection in a green light state, the simultaneous provision of signal information and remaining time can be used to prevent dangerous behavior wherein drivers recklessly cross the intersection. The vehicles that approached the intersection in a green light state had significantly different average speeds when the information with remaining time was provided than when it was not.

Notably, when the green remaining time information was provided, the speed of passing the intersection was 4.866 km/h higher than that when no information was provided. The average intersection approaching speed during the green signal time was 30.206 km/h, which was 4.866 km/h higher than when no information was provided (Figure 3a). In the case of providing remaining time, vehicle speed statistically increases on average by 5 km/h at a confidence level of 95%.

(**a**) Average vehicle speed (**b**) Speed of passing the intersection during yellow time

Figure 3. Vehicle speed by type of information provided at green and yellow light state.

Additionally, the speed of passing the intersection was 44.692 km/h when only the light condition information was provided and 42.142 km/h during the yellow signal time (Table 4); this was lower than that when information was not provided (42.768 km/h). This implied that when the remaining time information was provided, reckless entry into the intersection was reduced; conversely, the driver's driving speed could be maintained if information containing sufficient remaining time was provided (Figure 3b).

Table 4. ANOVA table for vehicle driving characteristics for the type of information at green light state.

	Information	N	Mean	Std.D	Std.E	Sum of Square	df	Mean Square	F	Sig.
V (kph)	None	463	25.340	9.316	0.433					
	Light condition	431	26.305	8.779	0.423					
	With remaining time	743	30.206	9.770	0.358					
	Between-group					8066.882	2	4033.441	45.748	0.000
	Within group					144,064.435	1634	88.167		
	Total					152,131.317	1636			
Std.V	None	463	1.202	0.882	0.041					
	Light condition	431	1.272	0.850	0.041					
	With remaining time	743	1.269	0.885	0.032					
	Between-group					1.512	2	0.756	0.987	0.373
	Within group					1249.898	1632	0.756		
	Total					1251.410	1634			

The biggest cause of the increase in speed is passing through the intersection without stopping or decelerating, and misusing the remaining time information. Furthermore, a case existed in which the participant accelerated through the intersection at high speed, based on the provided information.

4. Conclusions

Providing the driver with real-time information on the status of traffic lights and remaining time at an intersection was determined to help individual drivers to select vehicle speeds irrespective of the initial lighting condition. Moreover, when information on the traffic light status was provided, the selection of the speed resulted in a higher speed of passing through the intersection than when it was not. Therefore, future signal remaining time information studies should consider these characteristics and enhance the message format and communication method based on the requirements.

Author Contributions: Conceptualization, methodology, analysis, writing, E.C.; investigation and data curation, H.H.; validation, S.L.; supervision, O.J.; project administration, K.K. All authors have read and agreed to the published version of the manuscript.

Funding: This research was funded by Korea Road Authority.

Institutional Review Board Statement: Not applicable.

Informed Consent Statement: Not applicable.

Data Availability Statement: The datasets used/or analysed during the current study available from the corresponding author on reasonable request.

Conflicts of Interest: The authors declare no conflict of interest.

Reference

1. Bo, Y.; Shan, B.; Fred, F.; James, S. Examination and prediction of drivers' reaction when provided with V2I communication—Based intersection maneuver strategies. *Transp. Res. Part C* **2019**, *106*, 17–28.

engineering proceedings

Proceeding Paper

Quantifying and Reducing Uncertainty in Transportation System Resilience Assessment: A Dynamic Bayesian Network Approach †

Vishnupriya Jonnalagadda and Ji Yun Lee *

Department of Civil & Environmental Engineering, Washington State University, Pullman, WA 99164, USA; v.jonnalagadda@wsu.edu
* Correspondence: jiyun.lee@wsu.edu; Tel.: +1-509-335-3018

† Presented at the Second International Conference on Maintenance and Rehabilitation of Constructed Infrastructure Facilities, Honolulu, HI, USA, 16–19 August 2023.

Abstract: Transportation systems are complex, and due to their interdependence with other essential facilities, any damage to them would pose a significant threat to the well-being of communities. Given the frequent occurrences and grave consequences of natural disasters observed in recent years, research on the resilience assessment of transportation systems has received a great deal of attention. This paper develops a dynamic Bayesian network (BN)-based resilience assessment model for a highway network subject to seismic events that can explicitly quantify uncertainties in all phases of the model and investigate the role of inspection and monitoring in uncertainty reduction. The results from this study can be used as comprehensive decision-support information so that decision makers can better assess the resilience of a highway network and associated uncertainties.

Keywords: resilience assessment; Bayesian network; uncertainty quantification; transportation systems; monitoring and inspection; earthquakes

Citation: Jonnalagadda, V.; Lee, J.Y. Quantifying and Reducing Uncertainty in Transportation System Resilience Assessment: A Dynamic Bayesian Network Approach. *Eng. Proc.* **2023**, *36*, 29. https://doi.org/10.3390/engproc2023036029

Academic Editor: Hosin (David) Lee

Published: 11 July 2023

1. Introduction

Resilience of an engineered system is generally defined as its capacity to withstand, adapt to, and recover from disruptions [1,2]. In recent years, assessing critical infrastructure resilience has become crucial due to the growing number of natural disasters. Transportation systems, in particular, play a key role in providing access to affected regions during and after catastrophic events by rescuing people and transporting essential supplies. However, assessing transportation system performance is accompanied by significant uncertainties arising from aging and deteriorating infrastructure, the increasing complexity of networks, damage due to extreme events from natural hazards, budgetary constraints, and increasing operational demands. Failure to account for these factors and their associated uncertainties may impede a transportation agency's ability to achieve its predefined goals and objectives, leading to significant consequences for society.

While uncertainty quantification and reduction in transportation resilience assessment are essential to ensuring efficient resilience-enhancing strategies, many studies that propose a new resilience metric or resilience-based decision framework do not investigate the role of the inspection/monitoring process. One reason for this is that most metrics or frameworks are designed to measure scenario-based static resilience assuming that, at the time of hazard event occurrence, the full probabilistic descriptions of structural capacity and external loadings are known. However, there are substantial uncertainties in (a) the number, severity, and timing of hazard events, (b) time-dependent structural capacity, and (c) external/service loadings especially when climate change affects the performance of an asset or when traffic demands change significantly because of population growth and urbanization. If a resilience-based decision framework extends to a specified period and is intended to capture the time-dependent changes in structural performance and resilience, inspection/monitoring is necessary to increase the reliability of the predictions of structural capacity and/or external loadings in the future and, ultimately, determine transportation system resilience. To this end, this paper proposes a dynamic Bayesian network (BN)-based resilience assessment model for a highway network subject to seismic events. The model characterizes and quantifies uncertainties and investigates the role of inspection and monitoring in uncertainty reduction.

2. Methodology

Figure 1 shows the overall flowchart of the proposed dynamic BN-based seismic resilience assessment of a highway network. The model begins by measuring the time-dependent structural reliability of individual bridges exposed to corrosion and links this measure to post-earthquake traffic-carrying capacity. Then, combined with traffic demand data, the performance of individual bridges is aggregated through network analysis to evaluate the performance of the highway network in terms of total travel time prior to and following an earthquake event. By incorporating time-dependent restorative activities into the network analysis, the network's seismic resilience is assessed. In the meantime, the probability density functions (PDFs) of bridge functionalities and link traffic demands are updated based on inspection and monitoring data at every time step through the dynamic BN.

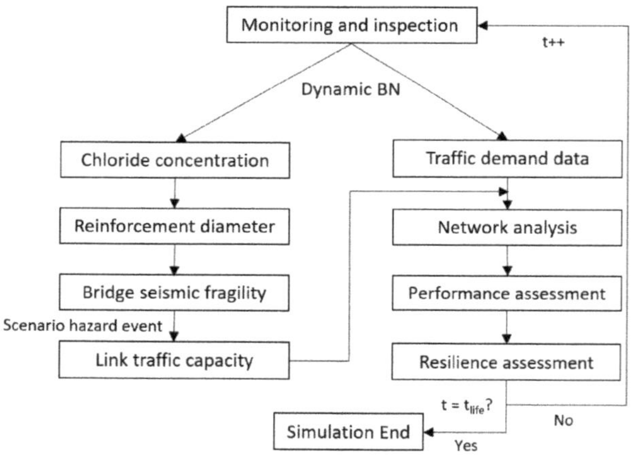

Figure 1. Proposed dynamic BN-based model procedure.

To examine the role of inspecting and monitoring major random variables, such as surface chloride concentration, earthquake events, road surface conditions, or traffic demand, in the seismic resilience assessment of highway networks, this study considers three cases: (a) Case A: the baseline case where network resilience is measured based on the prior distributions of random variables, which is consistent with most existing studies; (b) Case B: true transportation resilience which is never known in the real world; and (c) Case C: the case where transportation resilience is measured based on the updated distributions of random variables through the proposed model. More specifically, in Case C, it is assumed that major random variables that change over time are monitored and that their prior PDFs are updated over time through the dynamic BN. True time-dependent data of the major random variables (i.e., Case B) are never known in the real world due to epistemic uncertainties, and thus, in this study, synthetic true values are generated. Similarly, synthetic measurement data are also generated due to the absence of monitoring data for the major random variables. After generating the synthetic analytical values of the major random variables using their respective analytical models, the synthetic true values and measurement data are generated based on the following equations:

$$y_t(x) = y_a(x, \beta) + \varepsilon_a \tag{1}$$

$$y_m(x) = y_t(x) + \varepsilon_{obs} \tag{2}$$

where x = the independent variables; $y_t(x)$ = the synthetic true values of the major random variables; β = the model parameters; $y_a(x, \beta)$ = the output of the analytical model of the variables; ε_a = the modeling error; $y_m(x)$ = the synthetic measurement data of the variables; and ε_{obs} = the measurement error which is usually modeled as a Gaussian random variable. In general, the measurement error is smaller than the modeling error.

The synthetic measurement data of the major random variables are then used to update a prior PDF through the BN to reduce epistemic uncertainties. In Bayes' theorem, initial knowledge of a parameter (θ) is encoded in a prior PDF, $f(\theta)$. After incorporating the measurement data, a posterior PDF of the parameter, $f(\theta|y_m)$, is calculated by [3,4]:

$$f(\theta|y_m) = \frac{f(y_m|\theta)f(\theta)}{f(y_m)} \tag{3}$$

where $f(y_m|\theta)$ = the likelihood function that quantifies the likelihood of observing the data given θ; and $f(y_m)$ = the marginal probability of the data.

In the proposed model, it is assumed that uncertainties exist at all stages of the resilience assessment procedure. The functionality of the highway network, $Q(t)$, is expressed as a function of total travel time and is measured by [5]:

$$Q(t) = 100 - 100\left(\frac{TTT(t) - TTT_0}{TTT_0}\right) \tag{4}$$

where $TTT(t)$ = the total travel time at time t; and TTT_0 = the total travel time in the base model (measured without any disaster event). Once the time-dependent network functionality profile is constructed, the seismic resilience of a highway network is measured by the normalized area under the post-earthquake recovery trajectories and can be expressed as:

$$\boldsymbol{R} = \frac{1}{\boldsymbol{t_1} - \boldsymbol{t_0}}\int_{\boldsymbol{t_0}}^{\boldsymbol{t_1}} \boldsymbol{Q(t)dt} \tag{5}$$

where t_0 = the time of occurrence of an earthquake event; and t_1 = the time when a network is completely repaired. A higher value of R indicates a more seismically resilient network.

3. Benchmark Problem: A Highway Network in South Carolina

The proposed resilience assessment framework is illustrated with a highway network in South Carolina which was previously used as a case study in [6]. This network is selected as a benchmark problem in this paper because the surrogate seismic fragility curves of various types of bridges located in the network are available in [7]. The bridges in the network are also potentially exposed to marine chloride due to their proximity to the Atlantic Ocean, which makes these bridges more vulnerable to earthquakes over time. In this benchmark problem, two major random variables—chloride concentration and traffic demand—are monitored. After the synthetic analytical, true, and measurement values are generated (see Equations (1) and (2)), the measurement data on chloride concentration are used to update the fragility curves of individual bridges through the dynamic BN. Following a scenario earthquake event (i.e., the 1886 Charleston Earthquake), the failed bridges are realized using Monte Carlo simulation and incorporated in calculating link traffic capacities. On the other hand, analytical traffic demand for each link is forecasted over time using the ARIMA model and subsequently used to generate both synthetic true and measurement values. The total travel time of the network is estimated at every time step from the time of occurrence of the earthquake event to the time to full recovery through the CUBE voyager software program. Finally, the time-dependent performance and resilience of the network are calculated at every 10 years over a 50-year time period. Figure 2a shows the fragility curve of a specific bridge for the three different cases at Year 50: Case A (theoretical values), Case B (true values), and Case C (updated values through dynamic BN). As shown in Figure 2a, Case C is much closer to Case B as compared to Case A, which highlights the advantage of monitoring surface chloride concentration over time. From Figure 2b, a decrease in the resilience index of the network is observed over time mainly due to bridge deterioration and increased traffic demands. Another finding is that as time increases, the degree of overestimation of network seismic resilience by Case A increases. Conversely, Case C maintains a consistent and accurate estimate of network seismic resilience, even in the distant future. This also highlights the significance of monitoring and inspection in reducing uncertainties.

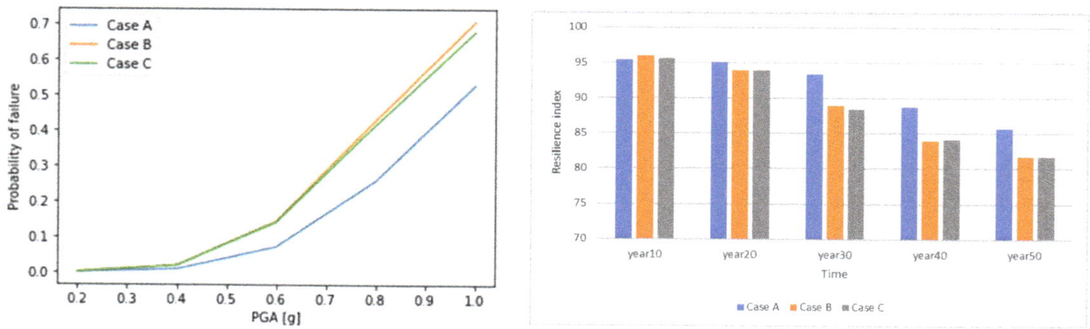

(a) (b)

Figure 2. Comparison of three cases: (**a**) fragility curve of a bridge at Year 50, and (**b**) time-dependent seismic resilience index.

4. Conclusions

Transportation systems are complex and are some of the highest-valued and largest public assets. To better understand transportation system performance during normal and disrupted conditions and ultimately mitigate consequences from hazards, it is critical to characterize uncertainties and reduce them through the efficient implementation of monitoring and inspection tools. To this end, this paper develops a dynamic BN-based resilience assessment model for a large-scale transportation system that can explicitly quantify uncertainties in all phases of the assessment and investigate the role of inspection and monitoring in uncertainty reduction. The benchmark problem results show that monitoring and inspecting major random variables reduces uncertainties in resilience assessment. The proposed model can help officials determine the optimal monitoring technologies and identify critical parameters to be monitored, thereby improving the prediction accuracy of the system resilience.

Author Contributions: Conceptualization, J.Y.L.; methodology, J.Y.L. and V.J.; simulation: V.J.; writing—original draft preparation, V.J. and J.Y.L.; visualization: V.J.; supervision, J.Y.L.; funding acquisition: J.Y.L. All authors have read and agreed to the published version of the manuscript.

Funding: This work was sponsored by the National Center for Transportation Infrastructure Durability & Life Extension (TriDurLE). This support is gratefully acknowledged. Any opinions, findings, conclusions, or recommendations expressed in this material are those of the authors and do not necessarily reflect the views of TriDurLE.

Institutional Review Board Statement: Not applicable.

Informed Consent Statement: Not applicable.

Data Availability Statement: The data that support the findings of this study are available from the corresponding author upon reasonable request.

Conflicts of Interest: The authors declare no conflict of interest. The funders had no role in the design of the study; in the collection, analyses, or interpretation of data; in the writing of the manuscript; or in the decision to publish the results.

References

1. Bruneau, M.; Chang, S.E.; Eguchi, R.T.; Lee, G.C.; O'Rourke, T.D.; Reinhorn, A.M.; Shinozuka, M.; Tierney, K.; Wallace, W.A.; Von Winterfeldt, D. A framework to quantitatively assess and enhance the seismic resilience of communities. *Earthq. Spectra* **2003**, *19*, 733–752. [CrossRef]
2. Chang, S.E.; Shinozuka, M. Measuring improvements in the disaster resilience of communities. *Earthq. Spectra* **2004**, *20*, 739–755. [CrossRef]
3. Zheng, X.W.; Li, H.N.; Lv, H.L.; Huo, L.S.; Zhang, Y.Y. Bayesian-based seismic resilience assessment for high-rise buildings with the uncertainty in various variables. *J. Build. Eng.* **2022**, *51*, 104321. [CrossRef]
4. Abimbola, M.; Khan, F. Resilience modeling of engineering systems using dynamic object-oriented Bayesian network approach. *Comput. Ind. Eng.* **2019**, *130*, 108–118. [CrossRef]
5. Kezhiyur, A.J. Analysis of Age-Dependent Resilience for a Highway Network with Aging Bridges. Master's Thesis, Pennsylvania State University, State College, PA, USA, 2015.
6. Rokneddin, K.; Ghosh, J.; Dueñas-Osorio, L.; Padgett, J.E. Seismic reliability assessment of aging highway bridge networks with field instrumentation data and correlated failures, II: Application. *Earthq. Spectra* **2014**, *30*, 819–843. [CrossRef]
7. Ghosh, J. Parameterized Seismic Fragility Assessment and Life-Cycle Analysis of Aging Highway Bridges. Doctoral Dissertation, Rice University, Houston, TX, USA, 2013.

engineering proceedings

MDPI

Proceeding Paper

Sun Damage on Roads: From UV Radiation to Bituminous Binders and the Protecting Effect of End-of-Life Tires †

Marie Enfrin [1], Jaffer Bressan Borinelli [2], Johan Blom [2], Cedric Vuye [2] and Filippo Giustozzi [1,*]

1 Civil and Infrastructure Engineering, Royal Melbourne Institute of Technology-RMIT University, Melbourne, VIC 3001, Australia; marie.enfrin@rmit.edu.au
2 Sustainable Pavements and Asphalt Research (SuPAR), Faculty of Applied Engineering, University of Antwerp, 2020 Antwerp, Belgium; jaffer.bressanborinelli@uantwerpen.be (J.B.B.); johan.blom@uantwerpen.be (J.B.); cedric.vuye@uantwerpen.be (C.V.)
* Correspondence: filippo.giustozzi@rmit.edu.au

† Presented at the Second International Conference on Maintenance and Rehabilitation of Constructed Infrastructure Facilities, Honolulu, HI, USA, 16–19 August 2023.

Abstract: Roads are exposed to solar radiation for the entire duration of their service life; depending on the season, the intensity and level of exposure can vary for different locations leading to differences in the magnitude of the damage. In Australia, many local road authorities have already started understanding that their roads are deteriorating faster due to non-load related effects (i.e., environmental damage). The present paper discusses the damage caused by UV radiation on bituminous binders and explores the protecting effect provided by the use of end-of-life tires in the form of crumb rubber. The results suggest that adding crumb rubber to bituminous binders has the potential to significantly reduce the non-load related environmental damage caused by the UV radiation of the Sun.

Keywords: asphalt; bitumen; crumb rubber; end-of-life tires; UV radiation

Citation: Enfrin, M.; Borinelli, J.B.; Blom, J.; Vuye, C.; Giustozzi, F. Sun Damage on Roads: From UV Radiation to Bituminous Binders and the Protecting Effect of End-of-Life Tires. *Eng. Proc.* **2023**, *36*, 30. https://doi.org/10.3390/engproc2023036030

Academic Editor: Hosin (David) Lee

Published: 10 July 2023

1. Introduction

Ensuring the durability of asphalt pavements is crucial to guaranteeing not only the performance of the road but also its safety as well as to reducing maintenance costs. The impact of traffic and extreme temperatures on the short- and long-term ageing of roads is well-known. Standard accelerated ageing tests are commonly used by the industry and the scientific community to assess the durability of binders and asphalt mixtures [1,2]. Weather events, such as heavy rains or sudden temperature changes, are expected to reduce the service life of roads; however, the impact of UV radiation has only been considered recently although roads are subjected to solar radiation during their entire service life.

Exposure to ultraviolet (UV) radiation is known to accelerate the ageing of a wide range of materials, such as polymers, by inducing chemical and physical changes which eventually lead to the deterioration of their mechanical properties [3]. A similar mechanism is known to occur in bitumen [4], but the complex composition of bitumen has contributed to delaying a comprehensive understanding of the real impact of UV on asphalt pavements. Polymers are known to be susceptible to oxidation and chain break-down, hence antioxidant additives are commonly used to prevent this phenomenon. For example, approx. 20–30% of carbon black (CB) is used in rubber tires to mitigate oxidation [5]. End-of-life tires can be recycled as crumb rubber (CR) and used in asphalt mixed as a bitumen modifier [6]; hence, the presence of CB in CR could prevent the photodegradation of bitumen and extend the service life of CR-modified roads, which would represent a significant economic benefit for road agencies.

Therefore, the aim of this study is to investigate the impact of crumb rubber, carbon black and its antioxidant potential on the ageing of CR-modified asphalt binders when exposed to UV radiation.

2. Materials and Methods

2.1. Bituminous Binders

Two base binders BB1 and BB2 were supplied by Australian and German suppliers, respectively. The crumb rubber (maximum particle size 0.5 mm) and carbon black used to modify the base bitumen were supplied from Rubber Recycling Overpelt (Pelt, Belgium) and Thermo Fisher Scientific (Belgium), respectively.

CR-modified binder (BB2 + CR) was produced by blending 15 wt% of CR per mass of binder in bitumen at 170 °C and 3500 rpm for 60 min. The same procedure was used to produce CB-modified binder (BB1 + CB) using 4 wt% of CB per mass of binder. A third blend was made by mixing both CR and additional 4 wt% CB in bitumen to produce a CB+CR modified binder (BB2 + CR + CB). In addition, the base binders BB1 and BB2 were also used as reference blends.

2.2. UV Ageing

Each of the five binders was poured into 50 × 50 × 1 mm moulds and exposed to solar radiation for 28 days using a SUNTEST XLS+ weathering chamber equipped with xenon lamps reproducing solar radiations at a controlled irradiance of 765 W/m^2.

2.3. Binder Characterisation

The surface of all binders was characterised before and after ageing with Fourier transform infrared (FTIR) spectroscopy using a Perkin Elmer Frontier instrument to assess their degree of oxidation. At least three samples were tested. The carbonyl index of each binder was calculated by considering the ratio between the area of the carbonyl (C=O) characteristic band of bitumen at ≈1700 cm^{-1} and the area of the aliphatic characteristic bands (C-C) at ≈1450 and 1370 cm^{-1} [7–9]. The viscoelastic properties of the binders were investigated before and after ageing using a Dynamic Shear Rheometer (Anton Paar, MDSC Rheometer) following AASHTO T315 and AASHTO T350 standard test methods. Two replicates were used for each parallel plate (8 mm and 25 mm). Master curves, black diagrams, and the rheological parameter (R) of the samples were assessed.

3. Results and Discussion

3.1. Impact of UV on Rheological Properties

To evaluate the effect of CB on the rheological properties of unmodified bitumen and crumb rubber modified binders, frequency sweep tests were performed. After UV ageing, it is expected that the complex modulus (G*) increases due to the hardening of the binder, particularly in the lower frequency domain. This trend can easily be seen in the master curves for all samples shown in Figure 1a. The presence of CB and CR in the binder reduced the relative increase in G* compared to the neat binder.

To further assess the activity of CB and CR on the binders, the Black diagram of the samples is presented in Figure 1b. The response of the samples became more elastic with UV ageing, which can be observed as a shift towards lower phase angle values. Blends containing CR and/or CB showed a smaller decrease in phase angle after UV ageing compared to the unmodified binders BB1 and BB2 exposed to the same ageing treatment, hence demonstrating a better resistance to UV ageing.

The evolution of UV ageing can also be analyzed by the increase in the rheological parameter (R), which is calculated as the difference between the log of the glassy modulus and the log of the crossover modulus (the point where the loss modulus coincides with the storage modulus). This rheological parameter is related to the width of the relaxation spectrum; hence, higher R values represent longer stress relaxation times and, therefore, a higher susceptibility to thermal cracking. As can be seen in Table 1, it is evident that all binders had an increase in R value after UV ageing but to a different extent. While the R values of BB1 and BB2 increased by 77 and 82%, respectively, the blends with CB and CR showed an increase of only 5 and 23%, respectively. Combining CB and CR led to a similar increase to that with CR only. Hence, the binder containing CB had the shortest relaxation time, followed by the binder modified with CR; thus, they outperformed the other blends in terms of resistance to UV. Adding CB to CR-modified bitumen did not seem to improve the UV resistance of the binder further, thus suggesting that an optimal content of CB could be used for anti-ageing purposes in bitumen.

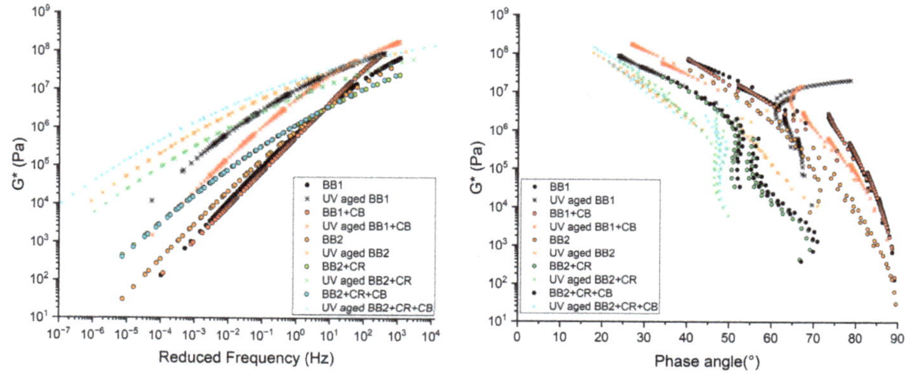

(a) (b)

Figure 1. (**a**) Complex modulus master curves and (**b**) Black diagram for a temperature range between +10 °C and +60 °C.

Table 1. Rheological parameter R of each binder before and after UV ageing.

Binder	R Parameter		Increase (%)
	Fresh	UV Aged	
BB1	1.37	2.43	77
BB2	1.68	3.05	82
BB1 + CB	1.32	1.38	5
BB2 + CR	2.74	3.38	23
BB2 + CR + CB	2.68	3.39	26

3.2. *Role of Crumb Rubber in UV Ageing of Bitumen*

ATR FTIR was employed to compare the oxidation levels of bitumen before and after UV ageing. It can be observed in Figure 2 that samples BB1 and BB2 had the highest values for I_{co} after UV ageing followed by the binders containing CR. The binder with CB has the lowest carbonyl index after UV ageing, confirming that the antioxidant properties of CB slowed down the formation of C=O groups in bitumen and, thus, the oxidative hardening of the binder. This is consistent with the rheological performance of the binder containing CB. CR slightly reduced the oxidation of the binder after UV ageing, most likely due to the presence of CB in CR material.

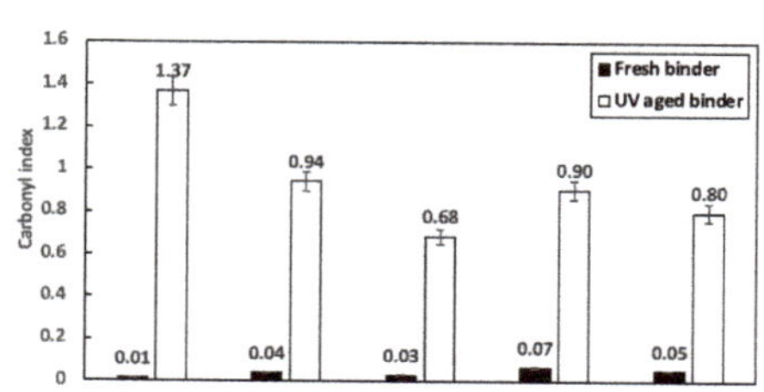

Figure 2. Carbonyl index of fresh and UV aged binders.

4. Conclusions

Crumb rubber and carbon black were used in this study as antioxidant additives to reduce the effects of ageing in bituminous binders when exposed to UV radiation. It can be concluded that CB showed a great potential to reduce the effects of ageing, slowing down the oxidation of the binder and making it less stiff and less susceptible to thermal cracking compared to the base binders. CR also improved the performance of the binder after UV ageing compared to non-modified binders, likely due to the presence of CB in CR. Hence, reusing CR in asphalt could improve the long-term resistance of roads against UV ageing.

Author Contributions: Conceptualization, F.G. and M.E.; methodology, M.E. and J.B.B.; formal analysis, M.E. and J.B.B.; writing—original draft preparation, M.E. and J.B.B.; writing—review and editing, F.G., M.E., J.B.B., J.B. and C.V.; supervision, F.G. and M.E.; funding acquisition, F.G., J.B.B., C.V. and J.B. All authors have read and agreed to the published version of the manuscript.

Funding: This research was funded by the Green.er funds managed by the King Baudouin Foundation, grant number KBS/2019-E1170430-212694, and by an FWO (Fund of Scientific Research Flanders) research stay, grant V429622N.

Institutional Review Board Statement: Not applicable.

Informed Consent Statement: Not applicable.

Data Availability Statement: The data presented in this study are available on request from the corresponding author.

Acknowledgments: The authors thank TotalEnergies Belgium N.V. and TotalEnergies Germany N.V. for providing bitumen for this study, and Rubber Recycling Overpelt (RRO) for its support by providing crumb rubber samples.

Conflicts of Interest: The authors declare no conflict of interest.

References

1. Shen, J.; AmirkhaniaN, S.; Tang, B. Influence of accelerated aging test temperature on the properties of binders. *Int. J. Pavement Eng.* **2006**, *7*, 191–198. [CrossRef]
2. Hu, Y.; Si, W.; Kang, X.; Xue, Y.; Wang, H.; Parry, T.; Airey, G.D. State of the art: Multiscale evaluation of bitumen ageing behaviour. *Fuel* **2022**, *326*, 125045. [CrossRef]
3. Yousif, E.; Haddad, R. Photodegradation and photostabilization of polymers, especially polystyrene: Review. *SpringerPlus* **2013**, *2*, 398. [CrossRef] [PubMed]
4. Petersen, J.C.; Glaser, R. Asphalt Oxidation Mechanisms and the Role of Oxidation Products on Age Hardening Revisited. *Road Mater. Pavement Des.* **2011**, *12*, 795–819. [CrossRef]
5. Bockstal, L.; Berchem, T.; Schmetz, Q.; Richel, A. Devulcanisation and reclaiming of tires and rubber by physical and chemical processes: A review. *J. Clean. Prod.* **2019**, *236*, 117574. [CrossRef]
6. Lo Presti, D. Recycled Tyre Rubber Modified Bitumens for road asphalt mixtures: A literature review. *Constr. Build. Mater.* **2013**, *49*, 863–881. [CrossRef]
7. Zhang, F.; Hu, C. The research for crumb rubber/waste plastic compound modified asphalt. *J. Therm. Anal. Calorim.* **2015**, *124*, 729–741. [CrossRef]
8. Ge, D.; Chen, S.; You, Z.; Yang, X.; Yao, H.; Ye, M.; Yap, Y.K. Correlation of DSR Results and FTIR's Carbonyl and Sulfoxide Indexes: Effect of Aging Temperature on Asphalt Rheology. *J. Mater. Civ. Eng.* **2019**, *31*, 04019115. [CrossRef]
9. Marsac, P.; Piérard, N.; Porot, L.; Van Den Bergh, W.; Grenfell, J.; Mouillet, V.; Pouget, S.; Besamusca, J.; Farcas, F.; Gabet, T.; et al. Potential and limits of FTIR methods for reclaimed asphalt characterisation. *Mater. Struct.* **2014**, *47*, 1273–1286. [CrossRef]

engineering proceedings

MDPI

Proceeding Paper

Deep Learning and Clustering-Based Analysis of Text Narratives for Identification of Traffic Crash Severity Contributors †

Cristian Arteaga and JeeWoong Park *

Department of Civil and Environmental Engineering and Construction, University of Nevada, Las Vegas, NV 89154, USA; arteagas@unlv.nevada.edu

* Correspondence: jee.park@unlv.edu

† Presented at the Second International Conference on Maintenance and Rehabilitation of Constructed Infrastructure Facilities, Honolulu, HI, USA, 16–19 August 2023.

Abstract: Crash narratives provide valuable information to understand traffic crashes and develop roadway safety countermeasures. However, manually reading long text narratives is time-consuming and error-prone. This study presents a deep-learning and clustering-based approach to identifying contributors to traffic crash severity in text narratives. We evaluate the approach using a dataset of narratives from Massachusetts and compare different deep-learning models for semantic similarity. The approach clusters semantically similar phrases in the narratives and provides an overview of frequent topics related to severe crashes, offering a valuable tool for roadway safety analysis and countermeasure development.

Keywords: crash narratives; clustering; deep learning; semantic similarity; severity contributors

Citation: Arteaga, C.; Park, J. Deep Learning and Clustering-Based Analysis of Text Narratives for Identification of Traffic Crash Severity Contributors. *Eng. Proc.* **2023**, *36*, 31. https://doi.org/10.3390/engproc2023036031

Academic Editor: Hosin (David) Lee

Published: 10 July 2023

1. Introduction

Understanding the factors that worsen the severity of traffic crashes is of paramount importance in prioritizing and implementing traffic safety countermeasures. Crash narratives, which describe in detail the context and circumstances of the crashes, are a valuable source of information to identify injury severity contributors. However, extracting insights from this unstructured text data is challenging, particularly when manual reading of thousands of narratives is required.

Past studies [1–3] have provided important advances in exploiting information from narratives for decision-making, but these studies have two main limitations. First, they focus on analyzing at the word level, which could suffer from ambiguous or incomplete insights. Second, their methods offer limited modeling of language semantics at the word level, which negatively impacts the quality of the revealed insights.

In a recent study, Arteaga et al. [4] proposed a method that addresses the limitations of past studies by identifying meaningful phrases (instead of individual words) from the narratives that describe potential crash severity contributors. However, for large databases of narratives, the method simply yields a large number of phrases, making the interpretation of the results challenging. To address this limitation, this study proposes a method that synthesizes numerous phrases that describe severity contributing factors in narratives, which facilitates analysis and decision-making for traffic safety.

2. Materials and Methods

To synthesize topics recurrently found in the narratives as correlated with severe crashes, this study integrates deep-learning techniques for semantic similarity and a clustering technique. The deep-learning techniques are Transformer-based models that provide enhanced semantic modeling capabilities by capturing interrelationships between words using an attention-based mechanism. Researchers [5] have pre-trained several types of Transformer-based models (e.g., MiniLM, MPNet, and DistilBERT) to excel in sentence similarity tasks by using large datasets of sentence pairs, such as the one billion sentence pairs dataset (1B), as well as datasets with paraphrasing information and question-answering sentence pairs (MultiQA). These Transformer-based models take a set of phrases as input and return a vector representation of the phrases. The vector representations enable the comparison of phrases based on their semantics (ideas expressed) instead of their individual words.

For the clustering task, this study uses the Agglomerative Clustering technique, which has shown promising results when applied in conjunction with Transformer-based models [5]. The proposed approach uses the vector representations provided by the Transformer-based models and clusters them based on their semantic similarity.

For evaluation, this study leverages a dataset of 1131 narratives provided by the Massachusetts Department of Transportation. By following the approach outlined in Arteaga et al. [4], we identified 5783 phrases correlated with severe crashes and synthesized them using the proposed approach. We evaluated four Transformer-based models for semantic similarity, two distance metrics for agglomerative clustering, and different numbers of clusters to evaluate their effect on the cluster distortion scores (the average of the squared distances from the cluster centers to each element in the clusters).

3. Results

Figure 1 shows the distortion scores for different numbers of clusters, similarity metrics, and Transformer-based models for semantic similarity.

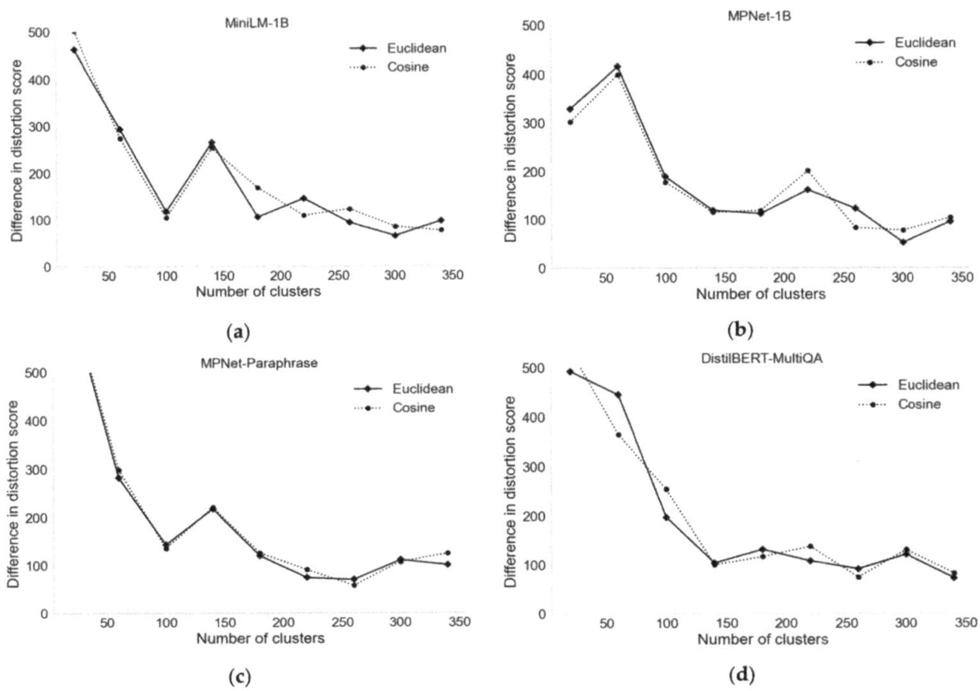

Figure 1. Distortion scores for different numbers of clusters and deep learning models for semantic similarity: (**a**) MiniLM-1B; (**b**) MPNet-1B; (**c**) MPNet-Paraphrase; and (**d**) DistilBERT-MultiQA.

These results show that all the deep-learning models exhibit similar patterns in the decrease of distortion scores as the number of clusters increases. The plots provide an indication of the points at which increasing the number of clusters yields diminishing returns in terms of reduction of distortion scores, which is equivalent to an analysis using the elbow method. Therefore, given that the goal is to have a low distortion score while keeping the number of clusters low, the value of 100 was selected as the number of clusters, and the MiniLM-1B model was used for subsequent analysis, as this model was the fastest to reach a low distortion score without excessively increasing the number of clusters. In terms of the distance metric, the Cosine and Euclidean distances yielded similar results for distortion scores, which indicates that both metrics are equally suitable for the clustering task.

Table 1 shows examples of the clusters returned by the developed approach. The table includes information about the most common words within a cluster to provide an indication of the type of contents that a cluster captures. For instance, the words "alcohol", "intoxicated", "bottles", and "marijuana" are top words in a cluster, and all the phrases are intrinsically related to the topic of driving under the influence of alcohol or drugs.

Table 1. Examples of clusters and phrases within clusters provided by the developed approach.

Top Words in a Cluster	Number of Phrases	Examples of Phrases in a Cluster
EMS, hospital, transported, and ambulance	2191	• "... EMS was requested, and he was later transported to the hospital for his injuries." • "... requested at least an ambulance." • "... needed medical attention."
Vehicle, lane, travel, and towing	1392	• "... partially in the travel lane." • "Wheels towing arrived on ... " • "... traveling west on Route 195 ... " • "... traveling in the middle travel lane ... " • "... blue hills towing ... "
Scene, arrived, officer, and trooper	324	• "The trooper arrived on scene ... " • "The scene was photographed by ... " • "... masters arrived on scene and ... " • "... arrived on scene ... " • "Also, on scene was ... "
Guardrail, crosswalk, pedestrian, and struck	217	• "A pedestrian was struck after exiting ... " • "... struck the guardrail, stopped and rolled ... " • "... communications reported a pedestrian who was hit by ... "
Head on, vehicle, pole, and crash	101	• "... strike the motorcycle head on and ... " • "... the utility pole was severed near the base." • "... reported head-on crash ... " • "... a head-on crash ... " • "... pole with heavy front end ... "
Speed, rate, high, and mph	91	• "... at a high rate of speed, presumably in ... " • "... very high rate of speed." • "... vehicle 1 began at a high rate of speed ... " • "... at a speed greater than 90 mph."
Alcohol, intoxicated, bottles, and marijuana	33	• "... round face and she appeared intoxicated." • "... bottles of Smirnoff ... " • "... appeared to be heavily intoxicated." • "... was heavily intoxicated. "

Table 1. *Cont.*

Top Words in a Cluster	Number of Phrases	Examples of Phrases in a Cluster
Control, swerve, lose, and lost	16	• "... unknown male lost control of ... " • "... she did not want to swerve ... " • "... lose control of basic ... " • "... then lose control and cross ... " • "... saw swerve abruptly ... "
Ice, weather, raining, and vehicle	7	• "... ice on the road." • "... with snow-covered/icy road conditions." • "... raining with low temperatures."

4. Discussion

The results in Table 1 indicate that the proposed approach effectively clusters phrases based on their semantic contents. Some clusters contain phrases that recurrently appear in narratives for severe crashes, although they are not necessarily severity contributing factors (e.g., EMS transporting people to hospitals and vehicles being towed from the scene). However, most of the clusters identified by the developed approach provide important insights about severity contributing factors, such as the involvement of pedestrians, speeding, the influence of intoxicating liquor, head-on crashes with utility poles, suspected marijuana use, and adverse roadway conditions. These phrases provide important insights to traffic safety analysts about the factors that require urgent attention for the implementation of countermeasures. Thus, the developed approach provides traffic engineers with a valuable tool to easily exploit the information in crash narratives for data-driven decision-making.

5. Conclusions

This paper presented an approach to synthesizing the results of an analysis of severity contributors in crash narratives. The developed approach addresses the limitations of past studies by integrating deep-learning-based semantic similarity and a clustering approach to provide an overview of frequent topics in the narratives associated with crashes of different severity levels (e.g., fatality, severe injury, minor injury, and property damage only). The insights returned by the approach can significantly help crash analyses as it enables the use of narratives as a valuable information source to compare and complement the insights derived from conventional statistical analyses of quantitative crash data, thereby facilitating a comprehensive diagnosis of traffic crashes. The identification of contributing factors based on the analysis of both text narratives and quantitative data can enhance analysts' confidence in the significance of such factors. Thus, by facilitating the extraction of insights from narratives, the proposed approach offers considerable value for the identification and prioritization of crash factors that need prompt attention.

Author Contributions: Both authors participated in the conceptualization, investigation, and writing of the paper. All authors have read and agreed to the published version of the manuscript.

Funding: This research was funded by the National Cooperative Highway Research Program (NCHRP) (IDEA Project #231).

Institutional Review Board Statement: Not applicable.

Informed Consent Statement: Not applicable.

Data Availability Statement: Data is unavailable due to privacy restrictions.

Acknowledgments: The authors express gratitude to MassDOT for supplying the data utilized in this investigation. The authors bear complete responsibility for the viewpoints, discoveries, and inferences presented in this work, which may not align with the positions of NCHRP or MassDOT.

Conflicts of Interest: The authors declare no conflict of interest.

References

1. Gao, L.; Wu, H. Verb-Based Text Mining of Road Crash Report. In Proceedings of the Transportation Research Board, 92nd Annual Meeting, Washington, DC, USA, 13–17 January 2013; pp. 5–16.
2. Arteaga, C.; Paz, A.; Park, J. Injury Severity on Traffic Crashes: A Text Mining with an Interpretable Machine-Learning Approach. *Saf. Sci.* **2020**, *132*, 104988. [CrossRef]
3. Rakotonirainy, A.; Chen, S.; Scott-Parker, B.; Loke, S.W.; Krishnaswamy, S. A Novel Approach to Assessing Road-Curve Crash Severity. *J. Transp. Saf. Secur.* **2015**, *7*, 358–375. [CrossRef]
4. Arteaga, C.; Park, J.; Paz, A. Enhanced Identification of Crash Severity Contributors from Text Narratives Using Natural Language Processing. In Proceedings of the 102nd Annual Meeting of the Transportation Research Board, Washintong, DC, USA, 8–12 January 2023.
5. Reimers, N.; Gurevych, I. Sentence-BERT: Sentence Embeddings Using Siamese BERT-Networks. *arXiv* **2019**, arXiv:1908.10084.

engineering proceedings

MDPI

Proceeding Paper

A Study of Artificial Neural Network-Based Real-Time Traffic Signal Timing Design Model Utilizing Smart Intersection Data †

Sang-Tae Oh and Jin-Tae Kim *

Department of Transportation Policy and Systems Engineering, Korea National University of Transportation, 157 Cheoldobangmulkwan-ro, Uiwang-si 16106, Republic of Korea; zilam94@naver.com

* Correspondence: jtkim@ut.ac.kr; Tel.: +82-31-461-8736

† Presented at the Second International Conference on Maintenance and Rehabilitation of Constructed Infrastructure Facilities, Honolulu, HI, USA, 16–19 August 2023.

Abstract: The smart intersection (SI) systems, as they are named in the Republic of Korea, are part of the ITS services implemented under local government projects with financial support from the central government. They collect real-time traffic data available at signalized intersections with advanced detection systems for surveillance purposes only. A traffic signal method utilizing such valuable data has been desirable but unavailable as yet in practice. This paper proposes a new approach to designing traffic signal timings, reflecting the demand changing in real time, by utilizing SI surveillance data. The proposed artificial neural network model generates suitable traffic signal timings trained to be near optimum based on surveillance data for each directional movement.

Keywords: traffic control; signal timings; deep learning; artificial intelligence; simulation; real time

Citation: Oh, S.-T.; Kim, J.-T. A Study of Artificial Neural Network-Based Real-Time Traffic Signal Timing Design Model Utilizing Smart Intersection Data. *Eng. Proc.* **2023**, *36*, 32. https://doi.org/10.3390/engproc2023036032

Academic Editor: Hosin (David) Lee

Published: 11 July 2023

1. Introduction

The traffic signal timing design process in practice has focused only on a certain period of hours—namely, a.m. and p.m. peak times—due to insufficient budgets. Through the intelligent transport system (ITS) deployment projects in the Republic of Korea, local governments have guided the improvement of traffic control systems. Despite such investments, it has been hard to achieve the long-term goal as intended due to the failure of sustainability in online data collection. While practitioners want to believe their ITSs are 'advanced', it has been hard to keep reacting to the ordinary changes in traffic demand in real time.

The smart intersection (SI) systems being deployed through national projects have recently allowed traffic engineers to use sustainable online data obtained from various service components, such as traffic flow surveillance, incident detection, etc. Despite the availability of real-time surveillance data, the SI project has excluded traffic control services from its core functionality. Therefore, it might be preferable to link the traffic control systems and SI surveillance data. This paper delivers an approach to designing traffic signal timings by utilizing SI surveillance data, reflecting the demand changing in real time.

2. Related Work

For the SI systems, Kim et al. [1] suggested using image detection data to ensure the highest level of reliability, depending upon the Korean ITS performance evaluation criteria. They concluded that the data from their detection systems are sufficient to support the real-time design of traffic signal timings. Kim and Kim [2] proposed a new approach to test the 'Smart Traffic' signal operation method, including various operations such as emergency vehicle preemption. They employed the 'Verkehr In Stadten Simulation (VISSIM)' model to test them by utilizing real-time external data from outside of the simulation model; it was like a hardware-in-the-loop system. It was found that the traffic signal timings designed using online data performed better in real time than those using offline data, which is traditionally used for time-of-day (TOD) operation.

3. Model Development

This study aims to (1) train an artificial neural network (ANN) model designing traffic signal timings based on real-time SI surveillance data and to (2) compare its performance against one using traditional Time of Day (TOD) operation. It employed two different computer programs to overcome the unnecessary bias in simulations: corridor simulation (CORSIM) for the model development and simulation of urban mobility (SUMO) for the validation. It should be noted that a set of training data was from the hypothetical intersection operation scenarios and that the validation data were directly from the SI systems at the isolated 'Baekwoun' intersection, Uiwang, Gyeonggi, from November 2021 to November 2022.

For the ANN training data, 1470 (= 7 × 5 × 6 × 7) different traffic conditions were prepared to support the hypothetical intersection scenarios. The variation includes (1) seven different traffic demand levels, (2) five different free-flow speeds, (3) six different percent left turns, and (4) seven different approach balances, as shown in Table 1.

Table 1. Various traffic conditions scenarios for the ANN training data.

Total Volume Levels (≈v/c)	Free-Flow Speeds (km/h)	Percent Left Turns (%)	Approach Balances
0.95	40	5	0.50:0.50
0.92	45	10	0.55:0.45
0.88	50	15	0.60:0.40
0.85	55	20	0.65:0.35
0.75	60	25	0.70:0.30
0.65	-	30	0.75:0.25
0.55	-		0.80:0.20
No. of variations = 7	No. of variations = 5	No. of variations = 6	No. of variations = 7

To find the acceptable control conditions for these 1470 traffic conditions one by one, the authors developed a computer program, which they named CORSIM optimal signal timing (COST), to increase the training-data-preparation efficiency. Such control conditions were prepared based on the control-delay-minimization strategy. COST found the optimal control conditions for each of the traffic conditions through an iterative heuristic searching method, which is named the hybrid hill-climbing method, by utilizing the authentic CORSIM evaluation technique and automatically shifting to the next when found. The hybrid hill-climbing method employed in this study utilized sixteen combination cases of each of the conventional dual-ring phase representation schemes and changed its step size based on conditions designed to ensure the global optimum search.

The ANN model was then developed based on the training data reflecting 1470 different intersection operation scenarios—each having their own best control conditions and the traffic signal timing set. It was expected that the proposed ANN model could find either some similarity with the ones previously built or a hidden trend desirably leading to liquidity among such data.

The model development considered various compositions and combinations of ANN inputs, outputs, and hidden layers. The prepared input layer contained twelve input neurons, reflecting twelve directional movement volumes at a four-lag isolated intersection. The prepared output layer contained eight neurons, reflecting each of the conventional dual-ring phase lengths. The proposed model utilized a single hidden layer with sixteen units of neurons.

The ANN model was developed based on the training data reflecting 1470 different intersection operation scenarios. Figure 1 shows the architectural configuration that utilized the ANN model through actually operated Smart Intersection(S.I) data. The state was calculated through monitoring, which was input into the ANN model. The optimal signal time was calculated through the existing learning method, and the input traffic volume was added as learning data. Through this process, the ANN model becomes more robust and can react to various situational data.

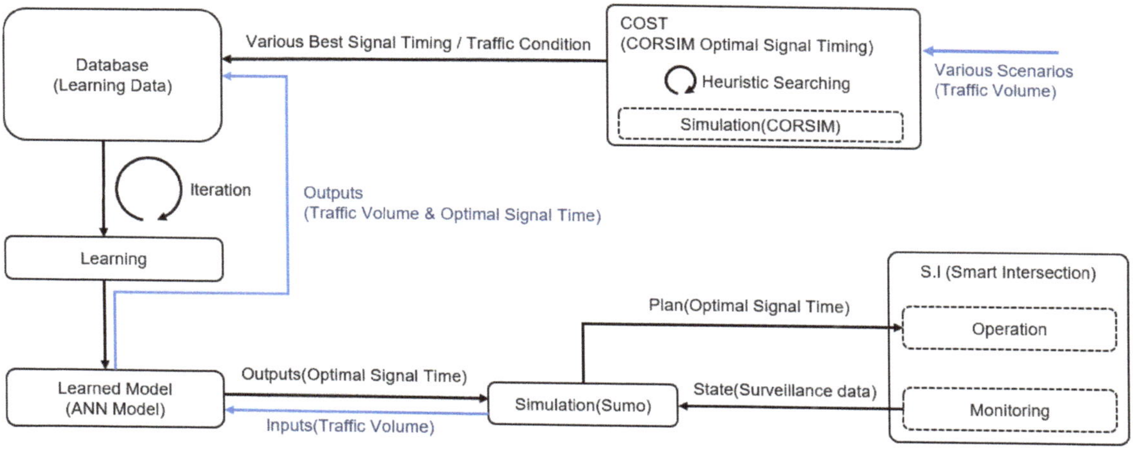

Figure 1. Study process.

4. Validation Test

To reflect the real-life demand variations in a day of twenty-four hours, the comparison test was done based on historical traffic volume counted for twenty-four hours on a weekday and a weekend from the SI surveillance systems. These data were obtained from the Uiwang SI traffic management center.

For the validation of the proposed ANN model, it was necessary to compare the performance of the isolated intersection with two different control conditions: the best-existing one in practice and the proposed hypothetical one utilizing SI surveillance data. The best-existing one was the TOD signal timings that have recently been updated in practice, whereas the proposed hypothetical one was from the ANN model developed in this study.

The authors employed weekday- and weekend-TOD signal timing plans updated through the official project of the Uiwang local government a month before this study. Table 2 presents a brief comparison of those control conditions in both TOD plans for weekday and weekend conditions.

Table 2. Existing/proposed signal operation method signal time setting.

TOD Classification	Existing		Proposed	
	Weekdays	Weekends	Weekdays	Weekends
00:00~06:00	TOD Plan 1	TOD Plan 5	The signal timings designed in real time with the proposed deep learning ANN model	
06:00~10:00	TOD Plan 2	TOD Plan 5		
10:00~17:00	TOD Plan 3	TOD Plan 8		
17:00~20:00	TOD Plan 4	TOD Plan 8		
20:00~21:00	TOD Plan 4	TOD Plan 5		
21:00~24:00	TOD Plan 1	TOD Plan 5		

The prepared traffic and control conditions were simulated with SUMO. Figure 2 illustrates the graphical comparison of the results. The horizontal axis represents the twenty-four hours of the day. The vertical axis represents the control delay estimated from the simulation.

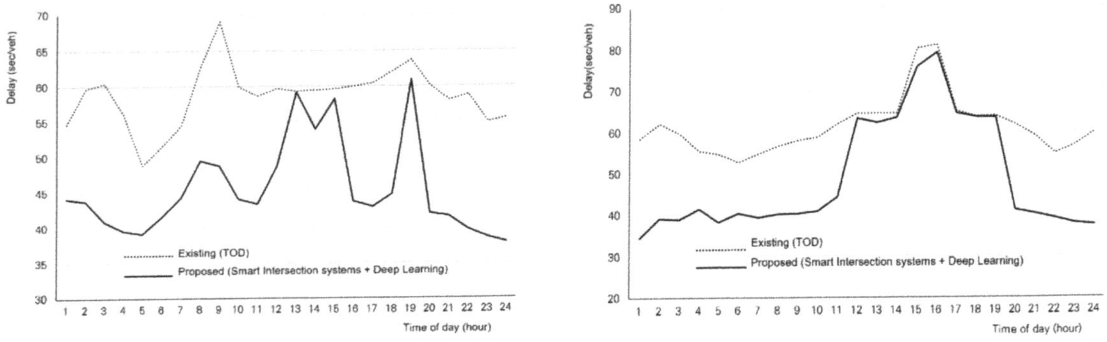

(a) (b)

Figure 2. Comparison of existing vs. proposed control conditions for (**a**) weekdays and (**b**) weekends.

It was found from the results that the existing signal timing design method performed well during peak hours but not for the rest of the day. It showed that the basic traffic volume data utilized for traffic signal design and the invested resources, such as money and manpower in the signal timing design procedure, might be limited in practice as long as it is done in an offline manner.

The proposed model showed that during peak hours, the performance level was close to the TOD values. In addition, it was able to provide better performance during non-peak hours. It should be noted that the proportion of non-peak hours in a day is significant. This is because the proposed method uses online data and updates the signal timings in real time. Although it cannot be concluded from the results that the proposed ANN model produced the best signal timings, it can be concluded that the proposed approach utilizing the ANN technique was able to find a hidden trend.

5. Conclusions

This paper presents an approach to designing traffic signal timings based on online SI surveillance data, reflecting real-time changes in traffic demand. The authors suggest that it is valuable to link SI surveillance services with traffic signal control systems. The performance of the proposed optimal signal timings showed that the level of the control delay was reduced by about 22% on average compared with the existing method widely used in practice.

Author Contributions: Conceptualization, J.-T.K. and S.-T.O.; methodology, J.-T.K.; software, S.-T.O.; validation, J.-T.K. and S.-T.O.; formal analysis, J.-T.K.; investigation, S.-T.O.; resources, S.-T.O.; supervision, J.-T.K.; project administration; J.-T.K. All authors have read and agreed to the published version of the manuscript.

Funding: This work was supported by a Korea Institute of Police Technology (KIPoT) grant funded by the Korea government (KNPA) (No.092021C29S02000, Development of on-site control technology for road traffic network control in the event of an incident or disaster).

Institutional Review Board Statement: Not applicable.

Informed Consent Statement: Not applicable.

Data Availability Statement: Not applicable.

Conflicts of Interest: The authors declare no conflict of interest.

References

1. Kim, D.W.; Kim, S.H. Smart Intersection and Smart Crosswalk Construction and Performance Evaluation. *J. Korean Soc. Transp.* **2022**, 429–430.
2. Kim, J.B.; Kim, J.T. A Study on the VISSIM Simulation Model for Real Time Signal Control. *Korea Inst. Intell. Transp. Syst.* **2019**, 601–603.

engineering proceedings

MDPI

Proceeding Paper

Prediction of Blow-Up Potential Due to Concrete Pavement Growth †

Youngkyu Kim [1], Huirak Ahn [2] and Seungwoo Lee [2,*]

1 Institute for Disaster Prevention, Gangneung-Wonju National University, Gangneung-si 24265, Gangwon-do, Republic of Korea; pave@gwnu.ac.kr
2 Department of Civil Engineering, Gangneung-Wonju National University, Gangneung-si 24265, Gangwon-do, Republic of Korea; ahr8338@naver.com
* Correspondence: swl@gwnu.ac.kr; Tel.: +82-33-640-2419
† Presented at the Second International Conference on Maintenance and Rehabilitation of Constructed Infrastructure Facilities, Honolulu, HI, USA, 16–19 August 2023.

Abstract: Concrete pavement growth can cause blow-ups and other pressure-related issues, such as concrete buckling and crushing at the transverse cracks or joints. In addition, these issues result in damaged to adjoining structures, such as bridge abutments, decks, and access structures in the pavement. However, available theoretical solutions and methods for evaluating pavement growth and predictions of concrete pavement blow-up are considerably limited. In this study, therefore, the pavement growth and blow-up analysis model was developed to predict the blow-up potential due to concrete pavement growth. This model considered factors such as the pavement structure and materials, climatic conditions, configuration of expansion joint, base friction characteristics, geometric imperfection, rotational stiffness of joints, and design reliability. In addition, the pavement growth and blow-up analysis model can be used to predict the service life of expansion joints and the blow-up occurrence time by considering the factors affecting it. Using the developed model, various sensitivity analysis was performed to investigate the factors affecting the potential concrete pavement growth and blow-up occurrence. It was found that the factors of substrate type, coefficient of thermal expansion, precipitation, and alkali–silica reaction had a significant effect on pavement growth and blow-up occurrence times.

Keywords: concrete pavement; pavement growth; blow-up; contraction joint; expansion joint

Citation: Kim, Y.; Ahn, H.; Lee, S. Prediction of Blow-Up Potential Due to Concrete Pavement Growth. *Eng. Proc.* **2023**, *36*, 33. https://doi.org/10.3390/engproc2023036033

Academic Editor: Hosin (David) Lee

Published: 11 July 2023

1. Introduction

Concrete pavement performance is typically impaired by pavement growth, which is a phenomenon that increases pavement length. Pavement growth is caused by several factors, such as increases in the pavement's temperature and moisture, infiltration of incompressible particles into cracks and joints, pumping of base material into cracks and joints beneath the concrete slab, and concrete expansion caused by the alkali–silica reaction (ASR) in concrete [1]. This issue can cause concrete pavement blow-up and other pressure-related issues, such as concrete crushing and damage to adjoining structures (e.g., bridge abutments, decks, drainage, manholes, and access structures in the pavement). The cumulative pavement growth in concrete pavements increases the probability of blow-up distress [1–3]. Ref. [3] presumed that most of these blow-ups were due to the generation of a high pavement pressure (this study took place before the use of pressure-relief joints was introduced). There is also a report that blow-ups occurred in the USA during a heatwave in the summer of 2016. In South Korea, numerous blow-ups and other pressure-related distresses occurred in many areas in the summer of 2018, when the air temperature and moisture levels were extremely high.

When a concrete pavement is newly constructed, the joint seals are intact and function well, reducing the probability of pressure generation in the pavement. However, as the pavement ages, the joint sealant fails owing to repeated joint movements, stiffening, traffic abrasion, etc. [4]. This failure accelerates the infiltration of incompressible particles into the joints, increasing the probability of pressure generation. This outcome is explained as a cyclical pressure-generation process caused by the infiltration of incompressible particles [2,5]. The process of a joint completely closing is referred to as joint closure (JC); in this case, the faces of adjacent slabs in the joint come into contact either with each other or the incompressible particles [6]. This contact generally occurs because of concrete slab expansion resulting from the increases in temperature and moisture. The temperature at which a joint closes is called the JC temperature (T_c); we note that *Tc* may be low if there are incompressible particles in the joint.

For years, pavement growth and blow-up have been major problems for highway and airport engineers, as well as for road users, because they directly impact the performance of concrete pavements and adjacent or nearby structures, such as bridge abutments. If the contact concrete slabs can move freely in the spacing of the expansion joints, blow-up does not occur. Therefore, controlling and maintaining the widths of functioning expansion joints during their service lives is crucial. The objective of this study was to develop a reproducible methodology for estimating pavement growth and predicting the time of blow-up occurrence. Blow-up and pavement growth are affected by the complex interactions of numerous factors, such as the climatic conditions, the quantity of incompressible particles infiltrating the joints, the pavement structure and materials, and the joint systems of the concrete pavement sections.

2. Descriptions of Pavement Growth and Blow-Up Analysis Model

In this study, a user-friendly tool called the pavement growth and blow-up analysis (PGBA) model was developed using MATLAB to estimate the service life of expansion joint and blow-up occurrence time. Figure 1 illustrates the overall framework for the PGBA model. For the input file, there are four groups of input data, which are climatic data, pavement structures and materials, configuration of the expansion joint, and design reliability. The backbone of the run file consists of the prediction of maximum concrete temperature (Tmax), the prediction of the trigger temperature for pavement growth (TTPG), the prediction of alkali–silica reaction (ASR), the prediction of the current width of expansion (W_EJ) and the service life of expansion joint, and the prediction of concrete pavement blow-up after the closing of expansion joints.

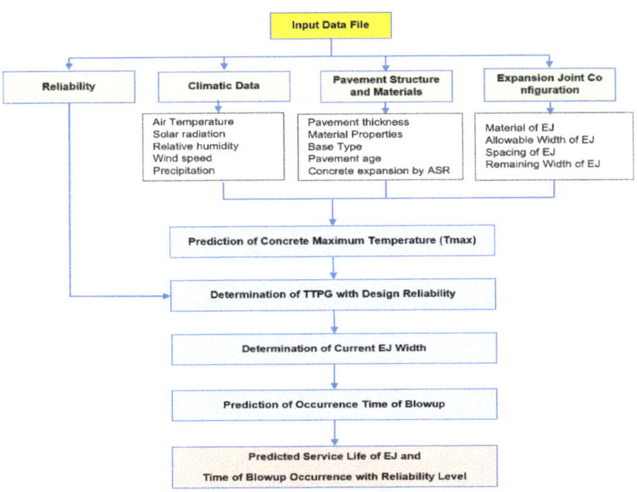

Figure 1. Overall framework for the PGBA program.

To predict the annual maximum concrete temperature (Tmax), a numerical predictive model for the concrete pavement temperature was established using the 1D finite difference method [7]. As previously mentioned, the TTPG is the temperature at which all transverse cracks and contraction joints among the expansion joints begin to come into contact and generate an axial compressive force in the pavement section. To estimate the magnitude of TTPG for a given concrete pavement section, [7] developed a statistical regression equation, which is in function with the concrete pavement age, transverse joint spacing, and maximum annual precipitation, using the Long-Term Performance Pavement Seasonal Monitoring Program (LTPP SMP) database.

3. Sensitivity Analysis of PGBA Model

Various sensitivity analyses were performed to investigate the factors affecting the potential of pavement growth and blow-up, as shown in Figure 2. In the case of the lean concrete base using a plastic sheet with low frictional resistance, the service life of expansion joint is less than 5 years (Figure 2a). The maximum temperature is reported as an influence factor in the thermal expansion of concrete. In the case of lean concrete using a plastic sheet, as shown in Figure 2b, there is a very short service life of less than 5 years for the expansion joint, even at normal temperatures of about 23 °C. In addition, if ASR does not occur, blow-up occurs within 40 years in extreme conditions with maximum air temperatures exceeding 40 °C. On the other hand, blow-up occurs within 20 years in ASR conditions. As shown in Figure 2c,d, the service life of the expansion joint decreased as the precipitation and CTE of concrete increased. The probability of blow-up occurrence increased when precipitation and CTE increased.

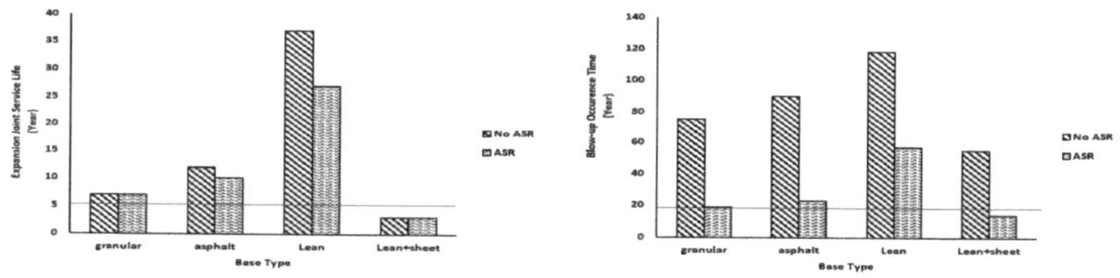

(a)

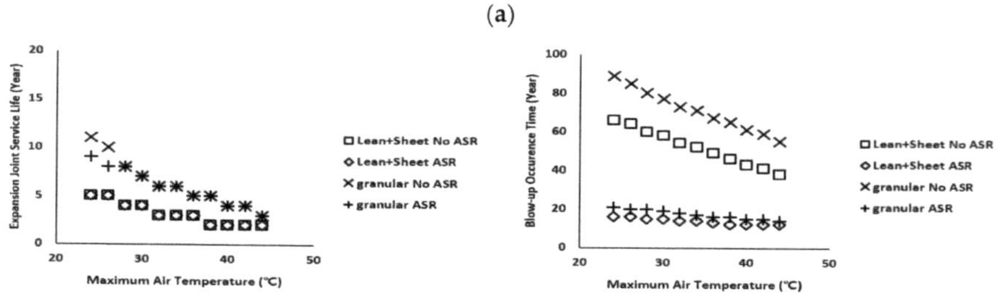

(b)

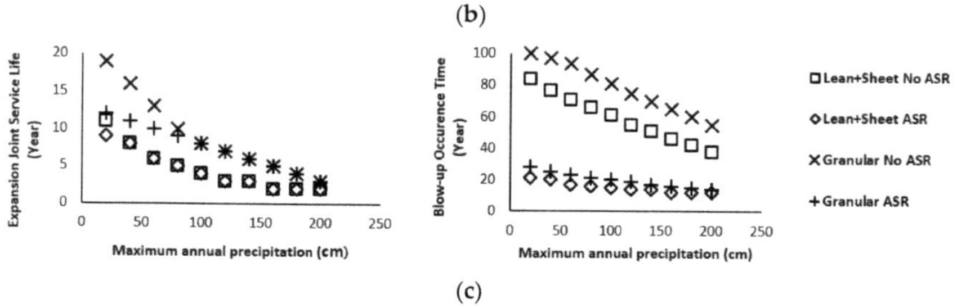

(c)

Figure 2. *Cont.*

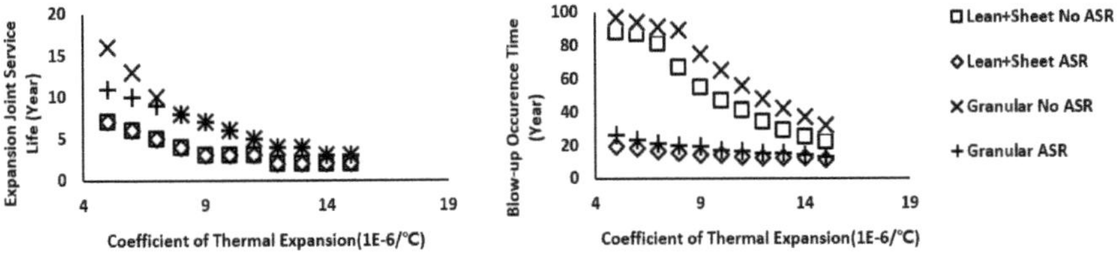

(**d**)

Figure 2. Sensitivity analysis results used to predict the service life of the expansion joint and blow-up occurrence times: (**a**) various base types, (**b**) the maximum temperature in the summer season, (**c**) annual average precipitation, and (**d**) the coefficient of thermal expansion.

4. Conclusions

In this study, the pavement growth and blow-up analysis model was developed to predict the blow-up potential due to the concrete pavement growth. This PGBA model considered factors such as the pavement structure and materials, climatic conditions, configuration of the expansion joint, base friction characteristics, geometric imperfection, rotational stiffness of joints, and design reliability. In addition, the PGBA model can be used to predict the service life of expansion joints and the blow-up occurrence time by considering the factors affecting it. Using the developed model, various sensitivity analyses were performed to evaluate the factors affecting the potential of concrete pavement growth and blow-up occurrence. It was found that the factors of substrate type, coefficient of thermal expansion, precipitation, and alkali–silica reaction had a significant effect on pavement growth and blow-up occurrence. In a further study, it will be necessary to verify the PGBA model by comparing it to previous research.

Author Contributions: Conceptualization, S.L. and Y.K.; methodology, H.A. and Y.K.; investigation and H.A.; original draft preparation, Y.K.; writing—review and edition, S.L. and Y.K. All authors have read and agreed to the published version of the manuscript.

Funding: This research was supported by the Basic Science Research Program through the National Research Foundation of Korea (NRF), which is funded by the Ministry of Education (2021R1A6A1A03044326).

Institutional Review Board Statement: Not applicable.

Informed Consent Statement: Not applicable.

Data Availability Statement: Not applicable.

Acknowledgments: The authors would like to thank the research team members at the Institute for Disaster Prevention, GWNU, for their guidance and support throughout the project.

Conflicts of Interest: The authors declare no conflict of interest.

References

1. Smith, K.D.; Snyder, M.B.; Darter, M.I.; Reiter, M.J.; Hall, K.T. *Pressure Relief and Other Joint Rehabilitation Techniques*; RES Consultants: Champaign, IL, USA, 1987.
2. Rogers, C.E.; Bouvy, A.; Schiefer, P. *Alleviating the Effects of Pavement Growth on Structures*; MDOT Division of Operation: Lansing, MI, USA, 2012.
3. Burke, M.P., Jr. Intergral & Semi-Intergral Bridges. In *Bridge Damage and the Pavement G/P Phenomenon*; Wiley-Blackwell: Hoboken, NJ, USA, 2009; pp. 21–39.
4. Yang, G.; Bradford, M.A. A refined modelling for thermal-induce upheaval buckling of continuously reinforced concrete pavements. *Eng. Struct.* **2017**, *150*, 256–270. [CrossRef]
5. Kerr, A.D. Blowup of a concrete pavement adjoining a rigid structure. *Int. J. Non-Linear Mech.* **1994**, *29*, 387–396. [CrossRef]

6. Lee, S.W.; Stoffels, S.M. Analysis of In Situ Horizontal Joint Movements in Rigid Pavements. *Transp. Res. Rec.* **2002**, *1778*, 9–16. [CrossRef]
7. Chhay, L.; Kim, Y.K.; Lee, S.W. Evaluation of temperature for pavement growth potential of concrete pavement. *Constr. Build. Mater.* **2021**, *278*, 121790. [CrossRef]

engineering proceedings

MDPI

Proceeding Paper

Analysis of LDWS Recognition Rate According to the Aging of Road Marking †

Soon-Yong Park * and Sung-Bum Yun

Seoul Institute of Technology, Seoul 03909, Republic of Korea; yunsb@sit.re.kr
* Correspondence: psy@sit.re.kr; Tel.: +82-2-6912-06959

† Presented at the Second International Conference on Maintenance and Rehabilitation of Constructed Infrastructure Facilities, Honolulu, HI, USA, 16–19 August 2023.

Abstract: In recent years, research on self-driving cars has been conducted in connection with road infrastructure in order to overcome the limitations of self-driving, such as the video detecting capability at day and night time. This study evaluates the operation of the lane departure warning system (LDWS) used in autonomous vehicles on real roads, as influenced by the aging of the road markings. The performance of the road markings was measured using the retro-reflection coefficient. Since there are restrictions on experiments on real roads, the experiment was conducted using the driving track of the Korea Institute of Civil Engineering and Building Technology while degrading the reflective performance of constructed road markings of different colors. The results indicated that there was no perceiving problem with the recognition rate of LDWS in day and night situations, but some situations occurred that depended on sun phantom or weather conditions.

Keywords: retro-reflection coefficient; lane departure warning system; connected autonomous vehicle; road marking performance

Citation: Park, S.-Y.; Yun, S.-B. Analysis of LDWS Recognition Rate According to the Aging of Road Marking. *Eng. Proc.* **2023**, *36*, 34. https://doi.org/10.3390/engproc2023036034

Academic Editor: Hosin (David) Lee

Published: 11 July 2023

1. Introduction

Lane departure warning systems (LDWS) were first introduced in the early 2000s, and initial studies found that the recognition rate was relatively low. For example, a study published in 2003 found that LDWS had a recognition rate of only 40–60% [1]. However, over the years, advancements in technology have led to significant improvements in the recognition rate of LDWS. A study published in 2016 found that the recognition rate had improved to over 90%, and more recent studies have reported even higher recognition rates [2]. It is worth noting that the recognition rate can depend on various factors, such as the type of road, weather conditions, and the quality of the road markings [3]. However, overall, it seems that the recognition rate of lane departure warning systems has been steadily improving over time [4]. However, there are no studies about performance changes of lane departure warning systems based on the aging of the lane makings. Therefore, in this study, we tried to examine the change in the recognition rate of the LDWS according to the aging of different colors of lane marking and variations in retro-reflectivity.

2. Experiment Method

2.1. Experiment Conditions

The experiment was conducted at the Yeoncheon SOC Demonstration Center of the Korea Institute of Construction Technology (KICT), where there is a weather reproduction demonstration test section that allows for the artificial expression of weather conditions, such as day and night. This section is crucial for testing autonomous vehicles. LDWS function tests were performed to meet the safety driving requirements of autonomous vehicles by the Ministry of Land, Infrastructure, and Transport (MOLIT).

The test conditions for the experiment included an air temperature range of 0–45 °C in the test furnace. The Lane Keeping Mode was set to a test-driving speed of 60 km/h, as recommended by MOLIT. These conditions were chosen to test realistic driving scenarios and to ensure accurate results.

2.2. Configure Autonomous Vehicle Speed and Trajectory

The vehicle speed was set to 60 km/h, which is a feasible speed for the LDWS to detect and respond to lane departures. The lateral lane departure of the vehicle was set to a departure angle of 5.01°, which was calculated using the tangent of 3.5 divided by 40. Additionally, the lateral speed of the vehicle was set to 1.458 m/s, which means that the vehicle was moving at a steady speed while staying within the desired lateral departure angle.

This indicates that the configuration of the autonomous vehicle's speed and trajectory took into consideration the painted and unpainted sections of the lane. Figure 1 describes the experimental test area and vehicle trajectory.

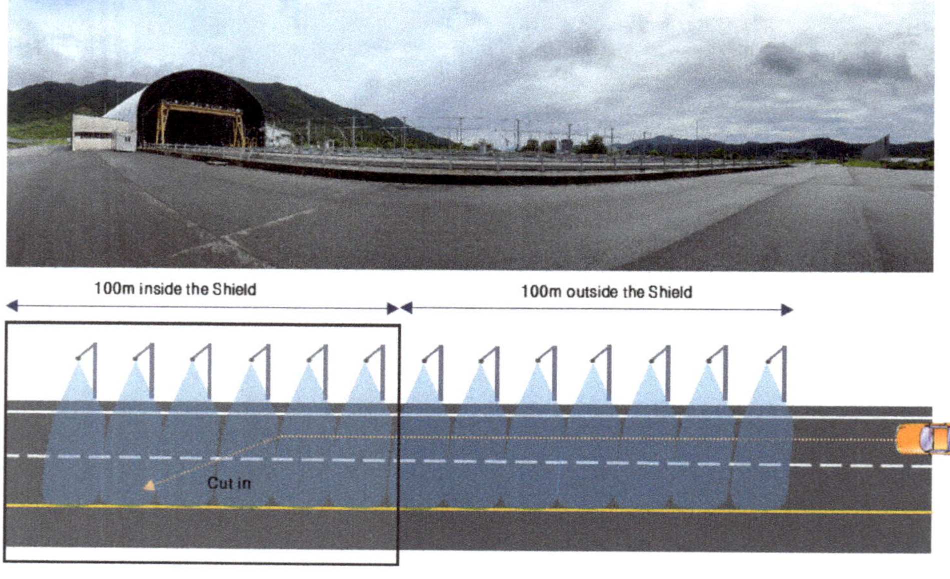

Figure 1. The Experimental Area of the Yeoncheon SOC Demonstration Center of KICT.

2.3. Construction of Road Lane Markings and Composition of Day/Night Experiments

The road surface marking construction was set according to the standards presented in the traffic road marking installation and management manual for the median line (yellow), lane (white), and exclusive lane (blue) using line types and specifications [5]. The following parameters were used for the construction:

1. Paint Length (L1): 300 cm
2. Empty Length (L2): 300 cm
3. Width (W): 15 cm

The experiment was conducted during the day from 12:00 to 15:00 and at night after 21:00 (with lighting on & off). Figure 2 shows the aging process, which was implemented using a grinding machine, and (b) describes the measurement of the reflection coefficient after aging.

(a) (b)

Figure 2. Experiment preparation: (**a**) aging machine and (**b**) retroreflection coefficient measuring machine (LTL-XL).

2.4. Construction of Roadlane Markings and Composition of Day/Night Experiments

Equation (1) is used to calculate the recognition rate (%) of LDWS for road lane markings based on the experimental results. The equation employs the number of road marking recognition counts and the total driving count to compute the recognition rate as a percentage.

$$\text{LDWS recognition rate} = (\text{No. of Warning}/\text{Total Exp. No.}) \times 100 \quad (1)$$

The experimental results were based on a sufficient number of driving scenarios to ensure the calculated recognition rate's reliability.

3. Test Results

LDWS experiments were conducted on road markings, and after the tests, the markings were ground to simulate road surface aging. The LDWS experiments were then conducted again to evaluate the impact of aging on the effectiveness of the LDWS system.

Based on the experimental procedure, the retroreflective coefficients of white, yellow, and blue road markings were initially measured as 365, 200, and 212 mcd/(m^2·Lux), respectively. To test the effect of aging on the retroreflective performance of the road markings, the markings were ground 2, 4, and 8 times, and aging was conducted. After aging, the retroreflective coefficients of white markings decreased to 253, 142, and 104 mcd/(m^2·Lux), respectively. For yellow markings, the retroreflective coefficient decreased to 164 and 76 mcd/(m^2·Lux) after two and four rounds of grinding, respectively, and it dropped to below 45 mcd/(m^2·Lux) after eight rounds of grinding. The retroreflective coefficient of blue markings decreased to 132, 62, and 47 mcd/(m^2·Lux) after two, four, and eight rounds of grinding, respectively.

On the demonstration tests conducted for white, yellow, and blue lanes, the following results were obtained:

For white lanes, 100% lane recognition was successful in daytime conditions. However, in nighttime conditions, there was a decrease in the lane recognition rate related to the decrease in retroreflection coefficient. Despite this decrease, a recognition rate of over 90% was still achieved regardless of whether there was night lighting or not, even in the situation of aging road markings.

For yellow lanes, a 100% lane recognition rate was achieved in all daytime conditions. However, it was confirmed that a recognition rate of 90% was secured at night (with or without lighting) when the retroreflective coefficient approached the repainting standard of 70 mcd/(m^2·Lux).

For blue lanes, a 100% lane recognition rate was shown in daytime conditions. In nighttime conditions, a 90% lane recognition rate was shown, regardless of the presence or absence of lighting.

Overall, the results suggest that the retroreflective coefficient of road markings is an important factor in determining the effectiveness of lane recognition systems, especially in nighttime conditions. However, the lane recognition rate can still be maintained at a high level even in situations where the retroreflective coefficient has decreased due to aging road markings.

4. Conclusions and Further Study

The study found that LDWS installed in autonomous vehicles resulted in a 100% recognition rate of road markings during the daytime and a 90% recognition rate at nighttime, where the recognition rate refers to the LDWS's ability to detect road markings. It was noted that the same 90% recognition rate was observed, regardless of the presence or absence of lighting at night time, which is thought to be due to the headlights of the vehicle. Currently, the minimum retroreflection coefficient is set based on the standard for a general driver, so it is not a problem for LDWS for driving support. However, future self-driving cars will require higher accuracy, and additional experiments will be needed to evaluate the performance of LDWS in adverse weather conditions, such as rain and fog, which were not tested in this study.

It is crucial to test the performance of LDWS under various weather conditions to ensure efficiency and safety in real-world scenarios. Therefore, experiments should be conducted in rainfall and low-to-moderate fog conditions to evaluate how the LDWS system responds to these challenging conditions and to identify potential areas for improvement. We plan to conduct experiments according to the conditions of rainfall intensity and medium-low visibility fog.

Author Contributions: Conceptualization, methodology, validation, S.-Y.P.; data curation, S.-B.Y. All authors have read and agreed to the published version of the manuscript.

Funding: This was supported by Korea Institute of Police Technology (KIPoT) grant funded by the Korea government (KNPA) (No.092021C26S02000, Development of Transportation Safety Infrastructure Technology for Lv.4 Connected Autonomous Driving).

Institutional Review Board Statement: Not applicable.

Informed Consent Statement: Not applicable.

Data Availability Statement: Not applicable.

Conflicts of Interest: The authors declare no conflict of interest.

References

1. Abdelgawad, H.; Abdel-Monem, M. Investigating the Effect of Lane Departure Warning System on Driving Performance: A Driving Simulator Study. *Transp. Res. Procedia* **2016**, *14*, 1263–1272.
2. Abdelgawad, H.; Abdel-Monem, M. Lane Departure Warning System: A Review. *IEEE Trans. Intell. Transp. Syst.* **2018**, *19*, 726–744.
3. Cheng, B.; Varma, R. Effectiveness of lane departure warning systems: A systematic review and meta-analysis. *Accid. Anal. Prev.* **2020**, *135*, 105325.
4. NHTSA. Evaluation of Lane Departure Warning Systems. *Tech. Rep. DOT. HS.* **2010**, *204*, 811.
5. NPA. *Manual for Installing and Managing Traffic Road Markings*; National Police Agency: Seoul, Republic of Korea, 2020.

engineering proceedings

MDPI

Proceeding Paper

Expeditionary Ground Rehabilitation for Military-Vehicle Traffic †

Haley Bell, Lulu Edwards * and John Rushing

U.S. Army Engineer Research and Development Center, Vicksburg, MS 39180, USA; haley.p.bell@usace.army.mil (H.B.); john.f.rushing@usace.army.mil (J.R.)
* Correspondence: lulu.edwards@usace.army.mil; Tel.: +1-601-634-3644
† Presented at the Second International Conference on Maintenance and Rehabilitation of Constructed Infrastructure Facilities, Honolulu, HI, USA, 16–19 August 2023.

Abstract: The research objective for this study is to identify and evaluate techniques for soil stabilization to support military-vehicle ground maneuver in contested environments. Various types of stabilizers mixed with silty sand are evaluated in the laboratory for their compressive strength at various soil moisture contents and in the field for their rutting performance. Field data are analyzed for the ability to withstand trafficking from a military ground vehicle by evaluating the rut depth and measured instrumentation data. The field testing shows that the rapid soil stabilization materials and techniques can produce repairs that withstand required traffic without traditional pavement surface materials.

Keywords: construction equipment; road repair; soil stabilizer; additive; silty sand

Citation: Bell, H.; Edwards, L.; Rushing, J. Expeditionary Ground Rehabilitation for Military-Vehicle Traffic. *Eng. Proc.* **2023**, *36*, 35. https://doi.org/10.3390/engproc2023036035

Academic Editor: Hosin (David) Lee

Published: 11 July 2023

1. Introduction

When the military's critical infrastructure assets do not meet mission load requirements, new materials, equipment, and techniques are required to alter the state of infrastructure and upgrade its capacity. The research objective of this project was to identify and evaluate materials for rapid soil stabilization to support military-ground-vehicle maneuvers in complex and contested environments. The target roads to be repaired are low-volume roads requiring approximately 1000 passes of military vehicles.

The chemical stabilization of soils improves quality most commonly by improving gradation, reducing plasticity index, or increasing durability or strength. Chemically stabilized layers can serve as a wearing course once curing has progressed to an adequate level for construction platforms, roads, or airfields where poor soils exist [1].

The typical design for stabilized soils in commercial applications includes detailed soil characterization such as gradation, Atterberg limits, and moisture–density relationships. Target strength and durability values are derived from detailed analysis of the number and magnitude of the loads using the structure. The design stabilizer content is the minimum dosage resulting in the desired properties. Specialized construction equipment available during construction results in thorough mixing and compaction to achieve desired properties. Military operations, however, are often limited in equipment and time to perform such tasks. Prescriptive methods for stabilizing problematic soils are needed that provide a standard recipe containing sufficient reliability to resolve the most common problems.

2. Method

Stabilization materials were studied in the laboratory and in the field. Unconfined compressive-strength (UCS) testing was completed on 13 chemical stabilizer products at moisture contents dry of optimum, optimum, and wet of optimum. The laboratory study also determined the appropriate moisture contents, dosage rates, and curing time requirements for use in field applications. Pavement repairs were then completed and subjected to military vehicle traffic in a full-scale concrete test section. Silty sand (SM) was chosen as the testing soil in the laboratory and field because it represents the most commonly found soil all over the world [2].

3. Field Testing and Results

Four downselected stabilizers and a control were tested with SM in a full-scale test section. The stabilizers included Type I/II cement, rapid-setting calcium-sulfoaluminate (CSA) cement, and xanthan gum. The rapid-setting CSA products are proprietary; no additional details are available. The control was pure SM with no additive and compacted at optimum moisture content.

The repairs were constructed within the interior of a 15-inch-thick concrete test section placed over natural lean clay subgrade. The natural subgrade had an average California bearing ratio (CBR) of 9% with a dry density of 108 lb/ft^3 and a moisture content of 18%. The repair location consisted of one confined slab, which was 12.5 ft wide by 15 ft long. Prior to the repair, the concrete slab was removed, and a geotextile was placed on top of the natural subgrade to prevent migration into the base. A 15-inch-thick heavy clay base with an in situ moisture content of approximately 35% and a CBR of 3 to 5% was compacted to a dry density of approximately 84 lb/ft^3 to provide a foundation for the repair material. Earth pressure cells (EPCs) were installed on top of the base in the wheel paths of each repair. Each of the approximate 7-inch-thick repair surfaces consisted of SM mixed with the stabilizers. The soil surfaces were compacted flush with the surrounding concrete.

The stabilizers were placed at 5% of the total dry weight of the SM. The only exception was Repair 2 where 3% of Type I/II cement was added to the SM. The moisture content of each SM stockpile before placement was approximately 10%. The preparation of each stabilizer-soil mixture occurred in batches by evenly spreading the stabilizer on top of a premeasured quantity of SM and mixing with a compact track loader (CTL) tiller attachment and flipping with a CTL bucket.

The SM mixtures were placed and compacted in two lifts approximately 4 to 5 inch thick. Compaction was completed with a vibratory drum compactor CTL attachment. The entire surfaces of each lift received two passes of the compactor—the vibrator was turned on for each second pass. Each stabilized surface material, including the control SM, was cured for 24 h after compaction before traffic was applied.

Each repair surface was subjected to nuclear-density gauge testing directly before trafficking. The nuclear gauge data showed that the moisture content of the stabilized surfaces ranged from 6 to 9%, and the dry density ranged from 101 to 118 lb/ft^3. Periodic rod and level surveys were completed to determine elevation changes with traffic applications.

Repair performance was evaluated with a four-axle common bridge transporter (CBT). The CBT weighed approximately 60,000 lb with tire pressures of 60 to 83 psi. Each repair was subjected to channelized traffic until failure, which was defined as 7 in. of rutting, or until 1000 passes. Often, repairs were trafficked beyond 7 in. of rutting.

Table 1 presents the field performance results in terms of rutting and EPC measurements. The number of passes at 7 in. of rutting is included for the north and south wheel paths. If a repair was trafficked beyond failure, the maximum rut shown in Table 1 was measured at the maximum passes applied. If the rut did not reach 7 in. in one of the wheel paths, the data were extrapolated to provide a number of passes for both wheel paths. The trafficking of each repair occurred over the course of one to two days after curing for 24 h.

As expected, the control (Repair 1) performed worst compared to the stabilized SM surfaces. The control had a rapid and steady increase in rutting with traffic in both wheel paths. Repair 5 with the rapid-setting C product had the least amount of rutting after 1000 passes of the CBT. Rutting did not exceed 1.20 in. for either wheel path.

Table 1. Field Performance results.

Repair	Additive	Passes at 7 in. of Rutting		Max Rut (in.)	Max Passes Applied	North EPC Peak (psi)	South EPC Peak (psi)
		N	S				
1	none	65	55	7.50	70	144	187
2	cement [a] (3%)	600 [b]	450	8.50	500	40	54
3	cement [a] (5%)	1050 [b]	n/a [b]	6.50	1000	23	38
4	rapid-setting cement A	385	425 [b]	7.25	400	47	48
5	rapid-setting cement C	n/a [c]	n/a [c]	1.20	1000	16	20
6	xanthan gum (3%) and cement (2%) [a]	500	450	8.25	500	--- [d]	40

[a] Type I/II cement; [b] Extrapolated data; [c] Rutting was not beyond 1.20 in.—extrapolation to 7 in. was not possible; [d] No data collected—EPC malfunctioned.

Type I/II cement (Repairs 2 and 3) was tested at different concentrations (3 and 5% cement). After 300 passes in Repair 2, a noticeable spike occurred in the rut depth data. Repair 3 performed significantly better except for the south wheel path. The rutting in the south wheel path steadily increased to a maximum of 6.50 in., while the rutting in the north wheel path remained relatively constant with rutting under 1 in. at 1000 passes.

Repair 4 failed at 400 passes. Citric acid had been added to the SM at a small dose (0.27% of the weight of rapid-setting cement) prior to mixing with the cement to help decrease the set time of the rapid-setting cement due to the high ambient temperatures. Repair 4 also had lower-than-expected dry density, and for unknown reasons, the moisture content of the SM mixture increased by 4% after trafficking.

Repair 6, consisting of 3% xanthan gum and 2% of Type I/II cement, failed at approximately 500 passes. The SM mixture appeared to be sticky to the touch and almost pliable. The dry density of Repair 6 was also lower than expected after 24 h of cure time.

The peak pressures measured from the EPCs typically occurred at or near failure for the repairs that performed poorly. The highest pressures were recorded in the control (Repair 1), which were at least three times as high as those of the other repairs. The lowest pressures were recorded in the repairs that did not develop 7 in. of rutting. This occurred in Repair 5 and the north wheel path of Repair 3. Repair 3 did achieve 1000 passes; however, the south wheel path had considerable rutting at 1000 passes. The peak pressure in the north wheel path was only 23 psi.

4. Discussion

Repairs 2 and 3, with Type I/II cement, showed that a small increase in stabilizer greatly increased the performance. Increasing the cement by two percentage points increased the performance enough to meet the objective of the study.

Some of the repairs had a much deeper rut on one wheel path compared to the other. Repair 3, stabilized with Type I/II cement, had a rut measurement difference of 5.5 in. with the south wheel path rut at 6.5 in. and the north wheel path rut at 1 in. The repairs with a large rut depth difference could be due to inconsistent mixtures or compaction.

Two of the repairs appeared not to have set; these had lower densities and worse performance. Repair 4, with the rapid-setting A product, failed earlier than expected for a cementitious product. The addition of a small amount of citric acid to decrease the set time may not have allowed the rapid-setting cement to reach final set even after curing for 24 h. Repair 6 may have had too much xanthan gum, as the material did not set after 24 h. After excavation of Repair 6, the soil-product mixture appeared to still be pliable.

It should be noted that when variability in repair performance occurred (i.e., varying rut depths with each wheel path or with each end of the wheel path), the maximum rutting occurred in the south wheel paths for each repair and on the west ends of each wheel path. This was particularly the case for Repairs 2, 3, and 6. The west end of each repair was the leading edge for trafficking and the trailing edge for compaction; traffic was applied west to east, and compaction was applied east to west.

5. Conclusions

The laboratory and field studies indicate that chemical stabilizers increase the load-carrying capabilities of repairs on low-volume roads. This was shown in the field when the rutting performance was improved from 70 passes to at least 400 passes of the CBT military vehicle. The field tests were also constructed wet of the optimum moisture content; therefore, the performance results are conservative. Additionally, chemical stabilizer types and dosages were identified that could achieve the 1000-pass target. Chemical stabilizers are an appropriate alternative for the military to use for rapid road repairs with and without surfacing with concrete or asphalt.

Author Contributions: Conceptualization, J.R.; methodology, L.E. and H.B.; validation, L.E., H.B. and J.R.; formal analysis, L.E. and H.B.; investigation, L.E. and H.B.; resources, L.E., H.B. and J.R.; data curation, L.E. and H.B.; writing—original draft preparation, L.E. and H.B.; writing—review and editing, J.R.; supervision, J.R.; funding acquisition, J.R. All authors have read and agreed to the published version of the manuscript.

Funding: This research was funded by the Assistant Secretary of the Army for Acquisition, Logistics, and Technology under Project Number 622114 BL7.

Institutional Review Board Statement: Not applicable.

Informed Consent Statement: Not applicable.

Data Availability Statement: The data presented in this study are available on request by emailing the authors at lulu.edwards@usace.army.mil or haley.p.bell@usace.army.mil.

Conflicts of Interest: The authors declare no conflict of interest.

References

1. *UFC 3-250-11*; Soil Stabilization and Modification for Pavements. Department of Defense: Washington, DC, USA, 2020.
2. Robinson, J.H.; Rabalais, C.P. *Performance of the Combat Engineer Vehicle with Mineplow Operating Worldwide and in Theaters of Operation*; Technical Report GL-93-23; U.S. Army Engineer Waterways Experiment Station: Vicksburg, MS, USA, 1993.

engineering proceedings

MDPI

Proceeding Paper

Infrastructure Measures to Protect the Unrecognized Vulnerable Road User: Motorcyclists †

Georgene M. Geary

GGfGA Engineering, LLC, Stockbridge, GA 30281, USA; ggeary@ggfga.com; Tel.: +1-470-443-2600

† Presented at the Second International Conference on Maintenance and Rehabilitation of Constructed Infrastructure Facilities, Honolulu, HI, USA, 16–19 August 2023.

Abstract: Motorcyclists are 28 times more likely to die in traffic crashes than occupants in passenger cars and they constitute 14% of the total traffic fatalities while being less than 3% percent of registered vehicles. Training and education of motorcycle operators has been an area of focus to improve these statistics, but, considering that motorcycle safety has become worse in the last 20 years, more must be done. This paper summarizes the needs and results of existing infrastructure-related improvements for motorcyclist safety, while also identifying areas of research needed to continue to advance motorcyclist safety through engineering-based infrastructure improvements.

Keywords: motorcycle; infrastructure; safety

Citation: Geary, G.M. Infrastructure Measures to Protect the Unrecognized Vulnerable Road User: Motorcyclists. *Eng. Proc.* **2023**, *36*, 36. https://doi.org/10.3390/engproc2023036036

Academic Editor: Hosin (David) Lee

Published: 11 July 2023

1. Introduction

A Vulnerable Road User (VRU) was recently defined in the VRU Safety Assessment section 23 U.S.C. 148(1) of the U.S. Bipartisan Infrastructure Law (BIL) as a "pedestrian, bicyclist, other cyclist . . . ", but motorcyclists were specifically noted as not included as a VRU [1]. Unfortunately, about the same time, the U.S. DOT National Highway Traffic Safety Administration (NHTSA) reported that more motorcyclists were killed in traffic crashes in 2020 than when data first was collected on fatal crashes back in the 1970s. They also reported that motorcyclists constitute 14% of the total traffic fatalities while being less than 3% of all registered vehicles [2], suggesting that they are indeed vulnerable.

This paper provides a brief history of motorcycle safety research and then describes the status of infrastructure design considerations for motorcycles. Next, several infrastructure-related improvements to safety are highlighted. The last section identifies future opportunities that could affect motorcyclist mortality.

2. Motorcycle Safety Historical Focus

With motorcycle fatalities reaching 5000 a year (see Figure 1), the NHTSA sponsored the first major U.S. motorcycle safety study in the 1970s. The "Hurt Study" (named for the lead investigator, Dr. Hurt) looked in depth at over 900 motorcycle crashes and identified the conspicuity of motorcycles, especially at intersections, as an important consideration to motorcyclist safety, but also recognized the potential contribution of infrastructure. The next major crash analysis study was performed in Europe (the MAIDS or Motorcycle Accident In-Depth Study) in the late 1990s. Both studies led to an increased focus on motorcyclist access to rider education safety programs and the use of personal protective equipment (PPE), such as helmets [3].

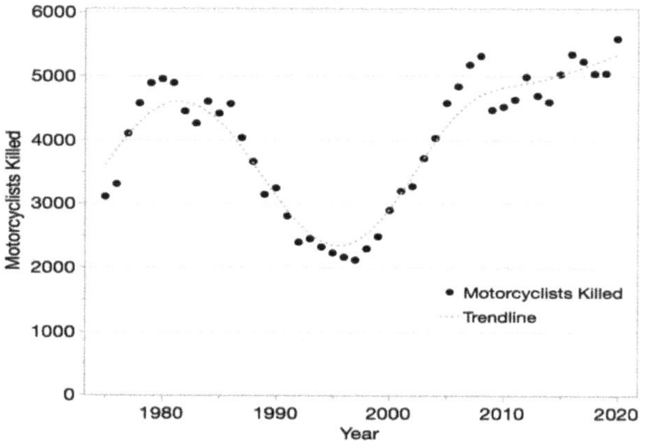

Figure 1. Motorcycle fatalities: 1975 to 2020 (Developed from FARS data [4]).

Since those studies, motorcycle fatalities have oscillated, as shown in Figure 1, and are now again above the "Hurt" study levels of 5000 [4]. It can be seen from the figure that U.S. motorcycle fatalities doubled from a low around 1994 to 2014, but it has also been reported that U.S. passenger car and light truck fatalities decreased by 34% in that same timeframe [5].

3. Motorcycle Safety and Infrastructure Design

Due to the NHTSA recommendations, currently, every state in the U.S. has a State Highway Safety Plan (SHSP) that identifies motorcyclists as a specific safety focus. The SHSPs predominately consider PPE education (i.e., helmets) or operator training efforts, with only limited descriptions of engineering methods or infrastructure changes used to improve motorcyclist safety. The main roadway design manual used by the U.S., the AASHTO Policy on Geometric Design of Highways and Streets (or "Green Book"), is relatively silent on motorcycles. In contrast, Australia's VicRoads, the United Kingdom's Institute of Highway Engineers and Transport for London all have guidance documents on roadway design specifically for motorcycles [6–8]. Even the World Health Organization (WHO) recognizes motorcyclists as VRUs and has guidelines specifically on how to design roads to be safer for motorcycles [9].

4. Specific Infrastructure Interests for Motorcyclists

The U.S. Motorcycle Advisory Council (MAC), which was authorized in the Fixing America's Surface Transportation (FAST) Act, identified 11 areas of infrastructure of concern to motorcyclists. As described in a recent Federal Highway Administration (FHWA) technical report, which was developed in response to the MAC recommendations, some noteworthy actions related to these areas of motorcycle safety in the U.S. have been identified [10]. These areas are described in more detail in the following sections.

4.1. Pavement Markings

The Minnesota and Florida Departments of Transportation (DOTs) are both addressing the friction of pavement markings in critical areas for motorcyclists. Minnesota has used high-friction markings on roundabouts and has research underway that is considering how to measure and address the effects of differential roadway friction. Florida has used high-friction pavement shields on interstate routes, starting with the interstates near Daytona Beach, the site of Annual Bike Week [10]. They are now expanding these requirements to the entire state.

Virginia Tech has also been testing continuous friction measurement equipment (CFME), which can identify pavement friction issues, including pavement marking friction, better than current U.S. equipment [3,11].

4.2. Work Zones

The Montana DOT has a motorcycle advisory sign policy that provides information for motorcycles in work zones that may be especially difficult for motorcycles to traverse (e.g., rough road or gravel-type conditions). This policy has been in place since 2013, and is aided by the Montana DOT's 511 website. Since it is not always possible to avoid all work zones, the National Work Zone Safety Information Clearinghouse also has a one-hour online training program for work-zone managers that offers guidance on how to ensure that their work zones address the specific concerns of two-wheeled vehicles (motorcyclists and bicyclists) [10].

4.3. Intersections

The proper detection of motorcycles for intersection approaches is critical to prevent trapping a motorcyclist in a dilemma zone, where they must choose between crashing due to braking too hard or crashing due to someone else in the intersection not seeing them enter on the end of a yellow. Only 10 states' DOTs specifically recognize the special requirements of detecting motorcycles at signalized intersections in their specifications (80% of states do not mention motorcycles at all) [3]. However, the Ohio DOT developed its own loop detector designed to specifically detect motorcycles. They also specify and test pole-mounted detectors to make sure they identify motorcycles accurately [10].

4.4. Roadside Barriers

Guardrails and other roadside barriers are designed to reduce the severity of crashes for motor vehicles. Although motorcyclists constitute about 14% of all traffic fatalities, they account for 40% of all vehicle–guardrail-related fatalities in the U.S. [12]. European Standard EN-1317-8, a full-scale vehicle–barrier crash testing standard for motorcycle-barrier impacts, has been in use for over a decade in Europe [13], but the U.S. does not have any similar standards. The Texas DOT has sponsored recent research using finite-element modeling and full-crash testing to develop barrier retrofits to improve motorcycle safety [14]. The North Carolina DOT, Utah DOT, and Caltrans have all tested motorcycle safety barriers on the roadway. Also called Motorcycle Protection Systems (MPS), they can be retrofitted on existing guardrails by placing a barrier rail below the main guardrail. This lower barrier prevents the motorcyclist from hitting the supporting poles of a guardrail, which is the kind of impact that is especially hazardous for motorcyclists. The Utah DOT documented a safety improvement with a MPS section that had a motorcycle injury every year in the five years before the installation, and none in the subsequent three-year period [13]. A recent U.S. cooperative research study recommended that MPS be further evaluated, and guidance be developed for their testing and use in the U.S. [12].

5. Crucial Infrastructure Opportunities of ITS (Intelligent Transportation Systems)

ITS is at the center of the future in transportation, tied in with automated and connected vehicles. The roadway, including traffic-safety appurtenances, and the vehicles themselves are all a part of the future of ITS. Even now, cars use different levels of automation or advanced driver-assistance systems (ADAS) to improve safety, such as blind-spot warnings in mirrors and the now-ubiquitous cruise control and anti-lock brakes. Automation is also available in motorcycles, and at least 14 different ARAS (advanced rider assistance systems) have been identified, the most common being anti-lock braking (ABS), but also including advanced curve warnings to identify if a rider is approaching a horizontal curve too fast [15]. The connected vehicles of the future also could improve the conspicuity of motorcycles and prevent hundreds of fatal motorcycle crashes per year [16].

6. Conclusions

Motorcyclists are vulnerable users of the roadway, as shown by their disproportionate number of fatal crashes as compared to other road users. Efforts to address these sometimes-forgotten VRUs can be found in the U.S., but more needs to be done. As part of the MAC effort, research problem statements have been developed related to measuring and specifying pavement marking friction and gathering information on how to modify other existing roadway guidance documents to reflect the particular needs of motorcyclists. One promising note: the AASHTO Guide for Pavement Friction recently added some language on the consideration of motorcycles as related to pavement friction [17]. Next, the "Green Book" and the Roadside Design Guide also need to recognize the vulnerable motorcyclist.

Funding: This paper received no external funding.

Institutional Review Board Statement: Not applicable.

Informed Consent Statement: Not applicable.

Data Availability Statement: Data sharing is not applicable.

Acknowledgments: The author would like to acknowledge the efforts of the research team that developed the reports which are described and referenced [3,10,13,15] in this paper and available at: https://highways.dot.gov/safety/other/motorcyclist/motorcycle-safety (accessed on 20 February 2023). The team was led by Mike Manser and Karen Dixon of the Texas Transportation Institute and included Karen Aspelin (Olsson Engineering) and the author as subcontractors.

Conflicts of Interest: The author declares no conflict of interest.

References

1. Walker, C. *Memorandum on Vulnerable Road User Safety Assessment Guidance*; Office of Safety Associate Administrator: Washington, DC, USA, 2022.
2. NHTSA. *Traffic Safety Facts: Motorcycles. DOT HS 813 306*; National Highway Traffic Safety Administration: Washington, DC, USA, 2022; p. 13.
3. Geary, G.M.; Dixon, K.; Aspelin, K.; Manser, M. *Addressing the Motorcyclist Advisory Council Recommendations: Synthesis on Roadway Geometry, Pavement Design, and Pavement Construction and Maintenance Practices*; FHWA Report SA-21-090; Federal Highway Administration, U.S. Department of Transportation: Washington, DC, USA, 2021; p. 80.
4. National Center for Statistics and Analysis (NCSA) Fatal Analysis Reporting System (FARS) Data. Available online: https://www.nhtsa.gov/crash-data-systems/fatality-analysis-reporting-system (accessed on 20 February 2023).
5. Nazemetz, J.W.; Bents, F.D.; Perry, J.G.; Thor, C.; Mohamedshah, Y.M. *Motorcycle Crash Causation Study: Final Report*; FHWA-HRT-18-064; FHWA: McLean, VA, USA, 2019.
6. VicRoads. *Making Roads Motorcycle Friendly: A Guide for Road Design Construction and Maintenance*; Victoria State Government Department of Transport: Melbourne, Australia, 2022; p. 32.
7. IHE. Institute of Highway Engineers. Guidelines for Motorcycling. Available online: https://motorcycleguidelines.org.uk/ (accessed on 20 February 2023).
8. TfL (Transport for London). Urban Motorcycle Design Handbook. 2014. Available online: https://content.tfl.gov.uk/tfl-urban-motorcycle-design-handbook.pdf (accessed on 20 February 2023).
9. WHO. *Powered Two- and Three-Wheeler Safety: A Road Safety Manual for Decision-Makers and Practitioners*; World Health Organization: Geneva, Switzerland, 2017; p. 108.
10. Geary, G.; Manser, M. *Motorcycle Safety Noteworthy Practices: Infrastructure and Engineering*; FHWA Report FHWA-SA-22-031; Federal Highway Administration, U.S. Department of Transportation: Washington, DC, USA, 2022; p. 32.
11. de León Izeppi, E.; Flintsch, G.; Katicha, S.; McCarthy, R.; Smith, K. *Pavement Friction Management Program Utilizing Continuous Friction Measurement Equipment and State-of-the-Practice Safety Analysis Demonstration Project*; FHWA-RC-20-0009; Federal Highway Administration, U.S. Department of Transportation: Washington, DC, USA, 2019; p. 137.
12. Gabler, H.C.; Daniello, A.; Tatem, W.; Tsoi, A.; Gabauer, D.; Stitzel, J.; Sink, J.; Barnard, R. *Motorcycle Crashes into Traffic Barriers: Factors Related to Serious and Fatal Injuries*; NCHRP Research Report 1005; National Academies of Sciences, TRB: Washington, DC, USA, 2022; p. 139. [CrossRef]
13. Silvestri Dobrovolny, C.; Geary, G.M.; Dixon, K.; Manser, M.; Chauhan, J. *Addressing the Motorcyclist Advisory Council Recommendations: Synthesis on Barrier Design for Motorcyclists Safety*; FHWA Report SA-21-069; Federal Highway Administration, U.S. Department of Transportation: Washington, DC, USA, 2021; p. 64.
14. Silvestri Dobrovolny, C.; Shi, S.; Kovar, J.; Bligh, R.P. Development and Evaluation of Concrete Barrier Containment Options for Errant Motorcycle Riders. *Transp. Res. Rec.* **2019**, *2673*, 14–24. [CrossRef]

15. Fowler, M.; Geiselbrecht, T.; Brydia, R.; Geary, G.M.; Manser, M. *Addressing the Motorcyclist Advisory Council Recommendations: Synthesis on Intelligent Transportation System Applications and Automated Technologies for Motorcyclists*; FHWA Report No. SA-22-033; Federal Highway Administration, U.S. Department of Transportation: Washington, DC, USA, 2022; p. 41.
16. Teoh, E.R. Motorcycle crashes potentially preventable by three crash avoidance technologies on passenger vehicles. *Traffic Inj. Prev.* **2018**, *19*, 513–517. [CrossRef] [PubMed]
17. AASHTO. *AASHTO Guide for Pavement Friction*, 2nd ed.; AASHTO: Washington, DC, USA, 2022.

engineering proceedings

Proceeding Paper

Re-Evaluating the Risk of Using Higher-Skid-Resistance Aggregates †

David Woodward *, Phillip Millar and Paul Sargent

Belfast School of Architecture and Built Environment, Ulster University, Belfast BT15 1AP, UK; p.millar@ulster.ac.uk (P.M.); p.sargent@ulster.ac.uk (P.S.)
* Correspondence: wdh.woodward@ulster.ac.uk; Tel.: +44-28-9536-5344
† Presented at the Second International Conference on Maintenance and Rehabilitation of Constructed Infrastructure Facilities, Honolulu, HI, USA, 16–19 August 2023.

Abstract: Aggregate with higher skid resistance is used in many countries around the world to improve road safety. It has been tested for skid resistance in the laboratory and measured in service for many years. Certain rock types and asphalt mixes offer more skid resistance. The laboratory test methods have not changed for many years. Road surfaces achieve a state of in-service equilibrium depending on many factors, and roads are now experiencing additional changes due to issues ranging from greater use of electric vehicles to climate change in its many forms. This paper considers whether the previous methods are still able to offer reliable prediction. The paper re-evaluates research performed in the period 1990 to 2010 to consider whether the testing of road surface aggregate needs to be updated for the early 21st century. The authors believe this paper is significant for all countries to optimize the performance of their road networks in line with changes in service conditions.

Keywords: re-evaluation; aggregate; skid resistance; performance

Citation: Woodward, D.; Millar, P.; Sargent, P. Re-Evaluating the Risk of Using Higher-Skid-Resistance Aggregates. *Eng. Proc.* **2023**, *36*, 37. https://doi.org/10.3390/engproc2023036037

Academic Editor: Hosin (David) Lee

Published: 11 July 2023

1. Introduction

Road surfacing asphalt used around the world contains approximately 92 to 95% aggregate. In-service performance depends on what happens at the tire/asphalt interface as a vehicle travels along a road. European aggregate, asphalt and concrete products used in roads are covered by the Construction Products Regulation (CPR) [1]. This regulation refers to performance over the life of the product, i.e., from initial design, through in-service life, to when it is recycled. This paper considers this statement and asks whether it is possible to improve the prediction of aggregate performance in the laboratory. The paper re-evaluates research projects carried out over a 20-year period between 1990 and 2010 with methods used in the laboratory to test aggregate. This re-evaluation has been prompted by developing 21st century issues, such as the move to semi-autonomous vehicles, the change from petrol/diesel power to electric and hydrogen, and the governmental response to climate change and all of its associated issues.

2. Why Re-Evaluate Old Research?

Much happened in the UK during the 1990s and early 2000s with regard to road surfacing products. Hot Rolled Asphalt (HRA) had previously been the dominant road surfacing material. However, a short period of higher-than-normal summer temperature caused the bitumen-rich HRA laid on the heavier trafficked networks to suffer widespread premature deformation failure. In response, the road industry looked to countries with high summer temperatures and the asphalt mixes they used. Coarse aggregate skeleton mixes, such as Stone Mastic Asphalt (SMA) and Porous Asphalt (PA), were researched for use in the UK. At this time, the UK had much higher requirements for aggregate, in-service skid resistance and texture depth than countries such as Germany and France. The high stone skeleton SMA and PA mixes were developed into proprietary Thin Surface Course

Systems (TSCS), which were optimized for in-service characteristics such as high-speed skid resistance, lower noise or improved rolling resistance.

A parallel factor during this period was the process of removing European trade barriers with the Construction Products Directive (CPD) published in 1998. This led to the first European Standards in 2004 and the Construction Products Regulation (CPR) in 2011 [1]. The CPR is based on the ideal of performance for the life of the product. It has seven generic requirements for all products used in construction, i.e., mechanical resistance and stability; safety in case of fire; hygiene, health and environment; safety and accessibility in use; protection against noise; energy economy and heat retention; and sustainable use of natural resources. These seven requirements were a response to how the world of construction was perceived to be moving into the 21st century.

In practice, the selection of European test methods for aggregate was not based on research to develop improved predictive test methods. Instead, they were existing methods that had been used for many years and had not changed since the early-to-mid-20th century when they were developed. The prediction of in-service risk based on laboratory testing was considered by Woodward [2]. This study concluded that standard test methods offer limited insight into structural and functional performance. Few aggregate test methods considered what happened during the life of the product and predict whether it would remain fit for purpose during the life of the surfacing.

On one hand was the expectation for performance for the life of the product and test methods which offer limited prediction. On the other hand was traffic growth, with new types of vehicle interfacing with the road surface in new ways and the impact of climate change in all of its different forms, from changing weather to government response. This perfect storm scenario forms the basis of this re-evaluation to determine whether anything can be learnt from the research carried out in the 1990s [2] to enhance roads in the early 21st century.

3. What Research Is Being Re-Evaluated?

It has been known for many years that not all aggregate is suitable for highway surfacing. The Travers Morgan Report in 1993 [3] and the Capita Symonds Report in 2004 [4] provided information regarding aggregates considered suitable for road surfacing. They remain a valuable resource for aggregates with higher skid resistance. A limitation of these reports is that they are based on data from standard aggregate testing. Recognizing this limitation, Woodward [2] investigated whether laboratory prediction of surfacing aggregate performance could be improved.

The subsequent SKIDPREDICT [5] and SKIDGRIP [6] projects considered whether the standard Polished Stone Value (PSV) test was a measure of the ultimate state of polish for an aggregate, and how skid resistance of asphalt mixes evolved in service, respectively. A wide range of rock types, asphalt mixes and in-service trafficking conditions were assessed. The projects evaluated the effect of test regimes simulating the conditions that aggregate experienced in different surfacing asphalt mixes in service; for example, increasing the load or induced stressing during PSV testing, or the effect of testing wet aggregate.

4. What Has the Re-Evaluation Identified with Respect to Current Trafficking?

A feature of the work by Woodward [4], SKIDPREDICT [5] and SKIDGRIP [6] was the deconstruction of standard test methods to determine whether a modified version offered better prediction of what happened in service. For example, increasing load during PSV testing found that the laboratory test result for skid resistance decreased. This deconstructed regime may be compared to the increasing use of electric vehicles, which tend to be heavier, more torque and smaller road/surface contact patches.

4.1. Deconstruction of Test Methods

Deconstructing the laboratory test regimes based on in-service conditions provided insight into how different aggregates in different types of asphalt mix react to different stress conditions around the road network. The tire/aggregate interface on the fast lane of a motorway is different to braking for a road junction or going around a bend. For each condition, the tire/aggregate creates an equilibrium condition that can, for example, be measured in terms of wet skid resistance.

4.2. Widening the Types of Aggregate That Would Typically Be Used

The UK has a wide range of aggregates. Not all are perceived as suitable for road surfacing. This has obvious issues with the greater use of local sources and other sustainable issues. Aggregates traditionally not used were considered. Selection was based on the hierarchy of rock types typically used in the UK as surfacing aggregate. Greywackes and sandstone would typically be used for the highest levels of skid resistance. Igneous rocks, such as basalts and granites, would be used on lesser trafficked roads. The projects also examined rock types such as limestones that would not be used in the UK because of their lower wet skid resistance. This questioned the traditional perception of what is considered a good surfacing aggregate. In simple terms, development in vehicle and tire technologies probably do not require the network use of high PSV aggregates, except in localized locations with the greatest skidding risk.

4.3. The PSV Test

With regard to road surface aggregate properties, skid resistance has always been considered the most important in the UK. The European test method to predict aggregate skid resistance in the laboratory is the Polished Stone Value (PSV). This has been a British Standard since 1960 before becoming the test method used across all European countries. Deconstructions of the standard 6 h test included the loss in skid resistance due to wetting the interface during pendulum testing for unpolished aggregate; how skid resistance changed during the standard 6 h PSV test; and what happened if fine emery polishing was extended up to an additional 30 h after initial 3 h polishing with coarse emery. Freeze/thaw cycles were used to investigate winter increase in measured in-service skid resistance. Sideways polishing was used to assess additional tire/surface interface stressing at bends of different radii.

These found that the standard PSV test offers limited insight into what happens at the tire/surface interface during the life of an asphalt mix. If the same aggregate is tested using deconstructed versions of the standard test, the result is different levels of equilibrium, agreeing with what happens in service.

4.4. PSV and Other Aggregate Properties

An important consideration was relating skid resistance data with other deconstructed aggregate test methods for measuring properties, such as strength, wear and durability. This confirmed the ranking that PSV depends on rock type. More importantly, for a given rock type, higher PSV was gained at the expense of all other properties, such as wet resistance to wear. Woodward [2] concluded that specifications that did not recognize rock type do not offer a sustainable solution.

4.5. Consideration of Rainfall

Surprisingly, it was found that a simple improvement in testing aggregate would be to determine what happens when the aggregate is wetted by rain. Irrespective of the extremes in climate change being experienced around the world, something as simple as rain fall is not considered. The most common aggregate tests used around the world are the ASTM or EN versions of the Los Angeles test. This is carried out on dry aggregate. Testing wet aggregate can, for some rock types, cause strength reductions greater than 50% compared to testing dry aggregate. Simple changes to accommodate climate change, such as more rainfall events, would be a significant addition to a risk-based sustainable optimization specification including dry and wet testing and a percentage change.

4.6. Aggregate Homogeneity

Many assume that an aggregate source is homogenous in composition and behaves in a predictable manner. However, many aggregate sources are heterogeneous, with the quarried product offering different responses to a tire at the interface, and to the many other climate-change-related factors such as rain. Woodward [2] gives examples illustrating how aggregate in the stockpile may be a mix of durable and less-durable particles. Under trafficking and environmental factors, such as rain, freezing, or application of deicing salt, the less-durable particles will perform differentially. This will affect the tire/surface interface and within the aggregate skeleton of the asphalt mix. Such problems may become more common due to more environmental stressors such as freeze/thaw, wetting and drying.

5. Conclusions

This paper has re-evaluated what may be perceived as old data dating back to projects carried out between 1990 to 2010. The authors believe this re-evaluation has highlighted the need to give more recognition to how road surfaces may be responding to changes in trafficking and climate around the world. Surfacing aggregate is not an inert material with constant engineering properties. There are complex interactions at the tire/aggregate interface which are primarily dependent on rock type. Deconstructing standard test methods to simulate what happens in service offers the opportunity to readdress how aggregate and asphalt mixes may better achieve the expectation of performance for the life of the product.

Author Contributions: Conceptualization, D.W.; writing—original draft preparation, D.W.; writing—review and editing, P.M. and P.S. All authors have read and agreed to the published version of the manuscript.

Funding: This research received no external funding.

Institutional Review Board Statement: Not applicable.

Informed Consent Statement: Not applicable.

Data Availability Statement: Data are available by contacting the lead author.

Conflicts of Interest: The authors declare no conflict of interest.

References

1. Regulation (EU) No 305/2011 of the European Parliament and of the Council of 9 March 2011 Laying Down Harmonized Conditions for the Marketing of Construction Products and Repealing Council Directive 89/106/EEC. Available online: https://eur-lex.europa.eu/LexUriServ/LexUriServ.do?uri=OJ:L:2011:088:0005:0043:EN:PDF (accessed on 1 July 2023).
2. Woodward, W.D.H. Laboratory Prediction of Surfacing Aggregate Performance. Doctoral Dissertation, University of Ulster, Coleraine, UK, 1995.
3. Thompson, A.; Greig, J.R.; Shaw, J. *High Specification Aggregates for Road Surfacing Materials: Technical Report. Report to the Department of the Environment*; Travers Morgan Limited: East Grinstead, UK, 1993; 270p.
4. Thompson, A.; Burrows, A.; Flavin, D.; Walsh, I. *The Sustainable Use of High Specification Aggregates in England: Report to the Office of the Deputy Prime Minister and the Minerals Industry Research Organisation*; Capita Symonds Ltd.: East Grinstead, UK, 2004.
5. Roe, P.; Woodward, W.D.H. *Predicting Skid Resistance from the Polishing Properties of the Aggregate (SKIDPREDICT) Final Report PR CSN/31/03*; University of Ulster: Coleraine, UK, 2004.
6. Woodward, W.D.H. *Predicting Early Life Skid Resistance of Highway Surfacings (SKIDGRIP). EPSRC Civil Engineering Responsive Programme GR/R09022/01*; University of Ulster: Coleraine, UK, 2003.

engineering proceedings

MDPI

Proceeding Paper

Safety Assessment of Cooperative Platooning in Mixed Traffic †

B. Brian Park [1,*], Hyejin Lee [2], Ilsoo Yun [3] and Jeehyung Park [4]

1 Department of Civil & Environmental Engineering, School of Engineering and Applied Science, University of Virginia, Charlottesville, VA 22903, USA
2 Department of Civil & Environmental Engineering, College of Engineering, Seoul National University, Seoul 08826, Republic of Korea; qwe3rty32@gmail.com
3 Department of Transportation System Engineering, Ajou University, Suwon 16499, Republic of Korea; ilsooyun@ajou.ac.kr
4 Center for Private Highway Studies, The Korea Transport Institute, Sejong 30147, Republic of Korea; jhpark@koti.re.kr
* Correspondence: bp6v@virginia.edu; Tel.: +1-434-924-6347
† Presented at the Second International Conference on Maintenance and Rehabilitation of Constructed Infrastructure Facilities, Honolulu, HI, USA, 16–19 August 2023.

Abstract: The safety impacts of cooperative platooning in mixed traffic consisting of human-driven, connected, and connected–automated vehicles were evaluated. The cooperative platooning, in the case of the mixed traffic control algorithm evaluated here, was cooperative adaptive cruise control with an unconnected vehicle (CACCu). Its safety and string stability were evaluated using a high-fidelity simulation based on real-world vehicle trajectories. An adaptive cruise control (ACC) algorithm was selected for comparison purposes. The results indicate that the cooperative platooning in mixed traffic control algorithm (CACCu) maintains string stability and operates with greater safety than the ACC.

Keywords: cooperative platooning; cooperative adaptive cruise control; surrogate safety; safety assessment; NGSIM

Citation: Park, B.B.; Lee, H.; Yun, I.; Park, J. Safety Assessment of Cooperative Platooning in Mixed Traffic. *Eng. Proc.* **2023**, *36*, 38. https://doi.org/10.3390/engproc2023036038

Academic Editor: Hosin (David) Lee

Published: 11 July 2023

1. Introduction

With the introduction of self-driving automated vehicles and connected vehicles, vehicles can benefit from automation and connectivity. One promising technology is cooperative platooning. Studies have shown that cooperative adaptive cruise control (CACC) can maintain a time headway of 0.6 s between vehicles [1]. Fully connected and automated vehicles can significantly improve mobility, safety, and sustainability by shortening time headway and ensuring string stability [2–4]. However, we do not expect that 100% fully connected automated vehicles will be available in the near future. A connected automated vehicle following another connected automated vehicle can engage in cooperative platooning using CACC. However, when it encounters a human-driven, unconnected preceding vehicle, it must fall back to adaptive cruise control (ACC), which can automatically control the vehicle's longitudinal movement to ensure a safe distance from the preceding vehicle. Although ACC performs better than simple cruise control and is safer than a human driver, it significantly downgrades the performance of the ego vehicle compared to that of CACC. The main reason why ACC does not perform well is the lack of cooperative control between vehicles.

Researchers have explored cooperative platooning control strategies with mixed traffic of connected automated and human-driven vehicles. These include the graceful degradation of CACC (dCACC) based on a preceding vehicle's estimated acceleration [5], connected cruise control exploring the benefits of communications with an out-of-sight preceding vehicle [6–8], and cooperative adaptive cruise control with unconnected vehicles (CACCu) that functions through a control strategy maximizing the string stability of the ego vehicle using the further connected preceding vehicle's feedforward signal [9,10].

The primary purpose of this paper is to assess the safety impacts of cooperative platooning in mixed traffic, including connected automated vehicles and human-driven vehicles. The remainder of this paper is organized as follows. Section 2 describes the control algorithm selected for this study and the safety assessment measures considered. Section 3 discusses the data used for this study, and Section 4 evaluates the data and presents the key findings. Finally, Section 5 discusses the conclusions and future research.

2. Cooperative Platooning in a Mixed Traffic Control Algorithm and Safety Measures

This section discusses the algorithm for cooperative adaptive cruise control with unconnected vehicles (CACCu) previously developed by Zheng and Park [9]. Figure 1 shows a framework of the CACCu. A key innovation over ACC is the use of a feedforward signal from a further connected preceding vehicle. Unlike ACC, which implements a feedback control using information from the immediately preceding vehicle based on its sensor, CACCu acts in the same manner as CACC by taking advantage of a connected further preceding vehicle.

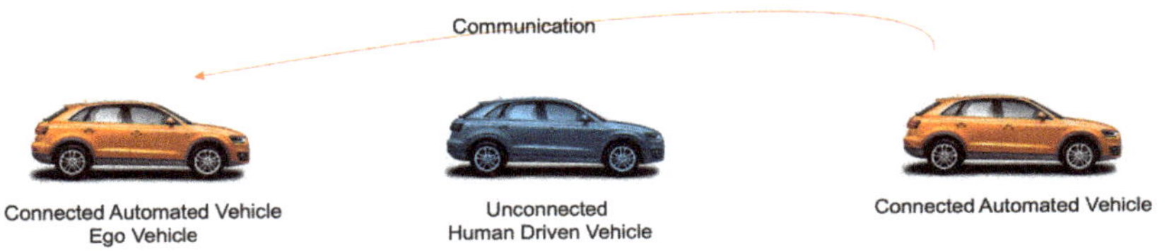

Figure 1. Cooperative adaptive cruise control with an unconnected vehicle framework.

The authors demonstrated that the virtual preceding vehicle filter attached to the original CACC feedforward filter helped the ego vehicle to stay string-stable at a gap significantly shorter than that in ACC. Both the high-fidelity simulation and field test concluded that CACCu outperforms ACC in regard to string stability, comfort, and sustainability based on the number of speed overshoots, acceleration root mean square, and fuel consumption, respectively [10,11].

As mentioned earlier, surrogate safety measures called time-to-collision (TTC) and the spacing error are used to assess safety impacts. TTC has been widely used in safety assessment in cases when the number of actual crashes is either low or not readily available [12,13]. The spacing error (SE) is determined by the difference between the summation of "ego vehicle speed x desired time headway + jam density" and the gap distance between the ego and preceding vehicles.

3. Vehicle Trajectory Data and High-Fidelity Simulation

The human-driven vehicle trajectories used in this study are from the Next-Generation Simulation (NGSIM) program [14]. Three representative trajectory data sets were carefully selected for analysis. The NGSIM data provide information on the position, speed, and acceleration of each vehicle every 0.1 s. It should be noted that the speed and acceleration data were smoothed to minimize noise and jerks caused by derivations from the position data.

As shown in Figure 1, the scenarios in this study are composed of a series of three or more consecutive vehicles from the NGSIM data. The performance of a human-driven vehicle is evaluated using the ego vehicle (i.e., the first vehicle), while the performance of ACC and CACCu vehicles is evaluated by applying the corresponding control algorithms to the ego vehicle. It is worth noting that the CACCu control algorithm utilizes the second preceding vehicle. For more details on the CACCu and ACC algorithms, please refer to [9]. The simulation was implemented with the MATLAB program, with three sets of three consecutive vehicles.

4. Simulation Results and Key Findings

When cooperative platooning is considered, the string stability is measured based on the number of speed overshoot events. This refers to the ability of an ego vehicle to smoothly follow its preceding vehicle without exceeding its speed. As noted, a desired time headway of 1.5 s is used in this evaluation to ensure safety, even during a field operation test. However, in the NGSIM data, human drivers do not necessarily maintain this time headway.

As shown in Figure 2, ACC resulted in five speed overshoot events for the ego vehicle, while CACCu resulted in zero overshoot events. This indicates that CACCu significantly improves ride comfort by maintaining string stability.

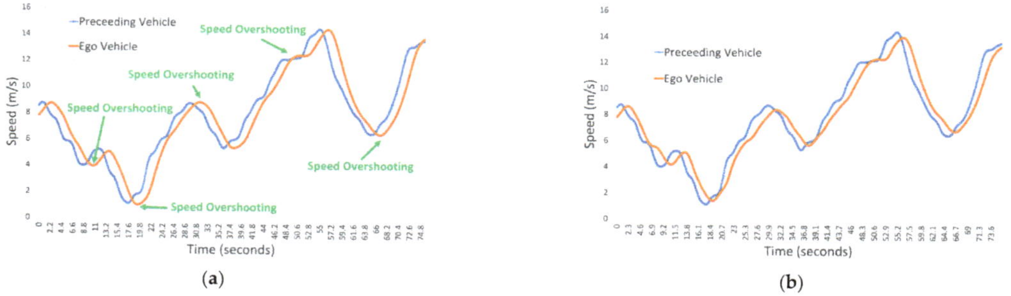

Figure 2. String stability evaluations using speed overshoot events. (**a**) ACC mode. (**b**) CACCu mode.

In addition to the overshooting speed events, safety was assessed using the time-to-collision (TTC) and spacing errors. With the desired time headway of 1.5 s, the TTC values are always higher than a safety concern. Thus, the spacing errors are considered for safety assessment. As shown in Figure 3, the spacing error distribution under ACC has far greater variability, often being higher than 3 m. On the other hand, CACCu shows a much tighter distribution, indicating a lesser likelihood of crashes. The spacing errors from the other two trajectories show similar trends.

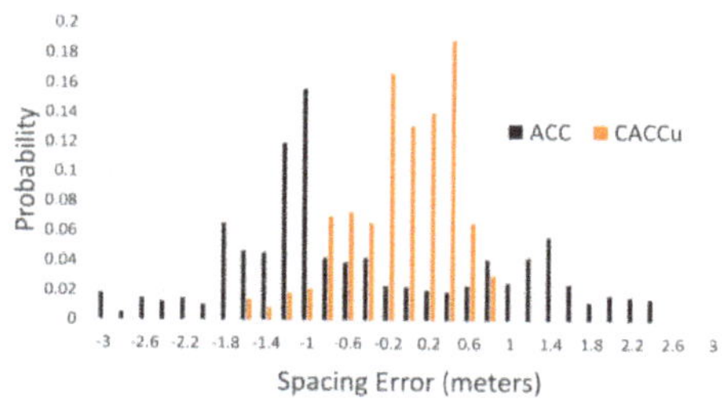

Figure 3. Spacing error histogram of ACC and CACCu.

5. Conclusions and Future Research

In this paper, the safety impacts of cooperative platooning in mixed traffic were assessed using spacing errors by comparing the performance of CACCu with that of ACC. The maximum spacing error was 1.6 m for CACCu, while results of up to 3 m were observed for ACC. The simulation results from this paper support the notion that CACCu, or cooperative platooning in mixed traffic, offers much safer control than the ACC. The results were consistent for three randomly selected real-world trajectories from the NGSIM.

Future research should consider the performance of field operation tests. It is recommended that additional simulations be performed using a vehicle-dynamics-based model, such as Carla or Carla-ROS. The latter would allow for the extension of a control algorithm to connected automated vehicles utilizing ROS.

Author Contributions: Conceptualization, B.B.P. and I.Y.; methodology and analysis, B.B.P.; writing—review and editing, B.B.P., I.Y., H.L. and J.P.; funding acquisition, B.B.P. and I.Y. All authors have read and agreed to the published version of the manuscript.

Funding: This research was supported by the National Science Foundation under Grant No. CMMI-2009342 and the Korea Agency for Infrastructure Technology Advancement (KAIA) grant funded by the Ministry of Land, Infrastructure, and Transport (grant No. 22AMDP-C161752-02).

Institutional Review Board Statement: Not applicable.

Informed Consent Statement: Not applicable.

Data Availability Statement: The data used in this paper are from the NGSIM program.

Conflicts of Interest: The authors declare no conflict of interest.

References

1. Milanés, V.; Shladover, S.E.; Spring, J. Cooperative Adaptive Cruise Control in Real Traffic Situations. *IEEE Trans. Intell. Transp. Syst.* **2014**, *15*, 296–305. [CrossRef]
2. Liu, H.; Kan, X.; Shladover, S.E.; Lu, X.Y.; Ferlis, R.E. Modeling impacts of Cooperative Adaptive Cruise Control on mixed traffic flow in multi-lane freeway facilities. *Transp. Res. Part C Emerg. Technol.* **2018**, *95*, 261–279. [CrossRef]
3. Lioris, J.; Pedarsani, R.; Tascikaraoglu, F.Y.; Varaiya, P. Platoons of connected vehicles can double throughput in urban roads. *Transp. Res. Part C Emerg. Technol.* **2017**, *77*, 292–305. [CrossRef]
4. Talebpour, A.; Mahmassani, H.S. Influence of connected and autonomous vehicles on traffic flow stability and throughput. *Transp. Res. Part C Emerg. Technol.* **2016**, *71*, 143–163. [CrossRef]
5. Ploeg, J.; Semsar-Kazerooni, E.; Lijster, G.; van de Wouw, N.; Nijmeijer, H. Graceful degradation of cooperative adaptive cruise control. *IEEE Trans. Intell. Transp. Syst.* **2015**, *16*, 488–497. [CrossRef]
6. Orosz, G. Connected cruise control: Modelling, delay effects, and nonlinear behaviour. *Veh. Syst. Dyn.* **2016**, *54*, 1147–1176. [CrossRef]
7. Ge, J.I.; Orosz, G. Data-driven parameter estimation for optimal connected cruise control. In Proceedings of the 2017 IEEE 56th Annual Conference on Decision and Control, CDC 2017, Melbourne, VIC, Australia, 12–15 December 2017; Volume 2018, pp. 3739–3744.
8. Ge, J.I.; Orosz, G. Connected cruise control among human-driven vehicles: Experiment-based parameter estimation and optimal control design. *Transp. Res. Part C Emerg. Technol.* **2018**, *95*, 445–459. [CrossRef]
9. Chen, Z.; Park, B.B. Cooperative Adaptive Cruise Control with Unconnected Vehicle in the Loop. *IEEE Trans. Intell. Transp. Syst.* **2020**, *23*, 4176–4186. [CrossRef]
10. Lee, D.; Lee, S.; Chen, Z.; Park, B.B.; Shim, D.H. Design and field evaluation of cooperative adaptive cruise control with unconnected vehicle in the loop. *Transp. Res. Part C Emerg. Technol.* **2021**, *132*, 103364. [CrossRef]
11. Cui, L.; Chen, Z.; Wang, A.; Hu, J.; Park, B.B. Development of a Robust Cooperative Adaptive Cruise Control with Dynamic Topology. *IEEE Trans. Intell. Transp. Syst.* **2021**, *23*, 4279–4290. [CrossRef]
12. So, J.; Dedes, G.; Park, B.B.; Alamdary, S.H.; Grejner-Brzezinsk, D. Development and evaluation of an enhanced surrogate safety assessment framework. *Transp. Res. Part C Emerg. Technol.* **2015**, *50*, 51–67. [CrossRef]
13. Park, B.; Chen, Y.; Hourdos, J. Opportunities for preventing rear-end crashes: Findings from the analysis of actual freeway crash data. *J. Transp. Saf. Secur.* **2011**, *3*, 95–107. [CrossRef]
14. Alexiadis, V.; Colyar, J.; Halkias, J.; Hranac, R.; McHale, G. The next generation simulation program. *ITE J. (Inst. Transp. Eng.)* **2004**, *74*, 22–26.

engineering proceedings

MDPI

Proceeding Paper

Unveiling the Benefits of Engineered Crumb Rubber for Asphalt Mixtures via Performance-Related Characterization: Rutting Behavior †

Usman Ghani [1], Silvia Milazzo [1], Gaspare Giancontieri [1], Chiara Mignini [1], Gabriella Buttitta [1], Fan Gu [2] and Davide Lo Presti [1,*]

1 Department of Engineering Ed.8, University of Palermo, Viale Delle Scienze, 90128 Palermo, Italy
2 School of Traffic & Transportation Engineering, Changsha University of Science and Technology, Changsha 410114, China
* Correspondence: davide.lopresti@unipa.it
† Presented at the Second International Conference on Maintenance and Rehabilitation of Constructed Infrastructure Facilities, Honolulu, HI, USA, 16–19 August 2023.

Abstract: Even though alternative paving materials, like rubberized asphalt, are sometimes present in specifications, these are still not widely adopted from road agencies mainly due to a lack of experience, reticence in changing work habits and, often, a lack of evidence of real gains in the change. Authors believe that performance-based laboratory characterization is a solution to highlight differences with conventional asphalt mixtures. Hence, this research wants to highlight the differences between designing asphalt mixtures modified with engineered crumb rubber (ECR) on the basis of conventional indirect tensile testing (ITS), as prescribed by Italian specifications, and by means of performance-related characterization. ECR allows to asphalt mixtures to be modified through a dry process without inconveniences such as uncontrolled swelling and the generation of fumes; on the other hand, performance-related characterization focuses on highlighting rutting behavior by using a basic approach, still based on ITS, and a more advanced viscoplastic methodology using the asphalt mixture performance tester (AMPT). As a result, performance-related characterization is necessary to highlight clear gains in the rutting behavior of the asphalt mixtures modified with ECR. Advanced methodology by means of AMPT provides a fine-tuned characterization; however, the basic approach by means of ITS already highlights the differences in performance. ECR could be widely used to improve the properties of dense mixtures for roads with low traffic; in fact, it solves many of the practical issues of adding crumb rubber through a dry process and greatly improves paving material properties compared to conventional asphalt mixtures, with an increase in cost of only 10%.

Keywords: rutting; permanent deformation; performance-related properties; tire rubber; modified asphalt mixtures; AMPT; ITS

Citation: Ghani, U.; Milazzo, S.; Giancontieri, G.; Mignini, C.; Buttitta, G.; Gu, F.; Lo Presti, D. Unveiling the Benefits of Engineered Crumb Rubber for Asphalt Mixtures via Performance-Related Characterization: Rutting Behavior. *Eng. Proc.* **2023**, *36*, 39. https://doi.org/10.3390/engproc2023036039

Academic Editor: Hosin (David) Lee

Published: 12 July 2023

1. Introduction

In recent years, the pavement industries have shown great interest in improving pavement construction practices, with the aim of increasing the service life and minimizing the efforts required for pavement maintenance and rehabilitation. Several research studies have investigated various ways of predicting the basic failure mechanisms for flexible pavements, such as rutting. [1]. In order to improve the rutting behavior, nowadays, technology developers are currently incorporating synthetic and/or recycled polymers into the conventional asphalt mixture since it has been proven to improve resistance to damage phenomena, including rutting. Despite having this specific advantage, together with other positive features, rubberized asphalt is still not fully implemented worldwide, including networks managed by Italian road authorities. Nevertheless, it has to be underlined that since 2021, Italian specification (ANAS, 2021) has included the use of rubber powder within their specifications; however, the required mechanical characterization of the mixes is still related to conventional parameters such as the indirect tensile strength test (ITS) at 25 °C [2]. It does not include any characterization related to assessing resistance to permanent deformation.

This study proposes a new approach for the performance-related characterization of asphalt mixtures complying with Italian specifications, which is key to discriminating the rutting behavior of a conventional asphalt mixture and a relative modified mixture with engineered crumb rubber (ECR). Along these lines, the experimental program used a multi-level approach, based on basic level testing, i.e., the high-temperature indirect tensile strength test (HT-ITS) at 54 °C and performance-based viscoplastic characterization based on advanced-level testing using an asphalt mixture performance tester (AMPT) was intended to be used [3]. For performance-based permanent deformation, a Stress Sweep Rutting (SSR) test was conducted. Based on the SSR test, the Rutting Strain Index (RSI) was calculated for climatic conditions in the Sicily region, Italy.

2. Methodology

2.1. Conventional Characterization

Conventional characterization included the indirect tensile strength (ITS). The ITS test was performed according to UNI EN 12697-23 at 25 °C on three replicates for each mixture investigated. Therefore, specimens did not need further temperature conditioning after curing. The indirect tensile strength is the maximum tensile stress calculated from the peak load applied at break specimen and the dimensions of the specimen, using Equation (1):

$$ITS = 2P/\pi DH \tag{1}$$

where P is maximum load, H is the height of the specimen and D is the specimen diameter.

2.2. Performance-Related Characterization—Basic Level

Like conventional testing at 25 °C, the basic level of the performance-related test was carried out through a high-temperature ITS test. The specimens were conditioned at the test temperature, 10 °C below the average as the 7-day maximum pavement temperature, 20 mm below the pavement surface at 50% reliability. The test was performed at 54 °C [4].

2.3. Performance-Related Characterization—Advanced Level

The performance-related characterization at the advanced level considered the study of rutting resistance using an Asphalt Mixture Performance Tester (AMPT) in accordance with AASHTO TP 79-13. The resistance to permanent deformation was evaluated with a Stress Sweep Rutting test (AASHTO TP 134-19), in which three cylindrical specimens were tested for each mixture.

Stress Sweep Rutting (SSR) tests at two temperatures (high TH: 54 °C; and low TL: 20 °C) were considered, and for each temperature, a vertical load was applied for 600 cycles at three deviatoric stress levels (200 loading blocks for each segment). Confining stress of 69 KPa was applied with a loading pulse of 0.4 s, followed by a rest time of 3.6 s for TH and 1.6 s for TL. The SSR test allowed us to determine the shift model for permanent strain using FlexMATTM rutting [4].

3. Results and Discussions

3.1. Conventional Characterization

Figure 1 represents the results obtained from the ITS test for the investigated mixtures. It can be observed that the examined mixtures have a similar ITS value with a percentage difference of 7%. Both mixtures fully meet the Italian road authority requirements [3].

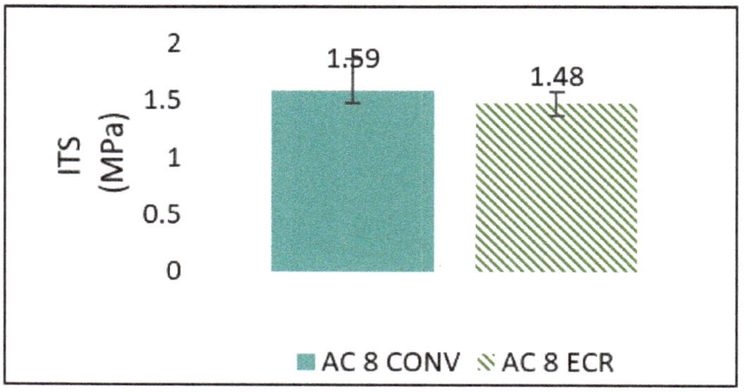

Figure 1. Indirect tensile strength test results.

3.2. Performance-Related Characterization—Basic Level

Figure 2 indicates the results of HT-ITS, and it can be observed that mix AC 8 ECR had the highest strength values with an average of 0.42 MPa, i.e., 420 KPa. Mix AC 8 was in the "Poor" rutting resistance tier, as shown in Figure 2, while AC 8 ECR was in the "Excellent" category. This difference between the two mixtures, in terms of percentage, is equal to 172%.

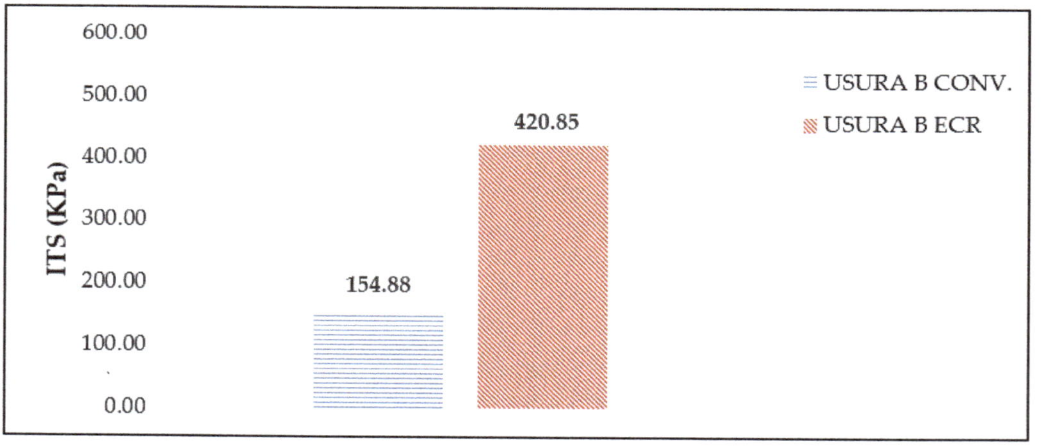

Figure 2. High-temperature indirect tensile strength test results with HT-ITS requirements.

3.3. Performance-Related Characterization—Advanced Level

The average SSR test results for 20 °C and 54 °C are illustrated in Figure 3a,b. The trend was permanent deformation accumulation with an increasing number of cycles [4]. At the high temperature of 54 °C, the AC 8 ECR mixture displayed a minimum achievement of 19,434 $\mu\varepsilon$ in the permanent strain. This confirms that the modified asphalt mixtures were less susceptible to rutting, with an average percentage difference of 50% as compared to the conventional AC 8 mix.

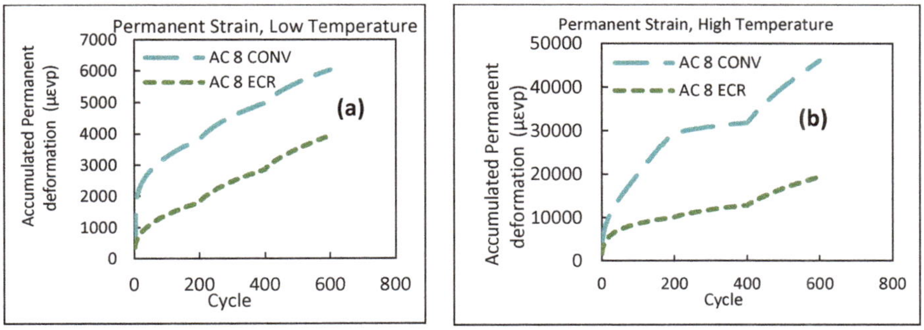

Figure 3. SSR test results (**a**) at 20 °C and (**b**) at 54 °C.

In addition, the micro strain obtained at 600 cycles for AC 8 was 43.185 µε, and for AC 8 ECR, it was 25.387 µε, showing better resistance as compared to conventional mixtures. In order to have a clearer understanding of permanent strain developed under the climate conditions of the Sicily region in Italy, the advanced level was extended to calculate the RSI.

The obtained RSI values for AC 8 and AC 8 ECR were compared with the standard specifications [4]. For conventional mixtures, the RSI value obtained was 71.52%, which is not recommended for any surface course, and the AC 8 ECR mixture value was approximately around 3.89%, which is recommended for heavy traffic on a surface course.

4. Summary and Conclusions

This research evaluated the prediction of permanent deformation using conventional and ECR mixtures. It was also intended to compare the methodology of basic and advanced-level testing performed on permanent deformation.

In summary, the research suggested that modifying conventional asphalt mixtures for wearing courses by adding ECR led to an improvement in terms of permanent deformation. Performance-related characterization is necessary to highlight differences due to modifications. Although advanced-level characterization for rutting resistance by means of AMPT is probably accurate and allows for further pavement design exercises, the basic approach by means of ITS already highlights the differences in rutting behavior.

As final recommendations for practitioners, the authors believe that ECR could be widely used to improve the properties of dense mixtures for urban and secondary roads; in fact, it solves many of the practical issues of adding crumb rubber through a dry process and greatly improves paving material properties compared to conventional asphalt mixtures, with an increase in cost of only 10%.

Author Contributions: Conceptualization U.G., S.M. and D.L.P.; data curation, U.G. and S.M.; formal analysis, U.G., S.M., D.L.P. and U.G.; funding acquisition, D.L.P.; investigation, S.M., G.G. and C.M.; methodology, S.M., D.L.P., G.G. and U.G.; project administration, F.G., S.M. and U.G.; resources, S.M. and D.L.P.; writing—original draft, U.G., D.L.P. and S.M.; writing—review and editing, U.G., S.M., D.L.P. and G.B. All authors have read and agreed to the published version of the manuscript.

Funding: Ghani is supported by the Italian Ministry MUR, with DM 1061 of 2021/08/10, under the financial resource of PON "Ricerca e Innovazione 2014–2020"Action IV.5 "Doctorates on green themes, while Milazzo is partially supported by SMACOM Srl.

Institutional Review Board Statement: Not applicable.

Informed Consent Statement: Not applicable.

Data Availability Statement: All the necessary data required for supporting this research work are included in this paper.

Acknowledgments: The authors would like to acknowledge Ecopneus scpa for the support provided through the project RUBBERAP and RubLab Srl and SicilBitumi Srl for their interest and support in-kind.

Conflicts of Interest: The authors declare no conflict of interest.

References

1. Meroni, F.; Flintsch, G.W.; Habbouche, J.; Diefenderfer, B.K.; Giustozzi, F. Three-level performance evaluation of high RAP asphalt surface mixes. *Constr. Build. Mater.* **2021**, *309*, 125164. [CrossRef]
2. Anas S.p.A. *Capitolato Speciale D'appalto-Norme Tecniche*; Azienda Nazionale Autonoma delle Strade: Roma, Italy, 2016.
3. Islam, M.R.; Hossain, M.I.; Tarefder, R.A. A study of asphalt aging using Indirect Tensile Strength test. *Constr. Build. Mater.* **2015**, *95*, 218–223. [CrossRef]
4. Kim, D.; Kim, Y.R. Development of Stress Sweep Rutting (SSR) test for permanent deformation characterization of asphalt mixture. *Constr. Build. Mater.* **2017**, *154*, 373–383. [CrossRef]

engineering proceedings

MDPI

Proceeding Paper

Combined Use of GPR and PMS Data for Composite Pavement Assessments †

Tae-Soo Kim, Chul-Ki Jung, Young-Mi Yoon, Byeong-Seok Kwak and Jung-Hun Lee *

Roadkorea Inc. Rm 1801~1805, The First Tower, 10, Dongtan-Daero 21-gil, Hwaseong-si 18471, Republic of Korea; aprilshower0@daum.net (T.-S.K.); ccg1231@gmail.com (C.-K.J.); transaromi@nate.com (Y.-M.Y.); kwak0730@hanmail.net (B.-S.K.)

* Correspondence: hun1347@gmail.com

† Presented at the Second International Conference on Maintenance and Rehabilitation of Constructed Infrastructure Facilities, Honolulu, HI, USA, 16–19 August 2023.

Abstract: The main objective of this study is to simultaneously evaluate the surface and existing layers of composite pavement using PMS data and GPR equipment. The distribution of dielectric constants according to the existing concrete conditions and the relationship between dielectric constants and surface distress was evaluated. As a result, the dielectric constant distribution of the existing concrete showed a significant difference depending on the AAR. In addition, the correlation between surface distress and the dielectric constants of the existing layers was low.

Keywords: GPR; dielectric constant; composite pavement; PMS; AAR

Citation: Kim, T.-S.; Jung, C.-K.; Yoon, Y.-M.; Kwak, B.-S.; Lee, J.-H. Combined Use of GPR and PMS Data for Composite Pavement Assessments. *Eng. Proc.* **2023**, *36*, 40. https://doi.org/10.3390/engproc2023036040

Academic Editor: Hosin (David) Lee

Published: 13 July 2023

1. Introduction

The total length of aged cement concrete pavement has continued to increase in Korea. As a way to extend the life of aged concrete pavement, an asphalt overlay is generally applied. However, if the existing pavement is in poor condition or the overlay design is inadequate, it can lead to various types of damage, such as reflective cracks, potholes, and efflorescence [1,2].

The Pavement Management System (PMS) includes data collection through Automated Pavement Condition Survey (APCS) equipment and analyzing the surface distress, International Roughness Index (IRI), and rut depth. For bridge decks with an asphalt overlay on Korean expressways, condition assessments through dielectric constants have progressed considerably [3,4]. In contrast, composite pavement is typically evaluated based solely on the surface layer's condition, and a standardized approach to maintenance is applied, which involves the use of uniform materials and thicknesses. Therefore, it is necessary to evaluate the surface layer and the existing layer in composite pavement simultaneously by using PMS data and Ground Penetrating Radar (GPR) equipment.

In this study, a case study was conducted on the correlation between surface distress and dielectric constants and the distribution of dielectric constants according to the existing concrete conditions in the pilot composite pavement section.

2. Pavement Assessment Method

2.1. Survey Equipments and Sections

The surface layer and existing pavement were evaluated concurrently using both APCS equipment and an air-coupled GPR system. PMS data (pavement image, IRI, rut depth, etc.) were collected through APCS equipment. The air-coupled GPR system includes four-channel 1 GHz air-coupled antennas with transmitter/receiver electronics (horn antenna type), a GPR control unit, a pulse encoder, and the distance measuring instrument. This GPR system was used to acquire the dielectric constant data of the existing layer.

All survey sections are currently in service, and an asphalt overlay was applied on existing concrete pavement. The Alkali-Aggregate Reaction (AAR) occurred in the existing concrete layer of Sections 1 and 2, but was not observed in Section 3 (as shown in Table 1).

Table 1. Survey sections.

Section No.	Route	Start Point (km)	End Point (km)	Direction	No. of Lanes	Condition of Existing Pavement
1	Gyeongbu Expressway	262.9	263.2	Seoul Bound	2	AAR
2		263.6	264.8		2	AAR
3		312.7	313.0		3	-

2.2. Assessment Procedure

In the assessment procedure, GPR data and PMS data (especially pavement surface images) were acquired and matched first. For analyzing and visualizing GPR data, the commercial software RADAN® [5] and Surfer® [6] were used. Additionally, surface distress was analyzed from pavement images by using image analysis software. After data analysis, the distribution of dielectric constants according to the existing concrete conditions and the relationship between dielectric constants and surface distress was evaluated.

3. Result and Discussion

3.1. GPR Survey

Figure 1 shows the dielectric constants obtained from GPS data analysis for each survey section. Generally, the dielectric constant of concrete ranges from 5 to 10 [7].

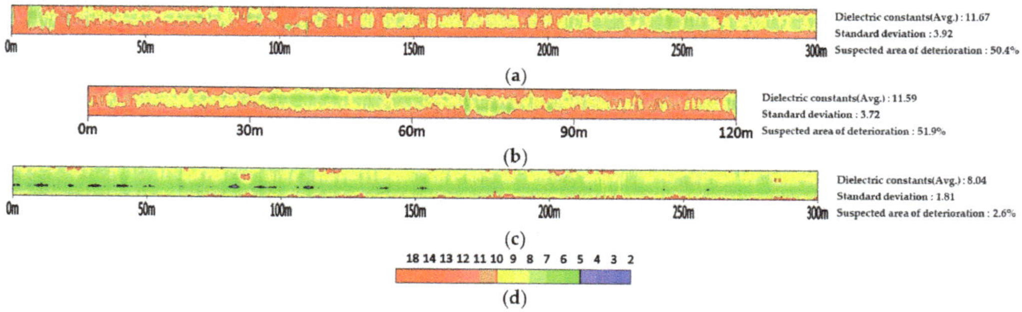

Figure 1. Condition maps of existing concrete obtained from air-coupled GPR measurements: (**a**) Section 1; (**b**) Section 2; (**c**) Section 3; (**d**) dielectric constants scale bar.

The average dielectric constants of existing pavements were 11.67 and 11.59 in Sections 1 and 2, respectively, which were very similar, and in the case of Section 3, it was 8.04, which was lower than those in Sections 1 and 2. The large dielectric constants values in Sections 1 and 2 may be due to the alkali-silicate gel caused by the AAR and the resulting concrete deformation [8]. The suspected deterioration area (with dielectric constants less than 5 and more than 10) in Sections 1 and 2 exceeds 50%, but in Section 3, it was very low at 2.6%.

Figure 2 shows a comparison of the distribution of dielectric constants from the surveyed sections. Sections 1 and 2 (AAR) show similar patterns and are evenly distributed in the direction of high dielectric constants. On the other hand, the dielectric constant distribution in Section 3 is concentrated in the range of about 4 to 13.

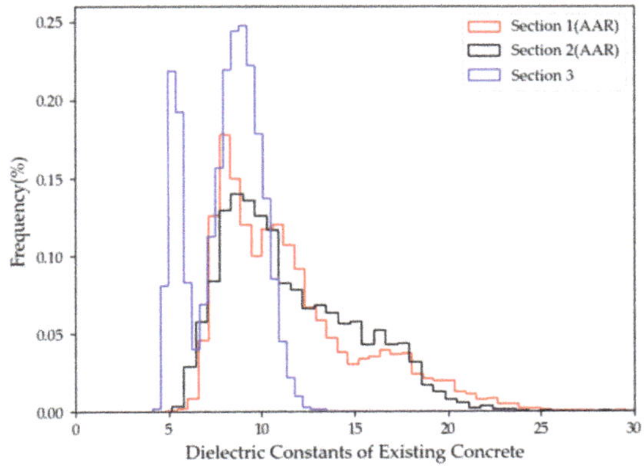

Figure 2. Distribution of dielectric constants of existing concrete pavement.

3.2. *Surface Distress*

Table 2 shows the results of a surface distress analysis for each section, and in the case of efflorescence, the number of observed locations is shown. Figure 3 provides an example of the surface distress analysis for a 10 m unit section. Efflorescence was found in both Sections 1 and 2, where the AAR occurred in the existing layer, but there was a large difference in the total cracks (%). In other words, the amount of distress to the surface layer may occur differently even with the same material, traffic volume, and environmental load. In the surface distress figures, the blue, purple, red line, and brown rectangle represent longitudinal cracks, transverse cracks, patching, and construction joints, respectively, and the values indicate the amount of distress (length or area). For a more accurate evaluation, it is judged that laboratory experiments through the coring are necessary. In the case of Section 3, the pavement condition was very good.

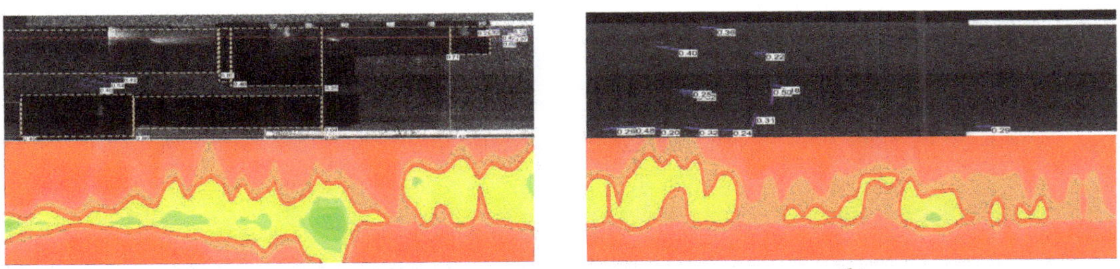

(a) (b)

Figure 3. Comparison of surface distress and existing concrete layer conditions: (**a**) Section 1 (104~114 m); (**b**) Section 2 (100~110 m).

Table 2. The result of a surface distress analysis.

Section No.	Total Area (m^2)	Total Cracks (%)	Major Distress		
			Patching (m^2)	Crack (m)	Efflorescence (Pcs.)
1	1044	14.2%	60.0	562.5	6
2	432	2.1%	4.4	31.6	4
3	1044	0.6%	5.9	3.7	0

3.3. Correlation Surface Distress and Dielectric Constants

Figure 3 shows the result of comparing the surface distress with the condition of the existing layer. In the case of Figure 3a, a wide range of patching and efflorescence occurred in the surface layer, and the dielectric constants of the existing layer were high. However, as shown in Figure 3b, the dielectric constants were high even in the section with little surface distress. In addition, as shown in Figure 4, as a result of comparing the dielectric constant of the existing layer and the total cracks (%) of the surface layer, it was found that the correlation between the two variables was low.

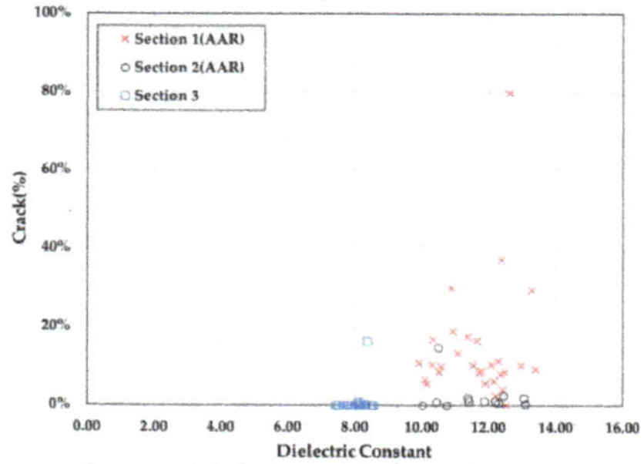

Figure 4. Comparison of the dielectric constant and the total cracks (%).

4. Conclusions

This investigation revealed a significant difference in the dielectric constant distribution of existing concrete depending on the presence of the AAR. Sections 1 and 2 (AAR) had wider dielectric constant distributions than Section 3, indicating possible deterioration in the existing concrete layer. In addition, the correlation between surface distress and the dielectric constants of the existing layer was low, so it was difficult to estimate the condition of the existing layer only by the surface condition. Therefore, it is necessary to introduce an investigation system capable of simultaneously considering the surface and existing layers in order to evaluate the performance of the composite pavement.

Author Contributions: Conceptualization, C.-K.J. and J.-H.L.; methodology, J.-H.L.; validation, Y.-M.Y. and B.-S.K.; formal analysis, Y.-M.Y.; investigation, T.-S.K., Y.-M.Y. and B.-S.K.; data curation, B.-S.K.; writing—original draft preparation, T.-S.K.; writing—review and editing, J.-H.L.; visualization, T.-S.K.; supervision, J.-H.L.; project administration, J.-H.L.; funding acquisition, C.-K.J. All authors have read and agreed to the published version of the manuscript.

Funding: This research received no external funding.

Institutional Review Board Statement: Not applicable.

Informed Consent Statement: Not applicable.

Data Availability Statement: Not applicable.

Conflicts of Interest: The authors declare no conflict of interest.

References

1. Kim, Y.K.; Lee, S.W. Performance evaluation of bonded concrete overlay. *Constr. Build. Mater.* **2013**, *49*, 464–470. [CrossRef]
2. Suh, Y.C.; Lee, E.J. Development of distress progress model for determining the appropriate time for overlaying durability cracks in concrete pavement. *KSCE J. Civ. Eng.* **2022**, *26*, 1773–1781. [CrossRef]
3. Rhee, J.-Y.; Park, K.-E.; Lee, K.-H.; Kee, S.-H. A Practical Approach to Condition Assessment of Asphalt-Covered Concrete Bridge Decks on Korean Expressways by Dielectric Constant Measurements Using Air-Coupled GPR. *Sensors* **2020**, *20*, 2497. [CrossRef] [PubMed]
4. Rhee, J.-Y.; Choi, J.-J.; Kee, S.-H. Evaluation of the depth of deteriorations in concrete bridge decks with asphalt overlays using air-coupled GPR: A case study from a pilot bridge on Korean expressway. *Int. J. Concr. Struct. Mater.* **2019**, *13*, 23. [CrossRef]

5. Geophysical Survey Systems, Inc. (GSSI). *RADAN 7 Manual*; GSSI: Nashua, NH, USA, 2020; Available online: http://www.geophysical.com (accessed on 27 February 2023).
6. Golden Software, LLC. Available online: https://www.goldensoftware.com (accessed on 27 February 2023).
7. *ASTM D6432-19*; Standard Guide for Using the Surface Ground Penetrating Radar Method for Subsurface Investigation. ASTM International: West Conshohocken, PA, USA, 2020.
8. Heifetz, A.; Strow, M.; Liu, Y.; Bevington, P.; Zapol, P.; Bakhtiari, S.; Bentivegna, A. Monitoring of dielectric permittivity in accelerated alkali-silica reaction concrete with microwave backscattering. *Mater. Struct.* **2020**, *53*, 130. [CrossRef]

engineering proceedings

MDPI

Proceeding Paper

Full Load Test for the Sheikh Jaber Al-Ahmad Al-Sabah Causeway Bridge (PSC Girder: 35 M) †

Kwangsoo Kim [1], Dooyong Cho [2,*], Raechul Lee [3], Sangcheol Lee [3], Joungyong Park [4] and Inbaek Hwang [1]

1 Civil Business Division, AI Safety Institute, Seoul 05818, Republic of Korea; whitesky68@gmail.com (K.K.)
2 Department of Convergence System Engineering, Chungnam National University, Daejoen 34134, Republic of Korea
3 Civil Business Division, SQ Engineering, Seoul 05818, Republic of Korea; leeraechul@sqeng.co.kr (R.L.); leesh111@naver.com (S.L.)
4 Civil Business Division, Korea Construction Disaster-Prevention Research, Seoul 05836, Republic of Korea; pjy6810@naver.com
* Correspondence: dooyongcho@cnu.ac.kr; Tel.: +82-10-9437-9238
† Presented at the Second International Conference on Maintenance and Rehabilitation of Constructed Infrastructure Facilities, Honolulu, HI, USA, 16–19 August 2023.

Abstract: The Sheikh Jaber Al-Ahmad Al-Sabah Causeway consists of an asymmetric cable-stayed bridge (340 m) and a prestressed concrete (PSC) box girder bridge (35.80 km) linking Kuwait City and Northern Kuwait. The full load tests were performed on the PSC box girder bridge and the load scale was set to 14,092 kNm which was 98.47% of the maximum design moment (14,310 kNm). A total of 12 individual 40 tonf dump trucks were exerted on the bridge for the tests. Based on the influence line of the target bridge, displacement sensors, and strain gauges were installed at the points where the maximum bending moment would occur. The collected deflection and strain data were compared with the finite element method analysis to analyze the change in stiffness of the bridge. From the analysis, it was found that higher stiffness behavior was identified compared to the design load.

Keywords: full load test; stress analysis; finite element method analysis; stiffness; deflection; strain

Citation: Kim, K.; Cho, D.; Lee, R.; Lee, S.; Park, J.; Hwang, I. Full Load Test for the Sheikh Jaber Al-Ahmad Al-Sabah Causeway Bridge (PSC Girder: 35 M). *Eng. Proc.* **2023**, *36*, 41. https://doi.org/10.3390/engproc2023036041

Academic Editor: Hosin (David) Lee

Published: 14 July 2023

1. Introduction

The target structure of this paper is a marine bridge with a total length of 36.14 km that connects Kuwait City to the new city of Subiyah, as shown in Figure 1.

Figure 1. Target facility location map.

The span composition of this marine bridge was constructed with an asymmetric cable-stayed bridge (340 m), which is the main bridge, and a PSC box girder bridge (35.80 km), and the initial value data, which are the basic foundation for the maintenance entity to carry out efficient maintenance work, were obtained.

Although the initial value data vary somewhat depending on the requirements of the ordering organization, which is the managing entity for the relevant bridge, the results of deflection and strain tests are basically obtained, and to that end, a full load test was performed.

Based on the acquired data, it is possible to quantitatively evaluate changes in the stiffness and sectional force of the bridge in use as of the time of completion, and it is expected that efficient maintenance work can be planned and executed based on the results [1,2].

2. Measuring Instrument Installation

In this load test, a total of three types of measuring instruments were used, which are LVDT, strain gauge, and total station. The linear variable displacement transducer (LVDT) was installed at the position of the elastomeric bearing at the supporting point to measure the amount of deflection of the bearing, and the strain gauges were installed inside the box girder to obtain the strain value as shown in Figure 2. In addition, considering that the space under the bridge is the sea, the amount of deflection was measured using the total station at the top of the bridge.

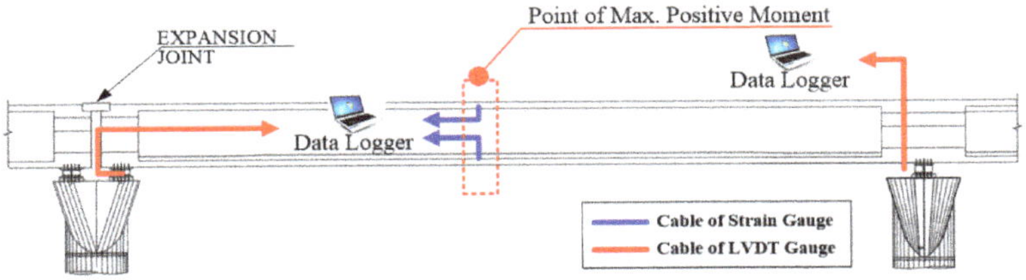

Figure 2. Diagram of the measurement system for data acquisition.

During the measurement, the data loggers were placed along the shortest moving line to minimize noise, and two data loggers were used to that end.

Measuring Instrument Installation Location

The location of the prism installed on the top of the bridge and the location of the strain gauge installed inside the box are shown in Figure 3, and the total station was located outside the expansion joint of the relevant span to eliminate the influence of loaded vehicles.

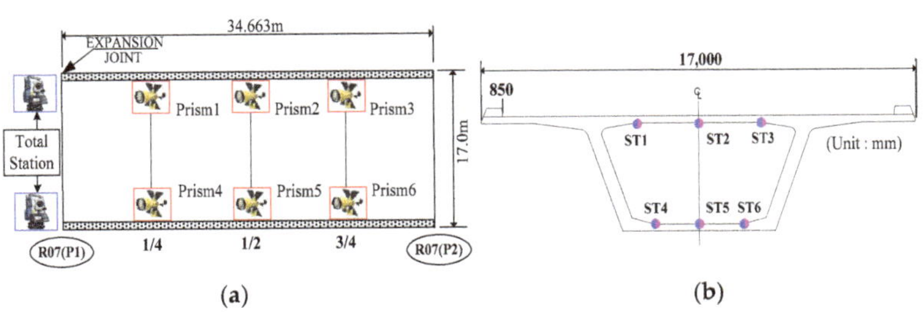

Figure 3. Diagram of the location where the measurement instrument was installed. (**a**) Locations of the total station and prism; (**b**) Location where the strain gauge was installed.

3. Load Test

3.1. Loading Vehicle Location Map

The target load for the load test was set as the design moment (14,310 kNm), and to that end, the number of trucks and the size of the load were determined through finite element analysis in advance. The sectional moment generated by the truckload used in this case was set as 14,092 kNm, which is 98.47% of the design load. Twelve 40 tonf dump trucks were loaded in the longitudinal and transverse directions as shown in Figure 4.

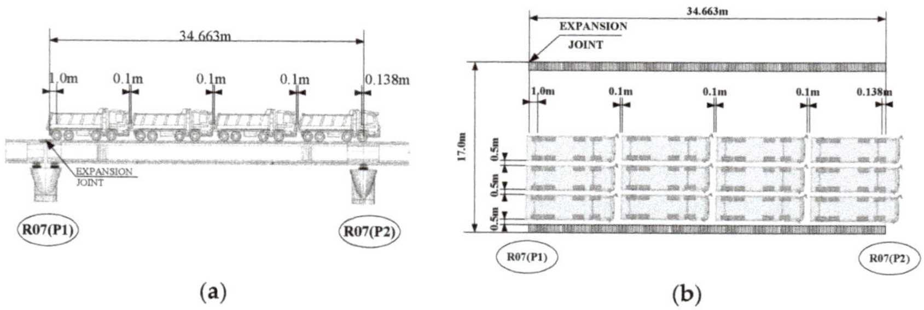

Figure 4. Diagram of longitudinal and transverse positions of the loaded vehicles. (**a**) Longitudinal positions of the loaded vehicles; (**b**) Transverse positions of the loaded vehicles.

3.2. Finite Element Analysis

It was determined to set the target load as close to the design moment as possible, and to that end, the load positions, and the load, which was 98.47% of the design moment, were determined through repetitive finite element analyses [3,4].

The deflection and strain values calculated as a result of the finite element analyses were compared and analyzed with the measured resultant values to verify the adequacy of the behavior of the theoretical structure and to quantitatively calculate the variation in the stiffness of the actual structure compared to the design, as shown as Figure 5.

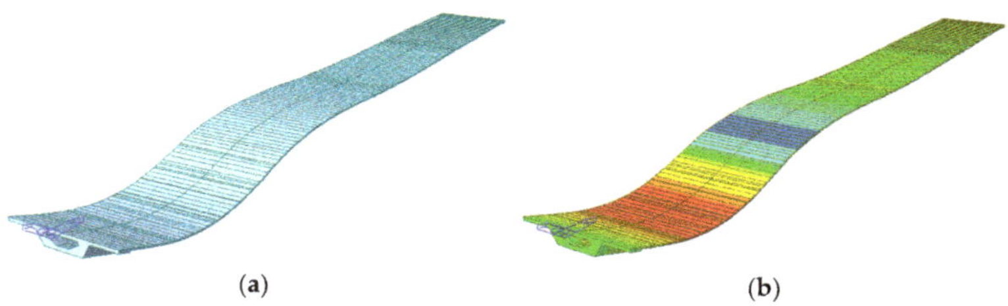

Figure 5. Results of finite element analyses. (**a**) Diagram of defection by the loaded vehicles; (**b**) Diagram of stress by the loaded vehicle.

3.3. PSC Results of Test of the Resilience of the PSC Girder

After the load test, additional amounts of defection were checked while maintaining the loaded vehicles for 24 h, but there was no change. In addition, it was identified that when all 12 loaded vehicles were removed, all the deflection was restored.

3.4. Results of Measurement of the Amount of Defection and Strain

The measured values of the amount of deflection and strain were compared with the values obtained analytically, and as a result of the comparison at the maximum point of the positive moment, a maximum response ratio of 1.12 was derived in the deflection, and a result exceeding the predetermined reference value was obtained in strain, as shown in Figure 6 and Table 1.

Table 1. Results of analysis and measurement of stress.

Classification		Analysis (MPa)	Measurement (MPa)	Limit Stress (MPa)	
				Compression	Tensile
PSC Girder (35 m)	Top	−0.94	−1.01	−22.50	1.767
	Bottom	1.37	1.13	−24.88	

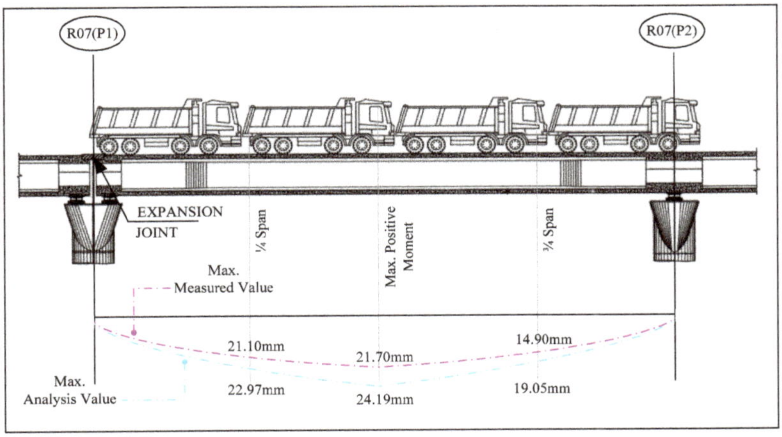

Figure 6. Results of analysis and measurement of defection.

4. Conclusions

This study was conducted to provide basic data for planning and execution of maintenance work in use hereafter by performing load tests at the time of completion of the bridge and analyzing the measurement results. By analyzing the results of the test performed in a limited range, it was confirmed that the stress range and mechanical behavior characteristics were similar to the results of the analysis.

Author Contributions: Conceptualization, K.K. and R.L.; methodology, K.K.; software, J.P.; validation, J.P., I.H. and S.L.; formal analysis, K.K.; investigation, S.L.; resources, R.L.; data curation, I.H.; writing—original draft preparation, K.K.; writing—review and editing, D.C.; visualization, D.C.; supervision, D.C.; project administration, S.L.; funding acquisition, R.L. All authors have read and agreed to the published version of the manuscript.

Funding: This research received no external funding.

Institutional Review Board Statement: Not Applicable.

Informed Consent Statement: Not Applicable.

Data Availability Statement: Not Applicable.

Conflicts of Interest: The authors declare no conflict of interest.

References

1. Bathe, K.-J. Formulation of the Finite Element Method. In *Finite Element Procedures*; Prentice Hall: Englewood Cliffs, NJ, USA, 1995; pp. 153–186.
2. Hambly, E.C. *Box Girder Deck. Bridge Deck Behaviour*, 2nd ed.; E & FN SPON: London, UK, 1991; pp. 135–156.
3. Logan, D.L. Development of Beam Equations. In *A First Course in the Finite Element Method*; PWS Publishing Company: Boston, MA, USA, 1993; pp. 151–187.
4. *Precast Prestressed Concrete Bridge Design Manual*; PCI: Chicago, MI, USA, 1997; Volume 1, pp. 354–387.

engineering proceedings

Proceeding Paper

Rheological and Aging Characteristics of Polymer-Modified Asphalt with Addition of Sulfur †

Ana Luiza Rodrigues *, Caio Falcão and R. Christopher Williams

Department of Civil, Construction, and Environmental Engineering, Iowa State University, Ames, IA 50011, USA; falcao@iastate.edu (C.F.); rwilliam@iastate.edu (R.C.W.)
* Correspondence: analuiza@iastate.edu
† Presented at the Second International Conference on Maintenance and Rehabilitation of Constructed Infrastructure Facilities, Honolulu, HI, USA, 16–19 August 2023.

Abstract: The polymer modification of asphalt binders was first introduced in Europe in the 1980s and has gained use, as lower-quality asphalt binders did not perform well under increasingly heavier traffic loading on pavements. The influence of chemical cross-linkers such as sulfur on the rheological, morphological, and aging characteristics of the polymer-modified asphalt (PMA) binder has been experimentally examined. The PMAs were prepared, blending different sulfur contents (0.03, 0.1, 0.3, and 0.5% by wt. of binder) with a neat binder. The samples were aged by a rolling thin film oven (RTFO) and a pressure aging vessel (PAV) and examined through rheologic investigations. Using models, including master curves, the Glover–Rowe parameter, and aging indexes, the effect of the aging resistance of the asphalt binder modified with sulfur was analyzed. The results indicate that adding sulfur up to 0.3% improved the performance grade range, elasticity, low-temperature cracking resistance, and rutting resistance of the PMA. Additionally, the introduction of sulfur improved the aging resistance of the PMA.

Keywords: asphalt binder; sulfur; polymer-modified asphalt; rheology; aging; performance

Citation: Rodrigues, A.L.; Falcão, C.; Williams, R.C. Rheological and Aging Characteristics of Polymer-Modified Asphalt with Addition of Sulfur. *Eng. Proc.* **2023**, *36*, 42. https://doi.org/10.3390/engproc2023036042

Academic Editor: Hosin (David) Lee

Published: 14 July 2023

1. Introduction

The polymer modification of asphalt binders improved their rheological and viscoelastic properties, allowing the pavement to achieve the performance needed for the current traffic demands [1,2]. The copolymer elastomer styrene–butadiene (SB) block is the most common polymer asphalt modifier. The styrene improves the strength of the material, and the butadiene component contributes to the elasticity of the material [3].

To reduce the phase separation between polymer and asphalt binder, crosslinking agents are used to improve their compatibilization [1,3]. Elemental sulfur is the most widely used crosslinking agent, which promotes the vulcanization of SB by chemically crosslinking the elastomer through the unsaturated bond of butadiene and chemically connecting polymer and asphalt molecules via sulfide and/or polysulfide bonds [4]. Most recent studies have placed more attention on the effect of sulfur on PMA's stability, and only a few [5,6] have addressed the improvement in the aging susceptibility and improved rheological properties of the binder.

During the mixing and service life, the binder is subjected to aging. Due to thermo-oxidation, the polymer and polymer network degradation can occur, displaying chain scission, causing the embrittlement of the asphalt [4,7,8]. Cuciniello [9] analyzed the anti-oxidative effect of sulfur in PMA and stated that crosslinked polymer-modified binders had a lower oxidative susceptibility than non-cross-linked binders.

An improvement in the aging resistance of the binder can increase pavement durability and decrease the maintenance cost over its life cycle. Thus, this paper aims to investigate the effects of adding elemental sulfur, an inexpensive, abundant, and available cross-linking agent, to PMA and its concentration on the rheological properties of the asphalt binder that affect pavement performance and durability. Additionally, the use of sulfur as an antioxidant in PMA and its effects on the aging susceptibility were evaluated through the Glover–Rowe parameter and dynamic modulus and phase angle master curves.

2. Materials and Methods

2.1. Materials

The neat asphalt binder used in this study was classified as PG 64-22S, according to Superpave performance grading (PG). For the polymer modification of the asphalt, a thermoplastic elastomer was used: a linear diblock copolymer composed of blocks of styrene and butadiene since it is one of the most effective ways of improving the binder performance [10]. The sulfur used as a crosslinking agent was a 100-mesh powder with 99.5% purity.

For this study, the properties of a control sample, a binder modified with sulfur, and five PMAs with different sulfur contents were compared. The first step consisted of slowly incorporating 3% of SB by weight into the neat, heated asphalt at 180 °C using a high-shear mixer at 4000 rpm and subsequently stirring for 60 min at 1500 rpm. Then, the PMA was mixed with a calculated ratio (0.075, 0.03, 0.1, 0.3, 0.5 wt.%) of sulfur using a mechanical agitator applying lower shear at 160 °C and 800 rpm for 4 h.

2.2. Methods

The asphalt binders were compared by their Superpave performance grades and rheological properties. The binder grading was performed following the AASHTO M 320 specification, which has a set of characterization tests that determines the temperature range in which the binder is suitable to be used in pavement construction. The Multiple Stress Creep Recovery (MSCR) measures the percent recovery and non-recoverable creep compliance of the binder, can be an indicator of rutting sensitivity, and was tested in accordance with AASHTO TP 70. The rheological master curve represents the asphalt binder characteristic in a viscoelastic region. It is a model based on a time–temperature principle that predicts the performance of the material over a range of temperatures and loading times or frequencies. In this work, dynamic shear modulus and phase angle master curves were used to describe the influence of the sulfur on the behavior of the binder.

Using a Dynamic Shear Rheometer (DSR), the magnitude of the dynamic shear modulus $|G^*|$ and the phase angle δ were measured at 60 °C at low strain before and after aging. The aging index was the ratio of both values, and lower values represented lower aging susceptibility. The Glover–Rowe (G–R) parameter evaluated the cracking performance of asphalt binders and could be used as an aging parameter by evaluating the durability of the binder regarding non-load cracking.

3. Results

As given in Table 1, the modification of the neat binder with polymer improved the high-temperature PG.

Table 1. Binder PG and MSCR results.

Sample	PG	Sample	PG
C	64-22	C + SB + 0.075 S	76-22
C + SB	72-22	C + SB + 0.1 S	76-22
C + 0.075 S	68-22	C + SB + 0.3 S	76-22
C + SB + 0.03 S	76-22	C + SB + 0.5 S	76-18

Figure 1 shows that up to 0.3% of sulfur blended increased the elasticity of the binder, improving recovery and decreasing Jnr, a parameter that was related to rutting susceptibility. The PMA with the addition of sulfur achieved the minimum % Recovery expected to its respective Jnr, according to the AASHTO TP 70 criteria.

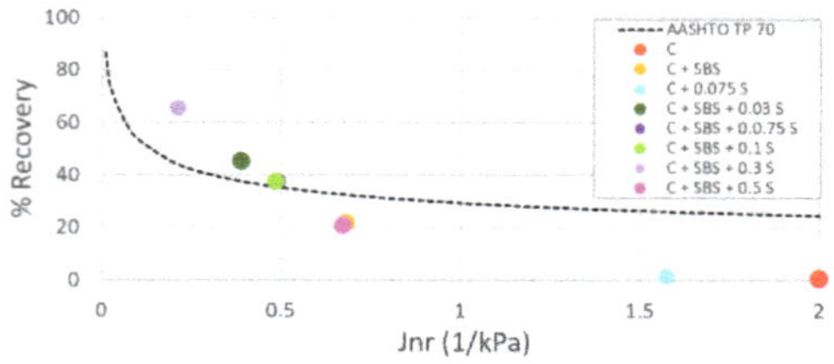

Figure 1. Elastomeric behavior of PMA.

Based on the Williams–Landel–Ferry (WLF) model and using 15 °C as a reference temperature, the master curves for complex shear modulus (G*) and phase angle (δ) were constructed and are shown in Figure 2. Lower frequencies represent slow traffic speeds, a condition that the binder is more prone to rutting, and higher frequencies correspond to shorter loading times. The addition of sulfur to PMA slightly decreased G* at lower frequencies, G* increased at higher frequencies, and the binder became marginally more viscous. Despite the outcome, the PMA with sulfur performed better than the neat binder.

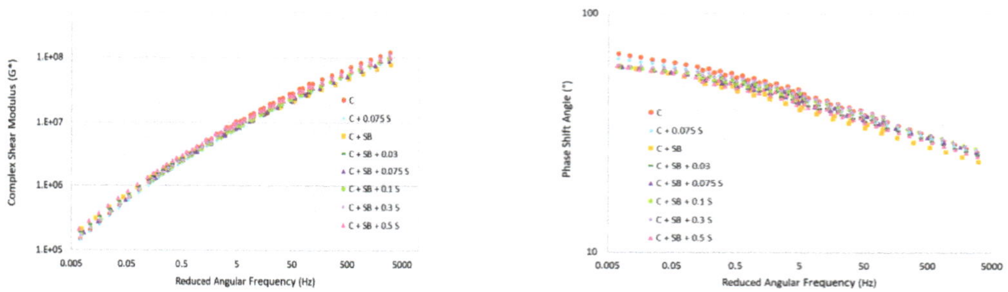

(a) (b)

Figure 2. Master curves (**a**) Complex Shear Modulus; (**b**) Phase Shift Angle.

The effect of sulfur as an antioxidant is suggested through the Glover–Rowe parameter and aging indexes, shown in Figure 3.

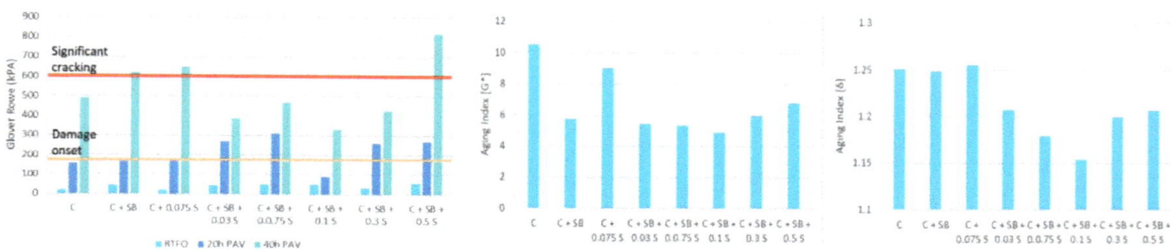

(a) (b)

Figure 3. Aging of PMA (**a**) Glover-Rowe parameter; (**b**) Aging Indexes (G* and δ).

Figure 3a shows that PMA increased susceptibility to cracking after long-term aging compared to the neat binder, and the addition of sulfur up to 0.3% could attenuate those effects. Figure 3b evaluates aging through the changes in G* and δ and shows that the aging susceptibility decreases when sulfur is added to the PMA blend, with the best formulation being with 0.1% of sulfur, complying with the Glover–Rowe parameter analyses.

4. Conclusions

This research investigated the effect of the addition of sulfur on PMA's rheological properties and aging mechanism. The following conclusions can be drawn.

- The content of sulfur has a significant impact on the extent of the changes in PMA, in which it has a positive effect of up to 0.3% by the total weight of the binder.
- Up to 0.3% sulfur improved the performance grade, elastomeric behavior, and low-temperature cracking resistance of PMA.
- The addition of sulfur decreased the aging susceptibility of PMA.

Therefore, the decrease in cracking and aging susceptibility can potentially increase the serviceability of the pavement, improve the ride quality, and require less maintenance over the pavement's life cycle.

Author Contributions: Conceptualization, A.L.R. and R.C.W.; methodology, A.L.R., C.F. and R.C.W.; validation, A.L.R., C.F. and R.C.W.; formal analysis, A.L.R.; investigation, A.L.R. and C.F.; resources, A.L.R.; data curation, A.L.R.; writing—original draft preparation, A.L.R.; writing—review and editing, A.L.R., C.F. and R.C.W.; visualization, A.R; supervision, R.C.W.; project administration, A.L.R. All authors have read and agreed to the published version of the manuscript.

Funding: This research received no external funding.

Institutional Review Board Statement: Not applicable.

Informed Consent Statement: Not applicable.

Data Availability Statement: No new data were created or analyzed in this study. Data sharing is not applicable to this article.

Conflicts of Interest: The authors declare no conflict of interest.

References

1. Sun, D.; Lu, W. Investigation and Improvement of Storage Stability of SBS Modified Asphalt. *Pet. Sci. Technol.* **2003**, *21*, 901–910. [CrossRef]
2. Imanbayev, Y.; Akkenzheyeva, A.; Bussurmanova, A.; Serikbayeva, A.; Boranbayeva, A. Preparation of Polymer Bitumen Binder in the Presence of a Stabilizer. *Processes* **2021**, *9*, 182. [CrossRef]
3. Sun, D.; Ye, F.; Shi, F.; Lu, W. Storage Stability of SBS-Modified Road Asphalt: Preparation, Morphology, and Rheological Properties. *Pet. Sci. Technol.* **2006**, *24*, 1067–1077. [CrossRef]
4. Airey, G.D. Rheological properties of styrene butadiene styrene polymer modified road bitumens☆. *Fuel* **2003**, *82*, 1709–1719. [CrossRef]
5. Oae, S.; Kim, Y.H.; Fukushima, D.; Takata, T. Oxidation of some biologically active and related sulfur-containing compounds. *Pure Appl. Chem.* **1977**, *49*, 153–162. [CrossRef]
6. Tang, N.; Huang, W.; Hao, G. Effect of aging on morphology, rheology, and chemical properties of highly polymer modified asphalt binders. *Constr. Build. Mater.* **2021**, *281*, 122595. [CrossRef]
7. Polacco, G.; Filippi, S.; Merusi, F.; Stastna, G. A review of the fundamentals of polymer-modified asphalts: Asphalt/polymer interactions and principles of compatibility. *Adv. Colloid Interface Sci.* **2015**, *224*, 72–112. [CrossRef] [PubMed]
8. Zaghdoudi, M.; Kömmling, A.; Jaunich, M.; Wolff, D. Scission, Cross-Linking, and Physical Relaxation during Thermal Degradation of Elastomers. *Polymers* **2019**, *11*, 1280. [CrossRef] [PubMed]
9. Cuciniello, G.; Leandri, P.; Presti, D.L.; Losa, M.; Airey, G. Investigating the Effect of Artificial Ageing on the Creep and Recovery of SBS-Modified Bitumen. *MATEC Web Conf.* **2019**, *271*, 03009. [CrossRef]
10. Aurilio, M.; Mikhailenko, P.; Baaj, H.; Poulikakos, L. Properties of Asphalt Binders with Increasing SBS Polymer Modification. In Proceedings of the 5th International Symposium on Asphalt Pavements & Environments (APE), International Society for Asphalt Pavements (ISAP), Padua, Italy, 11–13 September 2019; Springer: Cham, Switzerland, 2019; pp. 55–66.

engineering proceedings

MDPI

Proceeding Paper

Materials and Methods Used for the Expedient Repair of Concrete Pavements †

Jeb S. Tingle *, Charles E. Williams, Jr., William D. Carruth and Caitlin M. Tibbetts

U.S. Army Engineer Research and Development Center, Vicksburg, MS 39180-6199, USA; charles.e.williams@usace.army.mil (C.E.W.J.); william.d.carruth@usace.army.mil (W.D.C.); caitlin.m.tibbetts@usace.army.mil (C.M.T.)

* Correspondence: jeb.s.tingle@usace.army.mil

† Presented at the Second International Conference on Maintenance and Rehabilitation of Constructed Infrastructure Facilities, Honolulu, HI, USA, 16–19 August 2023.

Abstract: Many traditional methods for the repair and rehabilitation of concrete pavements require meticulous construction processes with specialized equipment and long material curing periods in order to develop adequate strength and durability prior to returning the pavement to service. This paper summarizes the results of research projects conducted by the U.S. Army Engineer Research and Development Center in order to develop innovative pavement repair procedures and evaluate numerous commercial repair materials that can produce fast long-lasting repairs that facilitate the rapid re-opening of critical pavement infrastructure to traffic. This paper summarizes methods used for the certification and selection of suitable concrete pavement repair materials. In addition, this paper outlines the key activities included in expedient concrete repair processes. Thus, this paper provides a valuable summary of state-of-the-art concrete repair procedures and materials for the rapid and effective repair and rehabilitation of concrete pavements.

Keywords: concrete pavement; concrete repair; concrete maintenance; concrete rehabilitation

Citation: Tingle, J.S.; Williams, C.E., Jr.; Carruth, W.D.; Tibbetts, C.M. Materials and Methods Used for the Expedient Repair of Concrete Pavements. *Eng. Proc.* **2023**, *36*, 43. https://doi.org/10.3390/engproc2023036043

Academic Editor: Hosin (David) Lee

Published: 14 July 2023

1. Introduction

Traditional concrete pavement repair methods can be characterized as meticulous procedures that require extensive specialized equipment and a lengthy return to service periods. Legacy materials and methods, such as Portland cement concrete placed with fixed forms or slipform pavers, require extensive planning and time commitments from maintenance crews, resulting in fewer repair completions for a given period of execution. In many instances, operational requirements require expedited repairs to either meet non-negotiable operational needs or to mitigate unacceptable user costs associated with extended delays. Examples include critical road and airfield pavements that support military missions, primary corridor highway repairs, and international airports. Figure 1 presents a photograph that illustrates the complexity associated with legacy concrete repair methods. Detailed planning and safety resources are required for the extended diversion of traffic for meticulous repair methods that effectively close the pavement to service for weeks or months.

The U.S. Army Engineer Research and Development Center (ERDC) has evaluated novel concrete pavement repair materials and developed expedient pavement repair processes (specifically for military operational needs). However, the same materials and methods can be applied to commercial applications, where the urgency of repair is of utmost importance. The pavement repair scenario often determines whether the utilization of expedient concrete pavement repair methods is suitable for a specific project. If suitable, then specific project needs, particularly the time allotted for repairs and the required time for return to service, will dictate what types of material solutions and repair processes are used. This paper summarizes the ERDC's concrete pavement repair material certification program and expedient construction practices in an effort to transfer these technologies to the public pavement repair industry.

Figure 1. Typical legacy concrete pavement repair in progress.

2. Repair Material Evaluation and Selection

As with any infrastructure engineering project, proper material selection is critical to achieving the required performance for an individual project. Thus, the selection of repair materials for concrete pavement repair is a critical aspect of the repair project that often dictates the equipment requirements necessary for compatibility with the material solution. The ERDC has developed and refined extensive laboratory test protocols that are used to certify concrete pavement repair materials. For small patches and spall repairs, the ERDC has developed two separate laboratory test protocols for cementitious and polymeric repair materials, as shown in Tables 1 and 2, respectively [1,2]. The required laboratory tests were selected to ensure that repair materials provide the required strength and environmental compatibility to provide expedient but long-lasting repairs. It should be noted that these protocols were developed specifically for expedient repairs with a focus on rapid return to service. These repair protocols were also adapted for the use of larger repairs, such as full-slab replacements [3].

Table 1. Laboratory test protocols for cementitious concrete pavement repair materials.

Tier 1 Test Requirements			
Property	**Test Method**	**Age**	**Criteria**
Compressive Strength	ASTM C39	2 hrs	≥2500 psi
		3 hrs	≥3000 psi
		1 day	≥4000 psi
		7 days	≥5000 psi
		28 days	≥5000 psi
Flexural Strength	ASTM C78	2 hrs	≥350 psi
		7 days	≥500 psi
		28 days	≥600 psi

Table 1. *Cont.*

Tier 1 Test Requirements			
Bond Strength (RS/RS)	ASTM C882	1 day	≥1000 psi
		7 days	≥1500 psi
Bond Strength (PCC/RS)		1 day	≥1000 psi
		7 days	≥1250 psi
Modulus of Elasticity	ASTM C469	2 hrs	$2 \leq x \leq 6$ Mpsi
		28 days	$2 \leq x \leq 6$ Mpsi
Time of Set	ASTM C403	Initial set	≥ 15 min
		Final set	15–90 min
Slump	ASTM C143	Within 5 min of added water	3–9 in. If ≥9 in., perform slump flow
Slump Flow	ASTM C1611	Within 5 min of added water	≥9 in.
Tier 2 Test Requirements			
Property	**Test Method**	**Age**	**Criteria**
Length Change	ASTM C157	28 days Stored in air	$-0.04\% \leq x \leq 0.03\%$ at 28 days Continue testing and report length change until 64 weeks
		28 days Stored in water	
Coefficient of Thermal Expansion	ASTM C531	-	≤7 (in./in./°F × 10^{-6})
Shrinkage Potential	ASTM C1581	14 days	Record microstrain but no pass/fail limits at this time
		28 days	Record microstrain and fail if any ring cracked

Table 2. Laboratory test protocols for polymeric concrete pavement repair materials.

Property	Test Method	Age	Criteria
Compressive Strength	ASTM C579	2 hrs	≥2500 psi @ 2 hrs
		3 hrs	≥3000 psi @ 3 hrs
		1 day	≥4000 psi @ 1 day
		7 days	≥5000 psi @ 7 days
		28 days	≥5000 psi @ 28 days
Flexural Strength	ASTM C78	2 hrs	≥350 psi @ 2 hrs
		7 days	≥500 psi @ 7 days
		28 days	≥600 psi @ 28 days
Bond Strength	ASTM C882	1 day	≥1000 psi @ 1 day
Repair Material to Repair Material		7 days	≥1250 psi @ 7 days
Bond Strength		1 day	≥1000 psi @ 1 day
Repair Material to Ordinary PCC		7 days	≥1250 psi @ 7 days

Table 2. *Cont.*

Property	Test Method	Age	Criteria
Modulus of Elasticity	ASTM C469	2 hrs	$2 \leq x \leq 6$ Mpsi
		28 days	$2 \leq x \leq 6$ Mpsi
Time of Set	ASTM C403	Initial set	≥15 min
		Final set	15–90 min
Thermal Compatibility	ASTM C884	Test to begin after 7 days cure	No Delamination
Chemical Resistance	ASTM C267	Test Method B-Fuel B Exposure Test Method B-JP-8 Exposure Test Method B-Oil-3 Exposure Test Method B-Air Exposure Test Method B-Water Exposure	≤20% strength loss and ≤10% weight change at 66 °C @ 1 day
Dynamic Mechanical Analysis	ASTM D5023	Sinusoidal 3-point bending-60 to 400 °F	>60 °C @ 7 days

3. Expedient Concrete Pavement Repair Processes

The overall pavement repair strategy must ensure that repair materials that meet project requirements are selected and that the pavement repair process is compatible with the required material solutions. In general, the concrete pavement repair process consists of several steps: defining the extent of damage, marking repair boundaries, cutting around the repair boundaries to maximum depth, demolishing and removing the damaged material, restoring the substrate, installing any dowels/tie bars, placing the repair material, and curing the repair material [3,4]. Suitable methods for sounding the concrete pavement, including the use of hammers or chains, should be used to identify the boundary between the damaged pavement and the sound intact pavement. This boundary should be clearly marked using paint or chalk lines to guide the pavement cutting activities. Cutting is ideally accomplished with walk-behind saws using diamond blades or skid-steer attachments equipped with diamond blades. In addition, wheel-saw attachments are successfully used for expedient repairs that provide a larger relief cut for the easier demolition of thick concrete pavements. Backhoes or excavators equipped with pavement breaker attachments and suitably sized buckets are also used to remove large debris. Hand tools and/or vacuum trucks are used to remove smaller debris. Any damage to the substrate resulting from the removal of the surface slab should be repaired either by compaction or the placement of rapid setting flowable fill. Once the foundation has been restored, dowels and/or tie bars can be installed. For expedient repairs of thick concrete pavements (PCC > 12 in.), the dowel bars and/or tie bars may be omitted, depending on the required service life. Once any reinforcement is placed, the rapid setting materials are placed in the repair and finished with a minimum of hand work. Rapid setting materials should be placed quickly from mixing buckets for small repairs or a volumetric mixer for large repairs. Care should be taken not to over finish rapid setting materials as excess paste will result in shrinkage cracks. Finally, the repair should be allowed to cure for the required period using either a topically applied polymeric curing compound or a wet burlap.

4. Recommendations/Conclusions

The laboratory test protocols and expedient repair processes described in this paper have been thoroughly evaluated through multiple projects by the ERDC and the Department of Defense for both road and airfield applications. Testing potential repair products according to recommended protocols ensures that the repair material will provide an adequate working time while meeting return to service requirements. In addition, strength testing will ensure that the repair materials are capable of withstanding applied loads, while environmental testing will ensure compatibility with the surrounding pavement under diverse environmental conditions. The expedient repair processes can be generally followed for most expedient concrete pavement repair scenarios with minor modifications to allow the use of existing equipment. The urgency of both repair and the service life required will dictate if any of the steps may be omitted for expediency while still meeting performance requirements.

Author Contributions: This project was conceived and executed by J.S.T. Laboratory testing and analyses were completed by C.E.W.J. Field testing for repair process development was performed by W.D.C. and C.M.T. The original draft was prepared by J.S.T. and C.E.W.J. The paper was reviewed and edited by W.D.C. and C.M.T. All authors have read and agreed to the published version of the manuscript.

Funding: This research received no external funding.

Data Availability Statement: The data are contained within this article. Additional details can be obtained by contacting any of the authors.

Acknowledgments: The authors would like to acknowledge Craig Rutland and the Air Force Civil Engineering Center for their general technical support.

Conflicts of Interest: The authors declare no conflict of interest.

References

1. Ramsey, M.A.; Tingle, J.S.; Rutland, C.S. *Evaluation of Rapid-Setting Cementitious Materials and Testing Protocol for Airfield Spall Repair*; ERDC/GSL TR-20-9; U.S. Army Engineer Research and Development Center: Vicksburg, MS, USA, 2020.
2. Tri-Service Pavement Working Group Manual 3-270-01.08-4 Testing Protocol for Polymeric Spall Repair Materials. 2017. Available online: https://www.wbdg.org/ffc/dod/supplemental-technical-documents/tspwg-m-3-270-01-08-4 (accessed on 12 April 2023).
3. Priddy, L.P.; Tingle, J.S.; McCaffrey, T.J.; Rollings, R.S. *Laboratory and Field Investigations of Small Crater Repair Technologies*; ERDC/GSL TR-07-27; U.S. Army Engineer Research and Development Center: Vicksburg, MS, USA, 2007.
4. Carruth, W.D. *Full-Scale Testing of Commercially Available Cementitious Backfill and Surface Capping Materials for Crater Repairs*; ERDC/GSL TR-20-17; U.S. Army Engineer Research and Development Center: Vicksburg, MS, USA, 2020.

engineering proceedings

MDPI

Proceeding Paper

Simplified Deterioration Modeling for Highway Sign Support Systems †

Myungjin Chae *, Lucas Voghell and Jiyong Choi

Construction Management, Central Connecticut State University, New Britain, CT 06050, USA; lucas.voghell@my.ccsu.edu (L.V.); jaychoi@my.ccsu.edu (J.C.)

* Correspondence: chae@ccsu.edu

† Presented at the Second International Conference on Maintenance and Rehabilitation of Constructed Infrastructure Facilities, Honolulu, HI, USA, 16–19 August 2023.

Abstract: Road sign support systems are not usually well managed, because bridges and pavements have budget and maintenance priority while the sign boards and sign supports are considered miscellaneous items. The authors of this paper developed a simple deterioration prediction model and a repair priority list for sign support systems. For asset management risk analysis, data were collected from the Connecticut Department of Transportation (CTDOT) asset management database. Asset ages, repair history, installation and repair costs, and other administrative information were collected. While there were many advanced and complicated structural deterioration prediction models, the Weibull function was used in this research because it is a simple probability distribution function that has been widely used in reliability theory. Assuming that the primary factors of the deterioration prediction are age and initial installation quality, the repair priority list was developed based on the risk analysis of the assets. Asset risks were calculated based on the deterioration prediction model and traffic impacts of the failure.

Keywords: sign support; asset management; deterioration modeling; preventive maintenance

Citation: Chae, M.; Voghell, L.; Choi, J. Simplified Deterioration Modeling for Highway Sign Support Systems. *Eng. Proc.* **2023**, *36*, 44. https://doi.org/10.3390/engproc2023036044

Academic Editor: Hosin (David) Lee

Published: 14 July 2023

1. Introduction

Sign support systems (Figure 1) are essential structures in the Connecticut Department of Transportation (CT DOT) bridge management system, and they are often neglected until they are in critical condition and fail. Periodic inspections and maintenance activities are needed as a long-term, cost-effective preventive maintenance strategy. Sign support systems are included in the Bridge Management System (BMS). The data collected from the BMS were analyzed using predictive deterioration curves with educated estimates from the literature review. This research developed the asset maintenance strategy using the deterioration prediction model and risk management for sign support systems.

Figure 1. Sign support 21311 on I-84, Hartford, CT.

2. Weibull Function for Deterioration Prediction Model

Usually, when machines or structures are newly built, during the operation's initial status, there is a high chance of failure or breakdown due to defective parts and installation. Once the initial status is settled, it goes into a stable service period until it reaches the end of its service life (Figure 2a).

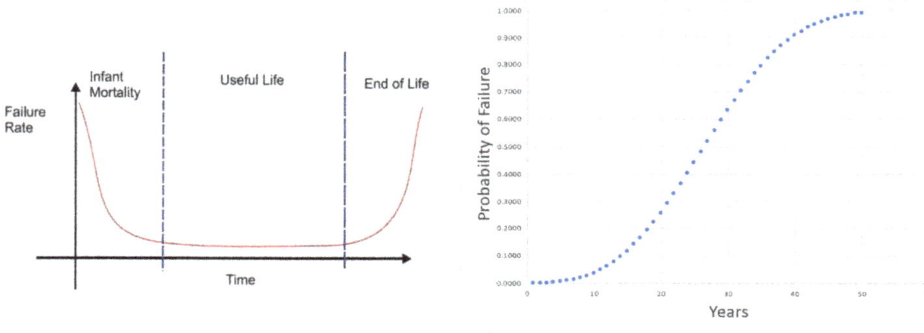

(**a**) (**b**)

Figure 2. (**a**) Bathtub curve for infrastructure failure rate; (**b**) Weibull function when MTTF β = 30 and shape factor γ = 3.

The Weibull distribution is perhaps the most widely used of all the failure time distributions, and it is noted for its flexibility as a failure prediction model. Reliability deals with reducing the frequency of failures over a time interval and is a measure of the probability of failure-free operation during a given interval, i.e., it is a measure of success for a failure-free operation. Reliability is quantified as MTBF (Mean Time Between Failures) for repairable products and MTTF (Mean Time to Failure) for the non-repairable product. In reliability theory, the Weibull function is widely used for its simplicity without losing too much accuracy. Figure 2b shows the typical failure rate over time by the Weibull function, where x is usage time and β and γ are the graph shape parameters. The shape parameters, γ and β, are assumed in this case to be 3 and 30, respectively, where β is the scale parameter that directly relates to the spread of the distribution curve. In reliability theory, β is either MTBF or MTTF. The cumulative distribution curve plots the cumulative probability of occurrence over time for a specific action. In this case, it plots the probability of failure over time.

$$f(x) = 1 - e^{-\left(\frac{x}{\beta}\right)^{\gamma}} \tag{1}$$

3. Infrastructure Asset Management Using Quantitative Risk Analysis

The objective of Infrastructure Asset Management is to produce the maintenance of priority and budget prediction of assets. In this paper, a quantitative risk analysis is performed. The risk assessment and asset deterioration and failure prediction model are created using the existing database provided by the DOT. The procedures of risk management plan development are (1) risk identification; (2) quantification of failure impact; (3) assessment and prediction of the probability of failure; and (4) risk quantification, calculated as *Probability of Occurrence (P)* $\times$ *impact (I), or risk* = $P \times I$. The estimated risk score can be considered as management and repair priority. The probability of occurrence (P) can be obtained by the *mean-time-to-failure (MTTF)* and the age of the infrastructure using the Weibull function or Normal distribution (Gaussian) function.

4. Development of Deterioration Prediction Model

MTTF (Mean-time-to-Failure) is calculated as an average of different adjustment factors relevant to each sign. The adjustment factors are as follows: (1) average daily traffic,

(2) lanes under the structure, (3) material type, and (4) structure type. These four factors are chosen as they most accurately represent the structures' average wear and tear from reviewing existing sign support literature. The factors of the sign support system failures were previously studied by Kipps [1], Barle [2], and Shboul [3].

All the factors below contain a variable that begins with an "s". These variables represent the years a structure would take to fail when no other influence is considered. For example, an adjustment value of 36 for a support constructed of A-36 steel is because the structure is made of A-36 steel where it is defined by ASTM [4]. The shape, location, and average daily traffic are not considered when assigning these scores, as they are all brought back together when averaging out.

This methodology of estimating *MTTF* would be to average these scores and then find the average number of years for a structural failure. Each structure is unique; therefore, coming up with an *MTTF* for each individual would take an incalculable amount of time and is beyond the scope of this paper. By assigning scores based on each factor's relative durability, we can assign a normalized *MTTF* score to each structure based on an average. These scores are assigned based on prior research, and each factor is present in at least one of each in any given sign support.

Material types of yield strength (*sMAT*) accurately represent how a sign support structure's material type plays into the general deterioration of the support over time. According to the ASTM 2019 standards reference documents, assuming all else are constant, the different materials and their yield strengths are shown in Table 1.

Table 1. Average daily traffic adjustment values and lanes under structure adjustment values.

Average Daily Traffic (ADT)	Adjustment Value	No. of Lanes under Structure	Adjustment Value
0–10,000	35	1	35
10,001–25,000	27	2	33
25,001–45,000	25	3	31
45,001–75,000	23	4	30
75,001–100,000	21	5	28
100,001–140,000	18	6	26
140,000+	16	7	24
		8	22
		10	20
		12	18

The formula to calculate the *MTTF* for any given structure is as follows:

$$MTTF = (sADT + sMAT + sTYP + sLNE)/4 \tag{2}$$

The *MTTF* score can be inserted into a probability distribution function to model the probability of failure. The Weibull function (Equation (1)) used to model this is as follows:

$$p(f) = 1 - e^{\left(-\left(\frac{Age\ of\ Structure}{MTTF}\right)^{\gamma}\right)} \tag{3}$$

where γ is the shape factor of the Weibull function. A shape factor of 3 is used to calculate *p*(*f*) because this shape factor relates to the failure rate behavior (increasing over time). This is known as a "wear-out failure", or a failure rate that increases over time. This is shown in Figure 2b.

The reliability factor (*r*(*f*)) can be found simultaneously alongside the *p*(*f*). The *r*(*f*) is a measurement to identify the measure of failures over a time interval. It measures the probability of failure-free operation during a given interval. *r*(*f*) is calculated using the following equation:

$$r(f) = e^{\left(-\left(\frac{Age\ of\ Structure}{MTTF}\right)^{\gamma}\right)} \tag{4}$$

The result is shown in Figure 3. The graph shows the assets, ages, and probability of failure. It shows only a few assets out of thousands in the database. Note that the numbers on the x-axis are the asset names. For example, the name of the first asset is "21231B", the second one is "20574", etc. As expected, and by the Weibull function's definition, age and probability of failure have a strong direct relationship.

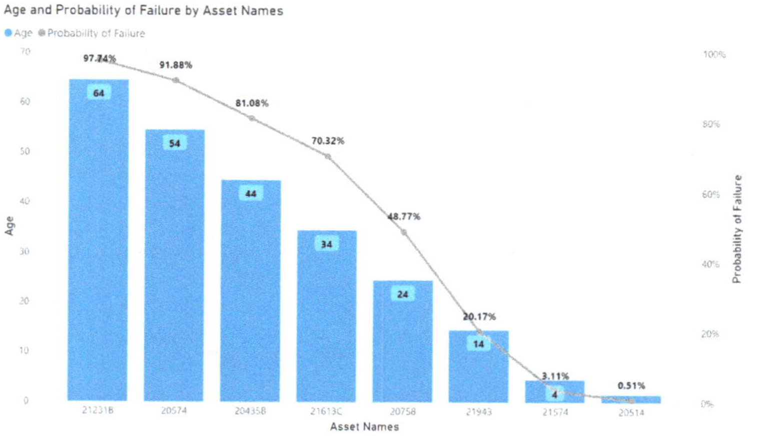

Figure 3. Age and probability of failure by assets.

Because the Mean-time-to-failure (MTTF) is defined by four variables such as (1) average daily traffic, (2) lanes under the structure, (3) material type, and (4) structure type, each sign support system has different MTTFs; however, the age of the structure is the main factor in estimating the probability of failure.

5. Summary and Conclusions

This paper presents a simplified deterioration model for highway sign support systems. Although there is much research to predict the deterioration of civil structures, because civil structures have a long life span, none of prediction models are proven. The deterioration prediction model is an important part of asset management. However, researchers and facility managers are often stuck in the deterioration prediction model development and fail to move forward to complete the asset management system development. This paper shows a simplified method of deterioration prediction model development.

The probability of the failure model adopts factors such as (1) average daily traffic (*sADT*), (2) lanes under the structure (*sLNE*), (3) material type (*sMAT*), and (4) structure type (*sTYP*). Then, the mean time to failure (*MTTF*) is calculated by averaging the four factors and it is assumed the reliability and the probability of failure are calculated using a simple probability distribution function, the Weibull function, which is widely used in failure prediction models of machine parts.

In conclusion, it is cost-effective to prioritize sign support repairs based on the findings. While the overall failure rate of these signs is low, preventative repairs should be considered, starting with the oldest supports and working toward the newest.

Author Contributions: M.C.: Conceptualization, methodology, formal analysis, and writing and editing paper; L.V.: investigation, data curation, visualization, writing draft; J.C.: review and editing. All authors have read and agreed to the published version of the manuscript.

Funding: CTDOT, SPR-PL-1(60) Part 2, SPR-2323.

Institutional Review Board Statement: Not applicable.

Informed Consent Statement: Not applicable.

Data Availability Statement: Not applicable.

Conflicts of Interest: The authors declare no conflict of interest.

References

1. Kipp, A.M.; Ehsani, M.R. Field Testing of Highway Sign Support Structures. *J. Struct. Eng.* **1987**, *113*, 4.
2. Barle, J.; Grubisic, V.; Vlak, F. Failure analysis of the highway sign structure and the design improvement. *Eng. Fail. Anal.* **2011**, *18*, 1076–1084. [CrossRef]
3. Shboul, K.W.A.; Rasheed, H.A.; Alshareef, H.A. Intelligent approach for accurately predicting fatigue damage in overhead highway sign structures. *Structures* **2021**, *34*, 3453–3463. [CrossRef]
4. *A794/A794M-18*; Standard Specification for Commercial Steel (CS), Sheet, Carbon (0.16% Maximum to 0.25% Maximum), Cold-Rolled. ASTM: West Conshohocken, PA, USA, 2018.

engineering proceedings

Proceeding Paper

Accelerated Bridge Construction on Maui's Hana Highway †

Sean Oroho [1,*] and Tom Kubicz [2]

[1] HDR Inc., Denver, CO 80202, USA
[2] Federal Highway Administration—Central Federal Lands Highway Division, Lakewood, CO 80228, USA; tomasz.kubicz@dot.gov
* Correspondence: sean.oroho@hdrinc.com
† Presented at the Second International Conference on Maintenance and Rehabilitation of Constructed Infrastructure Facilities, Honolulu, HI, USA, 16–19 August 2023.

Abstract: Through the partnership between the Hawaii Department of Transportation (HDOT) and the Central Federal Lands Highway Division (CFLHD) of the Federal Highway Administration, six of the bridge structures along the State-owned portion of the Hana Highway will be replaced. Due to the nature of this road, and the limited opportunities for detours and road closures, along with context-sensitive solution design and right-of-way considerations, these structures will be replaced utilizing an accelerated bridge construction (ABC) technique known as a lateral bridge slide. To our knowledge, this project will be the first implementation of this ABC technique.

Keywords: accelerated; bridge; construction; Hana; highway; lateral; slide; FHWA; ABC

Citation: Oroho, S.; Kubicz, T. Accelerated Bridge Construction on Maui's Hana Highway. *Eng. Proc.* **2023**, *36*, 45. https://doi.org/10.3390/engproc2023036045

Academic Editor: Hosin (David) Lee

Published: 14 July 2023

1. Introduction

The Central Federal Lands Highway Division (CFLHD), in partnership with the Hawaii Department of Transportation (HDOT), is preparing to deliver a construction project to replace six bridge structures along the Hana Highway on the northeast coast of Maui, Hawaii (See Figure 1). The Hana Highway is an iconic thoroughfare, stretching over 50 miles, with over 600 curves and nearly 60 bridges, many of which are one-lane bridges. Through extensive coordination with stakeholders, and outreach to the public, the bridges are being replaced in a context-sensitive manner, to continue to contribute to the character of the Hana Highway, which is a historically significant asset, subject to compliance with Section 106 of the National Historic Preservation Act, among other federal, state, and local regulatory compliance requirements.

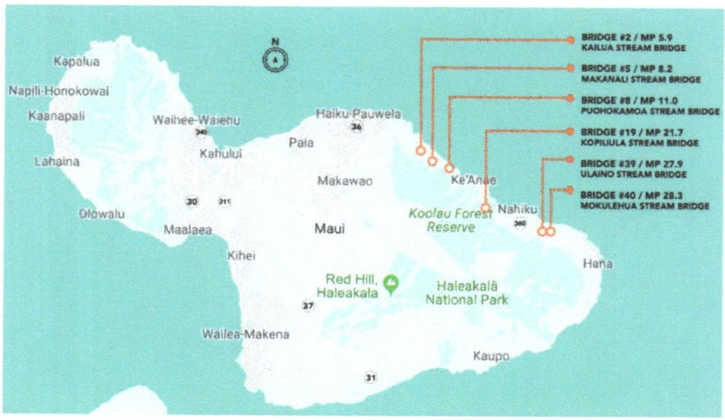

Figure 1. Hana Highway bridge replacement locations. (Source: FHWA-CFLHD).

In general, the existing reinforced concrete superstructures will be replaced by similar, one-lane bridge superstructures, capable of supporting modern AASHTO LRFD Bridge Design Specification loads, and including upgrades of key safety features, such as bridge railings, in a context-sensitive manner. The existing bridges are typically supported by either concrete rubble masonry (CRM) abutments, or reinforced concrete piers and abutments, flanked by CRM wingwalls. For each of the bridges, these abutments, piers, and wingwalls are considered to be key contributing elements to the character-defining features of the structures, and attempting to preserve these elements of the bridges was considered to be of paramount importance during design development. Because of this, five of the six bridges will be replaced by constructing a new superstructure, in a context-sensitive manner, which will be supported on micropile foundations located behind the existing structures. The bridge foundations will be hidden from view, and concealed behind the original bridge abutments and wingwalls, which are to remain, as shown in Figure 2.

Figure 2. Artistic schematic of a typical longitudinal section of the proposed Hana Highway bridges. (Source: FHWA-CFLHD).

The Hana Highway is the only viable route for the travelling public between Maui's central valley, including urban areas such as Kahului and Wailuku, and the community of Hana, on the east side of Maui. Additionally, there are communities along the Hana Highway that rely on this route for not only commuter travel, but also for the commerce the highway provides, as one of the top tourist destinations on Maui. There is no viable detour, and long duration closures of the Hana Highway would cut off each of these communities, and result in significant economic loss. These constraints necessitate the need for an accelerated bridge construction (ABC) solution, or a temporary bridge, during the construction.

Although temporary bypass bridges are commonly used in bridge construction throughout Hawaii, they were not considered to be a viable option for the bridges proposed to be replaced along the Hana Highway. This is largely due to the limited right-of-way (ROW) information available to determine property ownership along the Hana Highway. With limited ROW information available, the property research process to obtain an easement for a temporary bridge would be extremely onerous and detrimental to the project schedule, as it would take years of research, going back to the Great Mahele of 1848.

As a result of the constructability and alternatives analysis summarized above, ABC was considered to be necessary for this project. After consultation with CFLHD, HDOT, industry constructability professionals, and community stakeholders, the lateral bridge slide ABC technique was considered to be a potential construction methodology.

2. Lateral Bridge Slide—What Is It?

The lateral bridge slide technique, or slide-in bridge construction (SIBC), as titled by the FHWA Every Day Counts program, entails the construction of a new bridge superstructure immediately adjacent to the final alignment on a temporary support structure, and the structure being slid into place. In general, the structure is slid into place by placing the structure on rollers or a sliding surface, and either pushing the structure with hydraulic jacks or post-tension jacks, or pulling the structure through the use of winches or pulley systems. This allows vehicular traffic to continue to utilize the existing facility, while the new structure is constructed. Then, under a short-duration closure (typically 2 to 4 days), the new structure is slid into place, as shown in Figure 3, and traffic is re-established.

Figure 3. General schematic of the lateral bridge slide technique, and a view of the typical site conditions around the structures. (Source: https://www.youtube.com/watch?v=YzYCSw5mACQ, accessed on 7 July 2023, FHWA—CFLHD).

2.1. Lateral Bridge Slide—General Design Considerations

In general, the design procedure for a slide-in bridge is not very different from that of a conventional bridge. The permanent loading is nearly identical to a typical bridge design, although consideration should be given to the increased stiffness of the end diaphragms and regions near the bearing areas and/or sliding surfaces, in order to avoid secondary stresses and/or cracking due to fluctuations in support (i.e., the temporary loss of contact of a portion of the sliding surface) during the slide. Additional stiffness near the supports allows for the redistribution of temporary loading throughout the transverse width of the structure, and avoids concentrating these temporary stresses on the immediately adjacent elements.

Generally, in a design-bid-build procurement, the temporary works associated with supporting the new superstructure off-line, and the slide procedure itself, is designed by the contractor, with the parameters and special requirements provided in the contract specifications, to aid in the contractor's design. The design for the loading on the temporary support is generally in conformance with design guidelines, such as the AASHTO Guide Specifications for Bridge Temporary Works, as an example. For sliding forces, it is common to consider up to approximately 20% of the dead load as a lateral design force during construction on the temporary works and on the permanent structure. Further consideration, to determine the application of the force, is contingent on the method of anchoring the jack or pulley system [1].

2.2. Recent Examples of Lateral Bridge Slides

The lateral bridge slide technique has become fairly common in the United States. Some recent bridge slide examples within the last decade include the following structures:

- The Sellwood Bridge Replacement in Oregon
- The Mesquite Bridge Project in Nevada
- The I-84/Dingle Ridge Road Bridge Slide in New York
- The SR-201 Bridge in Utah

For more information related to the lateral bridge slide technique, below are a few technical resources which review design, construction, contracting, cost estimating, and key elements of the lateral bridge slide technique:

- FHWA Every Day Counts Slide-In Bridge Construction [2]
- Florida International University ABC University Transportation Center Project and Research Database [3]
- Iowa State University Institute for Transportation [4]

3. Hana Highway Bridges—Potential for the Lateral Bridge Slide Technique

The proposed superstructure will be slightly longer that the existing bridge, to preserve the existing substructure and foundation elements, as part of the historic mitigation strategy. The proposed structures along the Hana Highway are single-span structures supported on simple, semi-integral abutment caps over drilled micropiles. The construction of the substructure elements, behind each existing abutment, is anticipated to be performed utilizing overnight closures of the Hana Highway. During daytime hours, the work areas could be plated over, to allow traffic to operate normally.

The temporary supports for the construction of the new structure could be constructed adjacent to the existing structure, to allow the existing structure to remain in service. Due to the challenging, steep topography near the structures, foundation system(s) will need to be constructed as part of the temporary support structure.

Once the new superstructure is complete, and the on-alignment substructure is installed, the roadway will be shut down to allow for the slide installation procedure. It is anticipated that this shutdown will last for up to 4 days. Advanced communication and coordination with local emergency services is critical during the closure. Depending on the contracting and schedule, the coordination of the slide activities across several structures may be required, to minimize the closure times of the Hana Highway. To minimize traffic impacts, precast elements and high-early-strength concrete will be used to allow for one lane operations once the superstructure is slid into place. Temporary barriers placed along the travel lane will be used to allow for the cast-in-place operations to complete the exterior geometries of the slab, and the restoration of the site outside of the roadway, following the re-opening of the road. The generalized procedure for the potential use of the bridge slide technique for the Hana Highway bridges is presented in Figure 4.

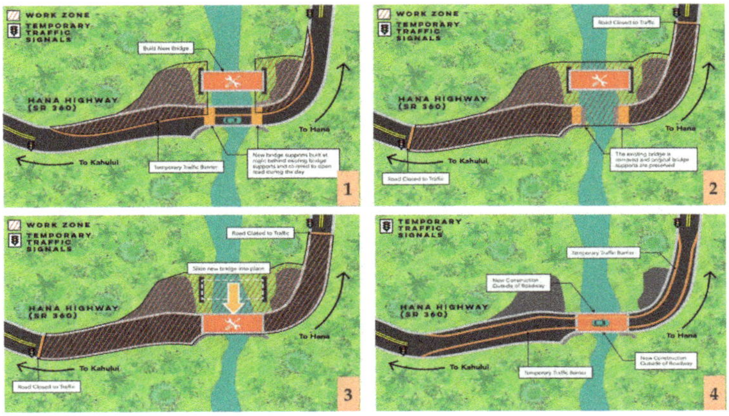

Figure 4. The generalized procedure for the potential Hana Highway bridges lateral slide. (Source: FHWA-CFLHD).

The construction is scheduled to commence in spring/summer 2025.

Author Contributions: Conceptualization, S.O. and T.K.; methodology, S.O. and T.K; writing—original draft preparation, S.O. and T.K.; writing—review and editing, S.O. and T.K.; visualization, S.O.; supervision, T.K.; project administration, T.K. All authors have read and agreed to the published version of the manuscript.

Funding: This research received no external funding.

Institutional Review Board Statement: Not applicable.

Informed Consent Statement: Not applicable.

Data Availability Statement: New data is not presented as part of this research.

Conflicts of Interest: The authors declare no conflict of interest.

References

1. Utah, D.O.T. *Michael Baker Corporation. Slide-in Bridge Construction Implementation Guide, Planning and Executing Projects with the Lateral Slide Method*; Federal Highway Administration: Washington, DC, USA, 2013.
2. U.S. Department of Transportation—Federal Highway Administration. Available online: https://www.fhwa.dot.gov/construction/sibc/ (accessed on 17 February 2023).
3. Florida International University—Accelerated Bridge Construction University Transportation Center. Available online: https://utcdb.fiu.edu/search/ (accessed on 16 February 2023).
4. Zhengyu, L.; Katelyn, S.F.; Justin, M.D.; Brent, M.P.; LaViolette, M. *Lateral Slide of Muti-Span Bridges: Investigation of Connections and Other Details—Phase 1*; Iowa State University Institute for Transportation: Ames, IA, USA, 2021.

engineering proceedings

Proceeding Paper

Benchmarking Material Use Efficiency for Building Projects †

Jiyong Choi *, Myungjin Chae and Namhun Lee

Department of Manufacturing and Construction Management, Central Connecticut State University, 1615 Stanley St., New Britain, CT 06050, USA; chae@ccsu.edu (M.C.); leen@ccsu.edu (N.L.)
* Correspondence: jaychoi@ccsu.edu
† Presented at the Second International Conference on Maintenance and Rehabilitation of Constructed Infrastructure Facilities, Honolulu, HI, USA, 16–19 August 2023.

Abstract: Reducing the quantities of engineered materials provides a significant opportunity to mitigate the environmental impacts caused by material production and processing. Although the efficient use of materials in building construction has been emphasized, there has been little attention given to measuring the material use efficiency (MUE) of a project. This research fulfills this gap by using data envelopment analysis (DEA) as a benchmarking tool to generate an overall perspective on the MUE and to further compare its efficiency with that of peer projects, thereby promoting enhanced efficiency through target setting. In this research, MUE was measured by adopting the quantities of a variety of materials consumed during construction as input variables and the floor area of a built facility as an output variable. To generate a reliable MUE performance, a stepwise variable selection process was applied and then the performance was ranked based on evaluating cross-efficiency. In addition, clustering analysis and DEA were fused to enable a more realistic target to be set for each input, thereby determining practical targets for each underperforming project. It is anticipated that the proposed MUE benchmarking model would enable projects to recognize the gap with the best-performing projects and help them determine the targets to focus on to become efficient.

Keywords: material use efficiency; data envelopment analysis; benchmarking; target setting

Citation: Choi, J.; Chae, M.; Lee, N. Benchmarking Material Use Efficiency for Building Projects. *Eng. Proc.* **2023**, *36*, 46. https://doi.org/10.3390/engproc2023036046

Academic Editor: Hosin (David) Lee

Published: 17 July 2023

1. Introduction

Recent research in the realm of the construction industry proposed possible measures that can generate a quantitative approximation of material use efficiency (MUE) by using efficiency measures that evaluate the ratio of useful input to output. These measures assume that an improved design that optimizes the required building size and functional areas can reduce the amount of building materials needed to construct a building [1]. To this end, some researchers used the quantities of the main building materials consumed for a project as inputs and used the area of floor space generated with the materials as the output. However, a robust approach for assessing the MUE based on factual data was not demonstrated in the study. Moreover, the research lacked a discussion on how to quantitatively assess the MUE at the project level and how to make continuous improvement in the efficiency of a project possible by cross-project comparisons.

One of the identified challenges with quantifying the MUE is that there are various types of materials used for building construction and they are generally quantified in diverse measurement units (e.g., tons for steel, cubic yards for concrete, or square footage for glazing). Moreover, different projects may operate in different environments (e.g., function, location, and size) and accordingly, the level (or degree) of MUE can vary from one project to another. In this regard, it would be desirable if all the materials could be considered together to measure the efficiency while taking into account differences in many project characteristics for credible cross-project comparisons and benchmarking. One modeling methodology that provides a flexible and powerful approach for measuring efficiency while facilitating cross-project comparison is data envelopment analysis (DEA), a linear programming (LP)-based relative efficiency evaluation methodology. Building on the existing body of knowledge on MUE and DEA, this paper proposes a MUE benchmarking model that can generate reliable MUE scores for multiple building projects, thereby allowing the scores to be compared for possible improvements through targeted design.

2. Background

2.1. Material Use Efficiency

Material efficiency includes a broad range of technical strategies aimed at pursuing sustainable building project delivery which are introduced by reducing environmental and economic impacts arising from material production and processing, as well as material consumption. Examples of such efforts include industrialized building construction [2], downsized building design [3], the flexible layout of building space [4], minimum quantity of materials used [5], more intensive use of a building (by reducing per capita floor area), extension of building lifespan, and reuse and recycling of building materials [6].

To alleviate the environmental impacts of construction materials and the consequences of increased material costs, one potential strategy is MUE which is rather focused on reducing the demand for construction materials needed to build the facility. When building designs use only the materials required, in the right place and without excess, the demand for materials is reduced for the same size of the facility which is often represented by the area of floor space or building gross square footage (BGSF) [6]. Likewise, ensuring each structural element is appropriately sized and working efficiently takes some additional design time but can result in substantial material savings during the construction phase.

2.2. Data Envelopment Analysis

DEA is a mathematical procedure that utilizes an LP technique and identifies a set of weights that individually maximizes each DMU's efficiency using the same weights for all DMUs [7]. There are two different types of models in DEA depending on the type of envelopment surface formed by the frontiers; (1) the CCR model developed by Charnes and colleagues and Cooper and Rhodes [8] and (2) the BCC model developed by Banker, Charnes, and Cooper [9]. The first model assumes that the increase in outputs is proportional to the increase of inputs at any scale of operation, and thus it is known as a constant returns-to-scale (CRS) model. The second one, however, allows the production to exhibit increasing or decreasing returns-to-scale so it is called a variable returns-to-scale (VRS) model. Both models are further classified by their orientation which indicates the direction that an inefficient DMU approaches the frontier; either an increase in its output levels while keeping the same level of input (i.e., output-oriented) or a decrease in its input while maintaining the same output level (i.e., input-oriented). In both input- and output-oriented models, the best DMUs are assigned an efficiency score of 1 (or 100%) and those of the others are less than 1. For brevity, the mathematical formulations for other DEA models are not presented here; instead, the reader is referred to a more comprehensive text [10].

3. MUE Benchmarking Model

In this study, an output-oriented CCR model was employed to answer the question; by how much can inputs be proportionally decreased while keeping the level of output constant? This way, the outcomes of the DEA analysis can be used to identify the savings or reductions in inputs and the most suitable direction to enhance inefficient DMUs. However, traditional DEA alone is not enough to set reliable benchmarks for MUE because of some critical limitations that the conventional DEA model exhibits. The MUE benchmarking model was designed to generate reliable benchmarking by improving the three main deficiencies of the traditional DEA model.

First of all, the DEA results depend heavily on the input and output variables used in the analysis, which means attention to variable selection is crucial for obtaining reliable outcomes [11]. This is mainly because the greater the number of input and output variables, the less constrained the model weights assigned to the inputs and outputs, resulting in less discriminating results [12]. In this regard, one of the main challenges in DEA application is to find a parsimonious model using as many variables as required but as few as possible. To address this issue, we employed a formal stepwise approach to prioritizing meaningful inputs. Such an approach involves sequentially minimizing the average change in the efficiencies as inputs are dropped from the analysis.

Moreover, each DMU selects its own most favorable input and output weights for computing its efficiency, instead of using the same weights for all DMUs. This flexibility in choosing the weights prevents DMUs from being compared on a common base. The same issue regarding this weighting scheme can also happen when some DMUs heavily weight a few favorable inputs and outputs while ignoring other variables to achieve a high efficiency score [13]. The MUE benchmarking model overcomes this limitation by evaluating the efficiencies based on a cross-efficiency method (see Table 1), such as a DEA extension tool.

Table 1. Generalized cross-efficiency matrix (CEM).

DMU	**Target DMU**				**Average Cross-Efficiency**
	DMU_1	**DMU_2**	...	**DMU_n**	
DMU_1	E_{11}	E_{12}		E_{1n}	$\sum_{k=1}^{n} E_{1k}/n$
DMU_2	E_{21}	E_{22}		E_{2n}	$\sum_{k=1}^{n} E_{1k}/n$
⋮	⋮	⋮	⋮	⋮	⋮
DMU_n	E_{n1}	E_{n2}	...	E_{nn}	$\sum_{k=1}^{n} E_{nk}/n$

Lastly, DEA presents the capability of determining a specific reference set for inefficient DMUs and deriving their potential improvements (or targets). Although this is a remarkable feature of DEA, the limitation of this feature is that an inefficient DMU and the corresponding reference set of the DMU may not be inherently similar in their operations or practices. This is because DEA assumes that DMUs are homogeneous and identical in operations or practices. Accordingly, the efficiency scores obtained from the DEA may cause the degree of improvement to become unattainable for an inefficient unit. To solve this problem, in the MUE benchmarking model, clustering analysis was adopted to cluster DMUs into a set of groups so that the best performing DMU in each cluster is utilized as the benchmark (or target) for other DMUs in the same cluster.

4. Summary and Conclusions

This study introduced a MUE benchmarking model that leverages the capability of DEA to measure the MUE performance and to identify useful benchmarks enabling inefficient projects to set potential improvement paths. To accomplish the goal, the model was designed to overcome the conventional application of DEA by selecting meaningful variables before running DEA analysis, ranking the efficiency scores on a common basis by measuring cross-efficiency, and integrating DEA with clustering analyses to determine practical references or targets from which inefficient DMUs can learn from. As a future direction, qualitative research will be conducted to validate the proposed model. With continuous data collection, it will be possible to conduct rigorous analyses to examine the relationship between a host of construction materials and different parameters of building products, which will help the authors update the model by including various input and output variables.

Author Contributions: J.C. carried out the experiment design and implementation and wrote the manuscript draft. M.C. provided technical support in building the model. N.L. provided supervision for the research work and edited the manuscript draft. All authors have read and agreed to the published version of the manuscript.

Funding: This research was funded by 2022–2023 CSU-AAUP Faculty Research Grant.

Institutional Review Board Statement: Not applicable.

Informed Consent Statement: Not applicable.

Data Availability Statement: No new data were created or analyzed in this study. Data sharing is not applicable to this article.

Conflicts of Interest: The authors declare no conflict of interest.

References

1. Ruuska, A.; Häkkinen, T. Material Efficiency of Building Construction. *Buildings* **2014**, *4*, 266–294. [CrossRef]
2. Kedir, F.; Hall, D.M. Resource Efficiency in Industrialized Housing Construction—A Systematic Review of Current Performance and Future Opportunities. *J. Clean. Prod.* **2021**, *286*, 125443. [CrossRef]
3. Wilson, A.; Boehland, J. Small Is Beautiful. *J. Def. Model. Simul.* **2005**, *9*, 277–287. [CrossRef]
4. Kumar Dhar, T.; Sk. Maruf Hossain, M.; Rubayet Rahaman, K. How Does Flexible Design Promote Resource Efficiency for Housing? A Study of Khulna, Bangladesh. *Smart Sustain. Built Environ.* **2013**, *2*, 140–157. [CrossRef]
5. Wu, G. A Multi-Objective Trade-off Model in Sustainable Construction Projects. *Sustainability* **2017**, *9*, 1929. [CrossRef]
6. Hertwich, E.G.; Ali, S.; Ciacci, L.; Fishman, T.; Heeren, N.; Masanet, E.; Asghari, F.N.; Olivetti, E.; Pauliuk, S.; Tu, Q.; et al. Material Efficiency Strategies to Reducing Greenhouse Gas Emissions Associated with Buildings, Vehicles, and Electronics—A Review. *Environ. Res. Lett.* **2019**, *14*, 043004. [CrossRef]
7. Sharma, M.J.; Yu, S.J. Performance Based Stratification and Clustering for Benchmarking of Container Terminals. *Expert Syst. Appl.* **2009**, *36*, 5016–5022. [CrossRef]
8. Charnes, A.; Cooper, W.W.; Rhodes, E. Measuring the Efficiency of Decision Making Units. *Eur. J. Oper. Res.* **1978**, *2*, 429–444. [CrossRef]
9. Banker, R.D.; Charnes, A.; Cooper, W.W. Some Models for Estimating Technical and Scale Inefficiencies in Data Envelopment Analysis. *Manage. Sci.* **1984**, *30*, 1078–1092. [CrossRef]
10. Charnes, A.; Cooper, W.W.; Lewin, A.Y.; Seiford, L.M. *Data Envelopment Analysis: Theory, Methodology, and Applications*; Kluwer Academic: Dordrecht, the Netherlands, 1994.
11. Wagner, J.M.; Shimshak, D.G. Stepwise Selection of Variables in Data Envelopment Analysis: Procedures and Managerial Perspectives. *Eur. J. Oper. Res.* **2007**, *180*, 57–67. [CrossRef]
12. Jenkins, L.; Anderson, M. A Multivariate Statistical Approach to Reducing the Number of Variables in Data Envelopment Analysis. *Eur. J. Oper. Res.* **2003**, *147*, 51–61. [CrossRef]
13. Sexton, T.R.; Silkman, R.H.; Hogan, A.J. Data Envelopment Analysis: Critique and Extensions. *New Dir. Progr. Eval.* **1986**, *32*, 73–105. [CrossRef]

engineering proceedings

Proceeding Paper

Application of Road Compaction Quality Control System to Road Pavement Construction for Advanced Quality Control †

Kei Sasaki [1,*], Hiroaki Aoki [2], Masakazu Jomoto [1] and Yasuhiro Mori [3]

1 Institute of Research and Development, Taiseirotec Corporation, Kounosu 365-0027, Saitama, Japan; masakazu_jomoto@taiseirotec.co.jp
2 Technology Center, Taisei Corporation, Yokohama 245-0051, Kanagawa, Japan; aoki-h@ce.taisei.co.jp
3 Machinery Department, Soil and Rock Engineering Corporation, Toyonaka 561-0834, Osaka, Japan; mori@soilandrock.co.jp
* Correspondence: kei_sasaki@taiseirotec.co.jp

† Presented at the Second International Conference on Maintenance and Rehabilitation of Constructed Infrastructure Facilities, Honolulu, HI, USA, 16–19 August 2023.

Abstract: Japan is undergoing a social transition with a decreasing workforce in the construction industry; therefore, improving productivity is an urgent issue for sustaining social infrastructure. This paper presents a quality control system for road pavement construction, relying on non-destructive and automatic measurement for compaction quality control. The system's applicability and measurement accuracy were verified in 2020 and 2021 for national highway pavement construction, commissioned by the Ministry of Land, Infrastructure, Transport and Tourism. The system is labor-saving and has equivalent accuracy to the currently used transmission RI for compaction quality control. This paper reports on an overview of the verification.

Keywords: productivity improvement; labor-saving; real time; quality control; RI

Citation: Sasaki, K.; Aoki, H.; Jomoto, M.; Mori, Y. Application of Road Compaction Quality Control System to Road Pavement Construction for Advanced Quality Control. *Eng. Proc.* **2023**, *36*, 47. https://doi.org/10.3390/engproc2023036047

Academic Editor: Hosin (David) Lee

Published: 17 July 2023

1. Introduction

Currently, road construction and base course construction ordered by the Ministry of Land, Infrastructure, Transport and Tourism are controlled by the degree of compaction as a quality control method. The Test Method for Soil Density by the Sand replacement method and Radio-Isotope(RI) instrumentation are commonly used to determine the degree of compaction.

These test methods are performed by setting several measurement points from the construction area where the compaction has been completed, which results in "point" control, and there is a concern that vulnerable points may be overlooked. In addition, human work is required due to the need to excavate or make boreholes. Therefore, at least one person in charge of quality control is required to conduct quality control tests. Since people often hunker down when working in the field, they may easily enter the blind spots of heavy machinery, such as compaction rollers, which may lead to serious accidents resulting from contact with heavy machinery.

In the Test Method for Soil Density by Sand Replacement method, the test results are not available until the next day at the earliest since the water content ratio test is conducted in a room after the on-site test. This time lag may lead to rework in the case of results that do not meet the specified values. The issues identified in response to the current situation described above are as follows.

1. Develop a method to measure not only "points" but also "surfaces" to prevent overlooking vulnerable points.
2. Eliminate human work, improve safety and reduce labor.
3. To grasp the measurement results in real time, to enable corrective measures to be taken during construction and to improve construction efficiency.

2. Technical Overview of Quality Control System for Ground Compaction

The ground compaction quality control system consists of a vibrating roller equipped with a rolling wheel-type RI instrument, as shown in Figure 1. The rolling wheel-type RI instrument is rolled on the measuring surface to measure the degree of compaction automatically and non-destructively continuously. The rolling wheel-type RI instrument can be raised and lowered by an actuator, and measurement is performed with the instrument grounded on the measurement surface. It can be raised and lowered with a tablet installed on the operator's seat. Since it is necessary to keep the rolling wheel RI instrument in close contact with the road surface during measurement, a mechanism has been developed to hold down the rolling wheel housing [1].

Figure 1. Vibratory roller with rolling wheel-type RI instrument.

As shown in Figure 2, the measurement results are linked to the GNSS location information and displayed on a tablet terminal installed in the driver's seat in the form of a heat map in real time, allowing the user to visually grasp the compaction level in the area of the measurement point. The measurement results uploaded to the cloud server can be processed into the required format and shared among the parties concerned at any time.

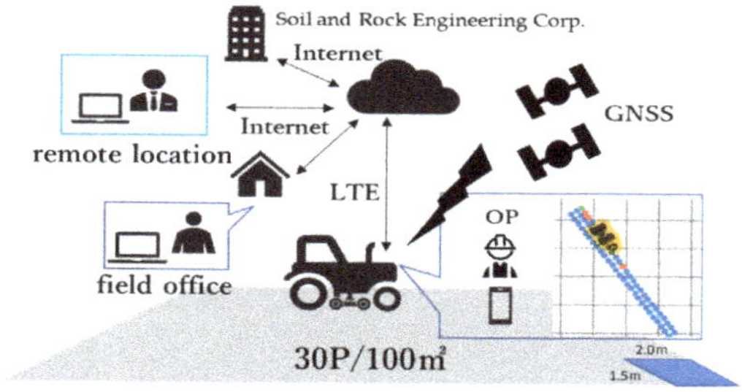

Figure 2. Configuration of the system.

The measurement principle of the rolling wheel-type RI meter is the same as that of the current scattering type RI meter. For the RI source used in the rolling wheel-type RI meter, both the density and moisture gauges are in compliance with the Japanese laws and regulations established in accordance with international standards. In addition, they have been registered with the Nuclear Regulation Authority of Japan (NRAJ) as a design-certified instrument.

As shown in Figure 3, a radiation source and a detector are built into the cylindrical density and moisture gauges, and the radiation emitted from the radiation source is scattered by the object to be measured and counted by the detector. The measured values of the degree of compaction by the rolling wheel-type RI meter are obtained by converting the gamma and neutron radiation readings. The wet density and water content ratio are calculated from the measured values using a calibration formula, and the degree of compaction is obtained from them. For the calibration equation, a calibration diagram showing the correspondence between the count rate ratio and the measured values was created by comparing the calibration equation with a reference measurement method, such as the Test Method for Soil Density by the Sand replacement method and the approximate curve was used.

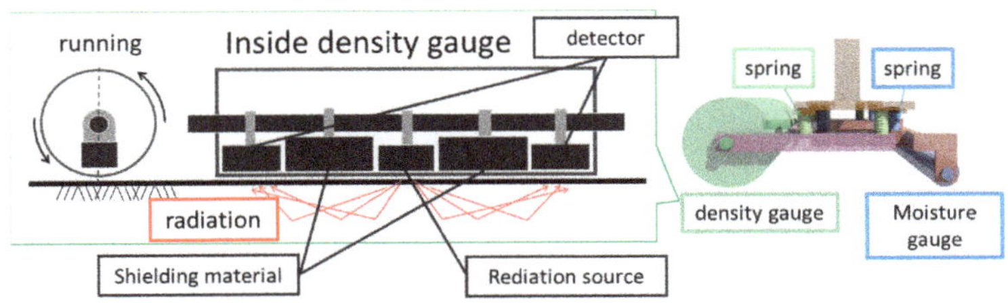

Figure 3. Structure of rolling wheel-type RI meter.

Based on past findings [2], the degree of compaction was calculated by counting the radiation emitted when the roller travels 2 m in the direction of travel as one measuring cell and estimating the density and moisture content based on the number of counts.

3. Verification of Applicability and Measurement Accuracy at Construction Sites

The Ministry of Land, Infrastructure, Transport and Tourism (MLIT) is promoting the "Project on the Introduction and Use of Innovative Technologies to Dramatically Improve Productivity at Construction Sites" using funds from the Public/Private R&D Investment Strategic Expansion PrograM (PRISM) promoted by the Cabinet Office. Currently, the project is being promoted, and the Ground Compaction Quality Management System has been adopted as a "technology to improve quality control in civil engineering works by utilizing data" in 2020 and 2021. Its applicability and measurement accuracy have also been verified at construction sites.

3.1. Results of Compaction Measurement of the Lower Base Course

The density and water content of recycled crushed stone 0–40 used in the subbase course were examined to see if they could be measured using the calibration diagram created using the previously measured values. The results are shown in Figure 4, which shows that the measurement accuracy of the rolling wheel-type RI meter used in this experiment was within ±0.1 [g/cm^3] for wet density and ±0.05 [g/cm^3] for water content compared to the transmission type RI meter at this experimental site. These values were found to be equivalent to the errors of commonly used transmission RI meters.

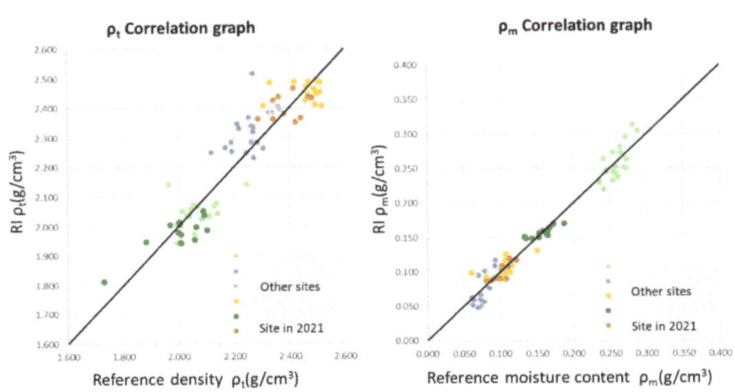

Figure 4. Correlation graph of transmission RI meters and rolling wheel-type RI instrument (lower base course).

3.2. Effects Related to Reduction of Work Hours and Manpower

The rolling wheel-type RI meter used in this trial can measure the degree of compaction during construction, so there is no need to set aside time for a separate field test, as is the case in the current system. Therefore, the time required for field density tests at conventional pavement construction sites, as shown in Table 1, can be reduced. Moreover, with this technology, there is no need for a person to conduct field density tests, a person to conduct laboratory tests, and a person to prepare forms, and all of these tasks can be performed by a single roller operator.

Table 1. Reduction of work time.

Type of Works	Base Course Construction
Assumed area	10,000 m^2
Time required for Test Method for Soil Density by the Sand replacement method	20 min/one place ... 10 place $\times$ 20 min
Reduction of work time	3.3 h

3.3. Effects on Quality Assurance and Improvement

The conventional management standard for base course construction is 1 point per 1000 m^2, but this technology enables management at approximately 333 points per 1000 m^2. Therefore, it is possible to prevent the overlooking of vulnerable areas and to more accurately determine the extent of ground improvement required. In addition, since the measurement results of the degree of compaction can be grasped in real time, corrective measures such as re-compaction can be taken promptly during construction to ensure and improve quality.

4. Conclusions

The use of the rolling wheel-type RI meter enabled us to grasp the ground conditions in real time and to control the surface, and the automatic measurement, which does not require human intervention, has confirmed the merits of advanced and labor-saving quality control. On the other hand, due to the characteristics of the rolling wheel-type RI meter, a problem in how to handle a counting error when the meter floats during measurement was also found. In the future, a verification process will be conducted to determine whether the errors are due to erroneous values or actual measurements.

Author Contributions: Conceptualization, H.A. and M.J.; methodology, H.A., Y.M. and M.J.; formal analysis, Y.M. and M.J.; investigation, K.S. and Y.M.; data curation, K.S., M.J. and Y.M.; writing—original draft preparation, K.S.; writing—review and editing, K.S. and M.J.; visualization, K.S.; supervision, M.J. All authors have read and agreed to the published version of the manuscript.

Funding: The demonstration at a construction site was funded by the "Project on the Introduction and Use of Innovative Technologies to Dramatically Improve Construction Site Productivity" promoted by the Ministry of Land, Infrastructure, Transport and Tourism under the Public/Private R&D Investment Strategic Expansion PrograM (PRISM) program promoted by the Cabinet Office. The funder for this project is the Ministry of Land, Infrastructure, Transport and Tourism.

Institutional Review Board Statement: Not applicable.

Informed Consent Statement: Not applicable.

Data Availability Statement: No new data were created or analyzed in this study. Data sharing is not applicable to this article.

Conflicts of Interest: The authors declare no conflict of interest.

References

1. Hiroaki, A.; Koichi, G.; Taichi, I.; Yasuhiro, M.; Sosuke, K. Automatic compaction degree measurement using a rolling wheel type RI density moisture meter—Application to an automated vibrating roller. In Proceedings of the 75th JSCE Annual Conference VI-1138, Online, 7–9 September 2020.
2. Koichi, G.; Hiroaki, A.; Taichi, I.; Yasuhiro, M.; Sosuke, K. Experiments on non-destructive measurement using a rolling wheel type RI density moisture meter—Toward application to an automatic traveling vibrating roller. In Proceedings of the JSCE 74th Annual Conference VI-356, Takamatsu, Japan, 3–5 September 2019.

engineering proceedings

MDPI

Proceeding Paper

Influence of Road Traffic Vibration on Micro-Dynamic Response of Precision Instrument Vibration Isolation Platforms †

Ying Zhang [1], Xihui Wang [1], Chenhu Lu [1], Kehui Liu [2] and Bo Song [1,*]

[1] School of Civil and Resource Engineering, University of Science and Technology Beijing, Beijing 100083, China; m202120166@xs.ustb.edu.cn (Y.Z.); xihuiw@126.com (X.W.); luchenhu@163.com (C.L.)
[2] Institute of Urban Systems Engineering, Beijing Academy of Science and Technology, Beijing 100089, China; lkh_2005@126.com
* Correspondence: songbo@ces.ustb.edu.cn
† Presented at the Second International Conference on Maintenance and Rehabilitation of Constructed Infrastructure Facilities, Honolulu, HI, USA, 16–19 August 2023.

Abstract: This paper analyzes the impact of environmental vibration caused by road traffic operations on precision instrument vibration isolation platforms. The research results show that there are directional differences in the speed response of the vibration isolation platform caused by vehicle load. The speed response is highest in the horizontal direction perpendicular to the vehicle's motion. The dominant frequency of the dynamic response of the isolation platform caused by road vehicle loads was concentrated at 2.5 Hz, mainly low-frequency vibrations. The monitoring results are basically consistent with the trend of structural vibration changes in the numerical simulations.

Keywords: traffic vibration; vibration isolation platform; numerical simulation; on-site monitoring; micro-dynamic response

Citation: Zhang, Y.; Wang, X.; Lu, C.; Liu, K.; Song, B. Influence of Road Traffic Vibration on Micro-Dynamic Response of Precision Instrument Vibration Isolation Platforms. *Eng. Proc.* **2023**, *36*, 48. https://doi.org/10.3390/engproc2023036048

Academic Editor: Hosin (David) Lee

Published: 17 July 2023

1. Introduction

Higher laboratory standards demand reduced environmental vibration, which, when transmitted through the laboratory platform, may introduce significant errors, thus hindering proper instrument functionality. Road vehicle-induced vibrations are among the primary factors influencing precision instrument performance.

Scholars worldwide have researched the characteristics of road vibration acceleration caused by vehicle driving, focusing mainly on amplitude and spectrum [1]. Chen et al. [2] measured ambient vibration on a soft soil site caused by heavy vehicles and found that the generated vibration exceeded the normal use standard of instruments, with heavy vehicles producing greater vibration than super-compact vehicles. Beben et al. [3] observed that the distance between buildings and fences played a vital role in reducing traffic vibration on damaged buildings. Cao et al. [4] identified that multi-peak ground vibration in a short time would impact the normal operation of precision instruments in a vehicle during monitoring of vehicle vibration on a trunk road in Beijing.

This study investigated the micro-dynamic response of a precision instrument vibration isolation platform to road traffic vibration using numerical simulations. Vehicle load parameter analysis was conducted on the platform at the Changping experimental base of the China Metrology Institute to provide a design and route planning reference for precision instrument vibration isolation platforms considering the impact of road traffic.

2. Three-Dimensional Dynamic Analysis Model Establishment

Engineering Background

The indoor vibration isolation platform level of a mechanical metrology laboratory in Changping Experimental Base is shown in Figure 1. The vibration isolation platform size was 13 m × 5.5 m × 3 m, with concrete strength class C30, modulus of elasticity E = 30 GPa, Poisson ratio μ = 0.2, and density ρ = 2500 kg/m^3, as shown in Figure 2.

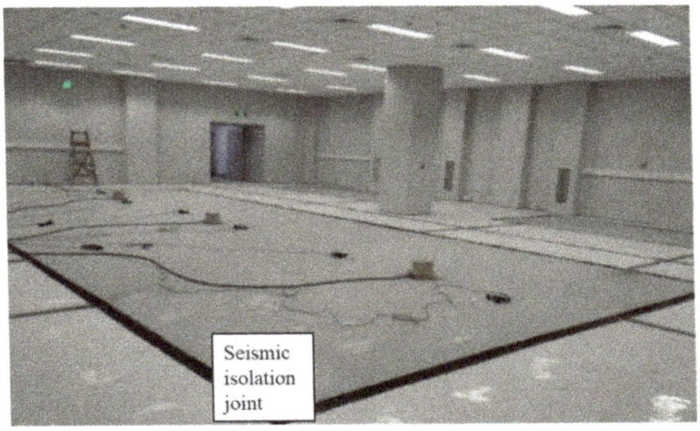

Figure 1. Site drawing of vibration isolation platform.

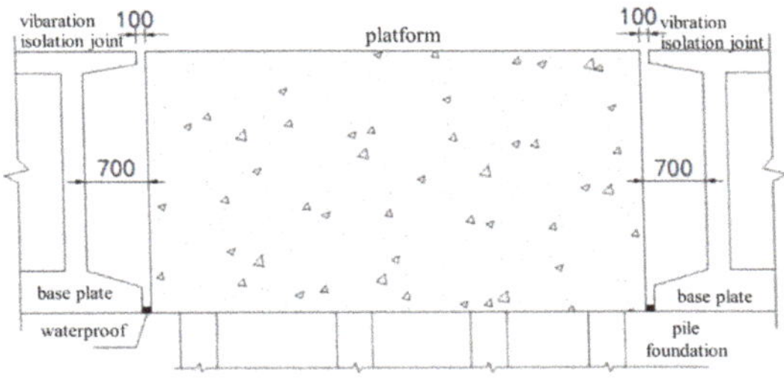

Figure 2. Section of vibration isolation platform (unit: mm).

In order to accurately reflect the vibration response of a vibration isolation platform, a finite element model including the Upper Structure–Vibration Isolation Platform–Soil Layer of the experimental building was established. The overall finite element model that has been established is illustrated in Figure 3.

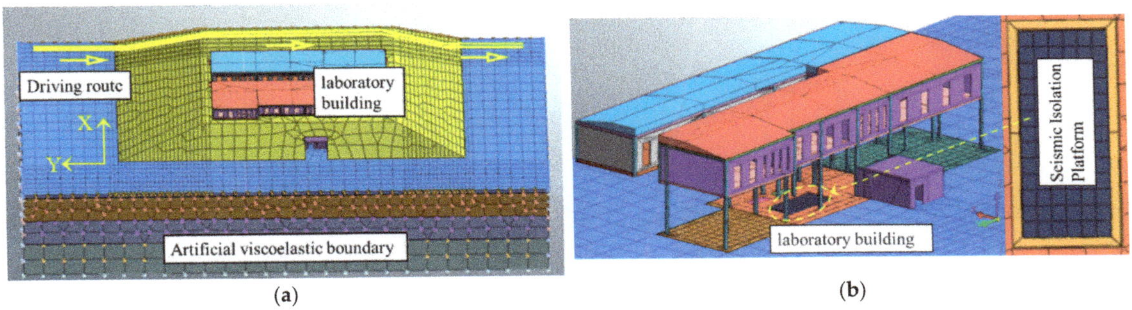

Figure 3. Three dimensional finite element model of superstructure–vibration isolation platform–soil. (**a**) Integral model; (**b**) upper structure and vibration isolation platform.

3. Micro-Dynamic Response Analysis

3.1. Analysis of Velocity Time-History Results of Vibration Isolation Platform

As shown in Figure 4, diagonal control points 1 and 2 were selected on the vibration isolation platform based on the platform requirements and micro-vibration observations. Figure 4 shows the velocity response time-history curves in the X, Y, and Z directions for these points. Under the action of a vehicle load on the road surface, the horizontal speed response of the vibration isolation platform perpendicular to the vehicle driving route (X) was the largest, followed by the horizontal response parallel to the driving route (Y) and the vertical speed response (Z) was the smallest. Therefore, precision instruments functioning in the horizontal direction can reduce the impact of traffic vibration by properly designing the platform direction.

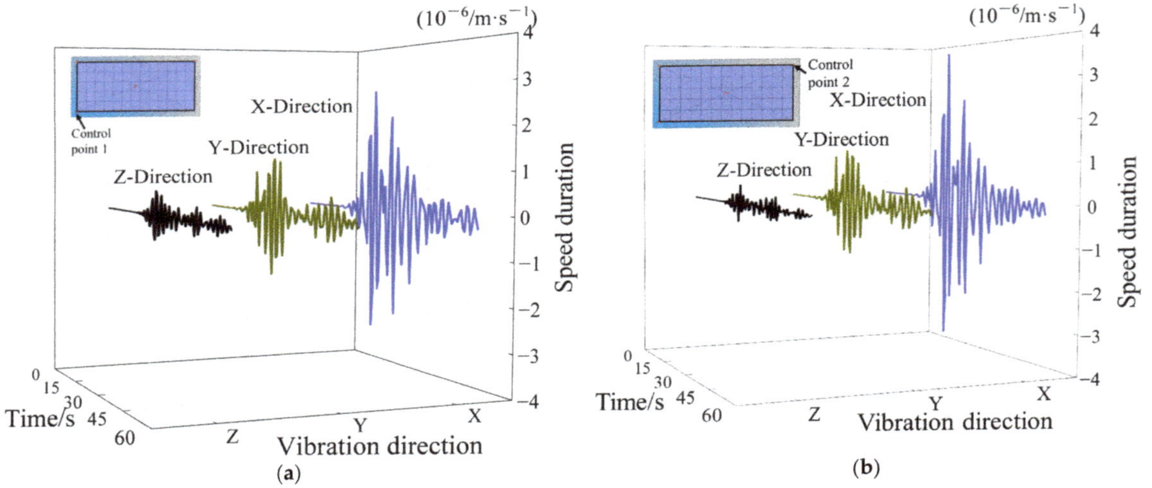

Figure 4. Speed response time history results. (**a**) Control point 1 speed history curve; (**b**) control point 2 speed history curve.

3.2. Vehicle Load Parameter Analysis

When the distance between the vibration sources was constant (35 m), the peak vertical vibration speed of the vibration isolation platform was extracted as shown in Figure 5. As can be seen from Figure 5, when the vibration source distance was constant and the vehicle speed was 20 m/s, the peak speed increased with the increase in vehicle load.

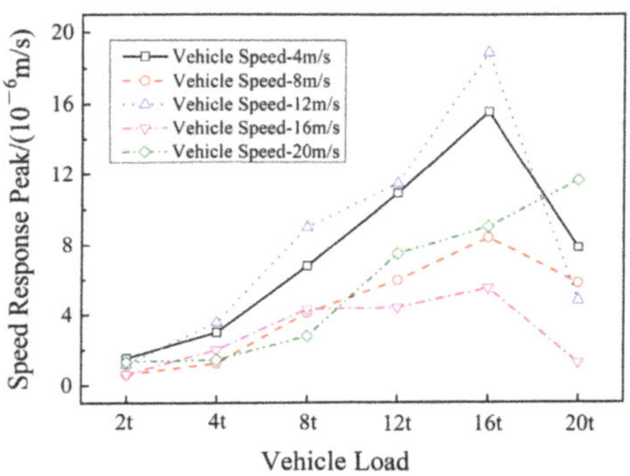

Figure 5. Relationship of response to load.

For a vehicle speed below 20 m/s, when the load exceeded 16 t, the peak curve showed a downward turning point. It can be seen that the vibration isolation platform had the largest dynamic response when the vehicle speed was 12 m/s and the load was 16 t.

When the vehicle load was constant (16 t), the variation of the peak vertical vibration speed of the vibration isolation platform with the source distance were extracted at different vehicle speeds, as shown in Figure 6. It can be seen from Figure 6 that when the vehicle load was constant, the peak vibration speed increased as the vibration source distance decreased. At a speed of 12 m/s and a vibration source distance of 35 m, the vibration isolation platform had the largest dynamic response.

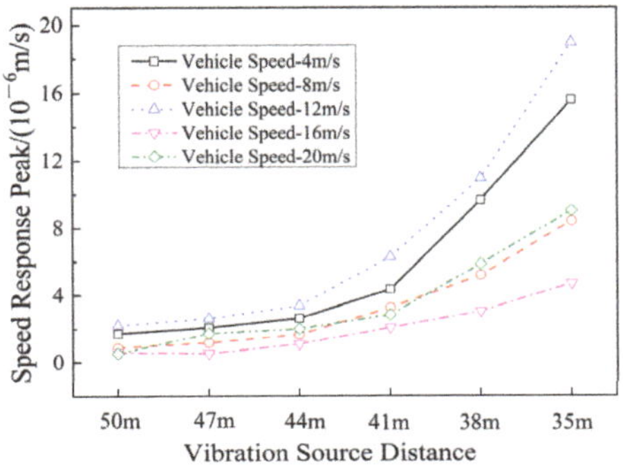

Figure 6. Relationship between velocity response and distance.

In summary, through parameter analysis of three key influencing factors, i.e., vehicle load, horizontal distance of vibration source, and driving speed, it can be concluded that the vibration response of vibration isolation platform is the most violent when driving at a medium speed with a vibration source distance of 35 m, load of 16 t, and speed of 12 m/s, which can be considered as the most unfavorable calculation conditions.

4. Comparison and Analysis of Field Vibration Measurement

High-precision sensors are required for monitoring the dynamic response of structures affected by traffic vibration due to the very small amplitude. Two high-precision sensors, VSE-355G3 from Tokyo and 941B vibration pick-up instrument from China Seismological Bureau, were utilized in the field as illustrated in Figure 7.

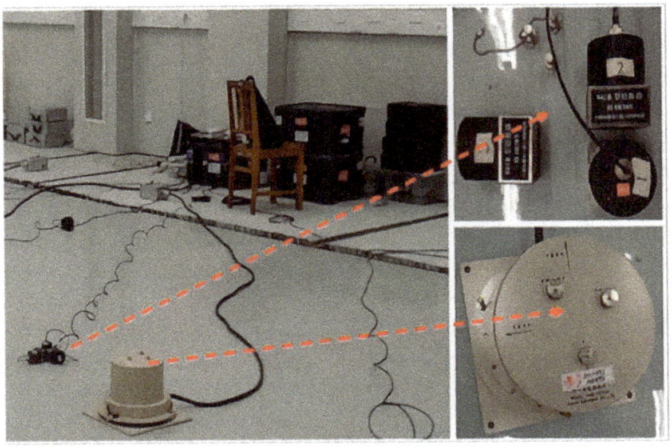

Figure 7. On-site monitoring.

To compare the measured and simulated results, a common mini-bus with a 2-ton weight, driving at 12 m/s and at a horizontal distance of 35 m from the vibration isolation platform was used. The vertical vibration speed time history was extracted and converted to 1/3 times frequency range results, as shown in Figure 8. The dynamic response dominant frequency of the vibration isolation platform caused by road traffic vibration was mainly concentrated at 2.5 Hz, which is a low frequency. The measured results were generally larger than the numerical simulation results due to the influence of on-site background vibrations. However, the results measured in the field match the numerical simulation results in terms of the change trend, verifying the accuracy of the numerical model.

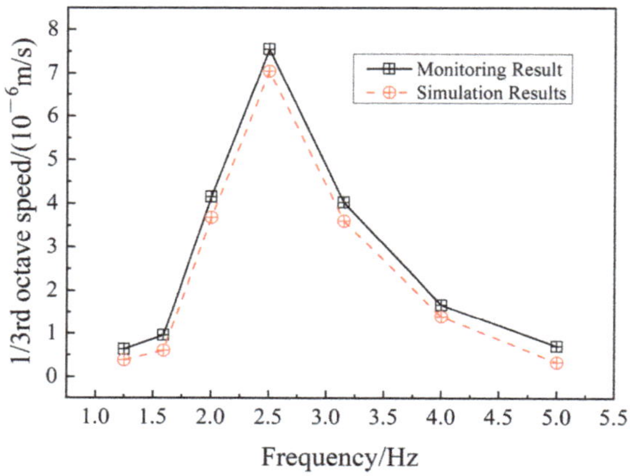

Figure 8. Comparison of measured and simulated results.

5. Conclusions

(1) There is a directional difference in the peak velocity response of an isolation platform caused by vehicle load. The horizontal velocity response perpendicular to the vehicle's driving direction is the largest, followed by the horizontal velocity response parallel to the driving direction, and the vertical velocity response is the smallest. Therefore, the impact of traffic vibration can be reduced by designing the platform direction.

(2) We conducted parameter analysis on three key influencing factors: vehicle load, vibration source horizontal distance, and driving speed. The vibration response of the isolation platform was the most severe when the vibration source distance was 35 m, the load was 16 tons, and the vehicle speed was 12 m/s. This set of parameters can be used as the most unfavorable working conditions for evaluating the micro-dynamic response of precision instruments.

(3) The dominant frequency of the dynamic response of the isolation platform caused by road vehicle loads was concentrated at 2.5 Hz, with low-frequency vibrations being the main factor. Due to the influence of on-site background vibration, the overall monitoring results were greater than the numerical simulation results. The variation trends of the on-site measurement results and numerical calculation results are basically consistent, verifying the correctness of the numerical model. Use of specifications that exceed their limits may affect the normal use of the instrument.

Author Contributions: Methodology, B.S. and X.W.; Software, C.L. and X.W.; Validation, C.L. and Y.Z.; Formal analysis, B.S. and X.W.; Investigation, K.L.; Writing—review and editing, Y.Z. All authors have read and agreed to the published version of the manuscript.

Funding: This research was funded by the National Natural Science Foundation of China, No. 52078038, and the USTB Research Center for International People-to-People Exchange in Science, Technology, and Civilization, No. 2022FKYB012.

Institutional Review Board Statement: Not applicable.

Informed Consent Statement: Not applicable.

Data Availability Statement: The data are not publicly available due to privacy restriction.

Conflicts of Interest: The authors declare no conflict of interest.

References

1. Zhang, F.; Feng, D.; Ling, X.; Wang, Z.; Zhu, Z. Vertical dynamical analysis model of heavy vehicle-pavement-subgrade during spring-thawing period. *China J. Highw. Transp.* **2011**, *24*, 7–14.
2. Chen, X.; Jiang, J.; Hu, Y.; Sheng, T.; Tang, K. Experimental Study on The Influence of Precision Instruments Caused by Heavy Vehicles Vibration. *J. Phys. Conf. Ser.* **2021**, *2044*, 012113. [CrossRef]
3. Beben, D.; Maleska, T.; Bobra, P.; Duda, J.; Anigacz, W. Influence of Traffic-Induced Vibrations on Humans and Residential Building—A Case Study. *Int. J. Environ. Res. Public Health* **2022**, *19*, 5441. [CrossRef] [PubMed]
4. Cao, Y.; Li, Z.; Yang, C. On-Site Experiment and Characteristics Analysis of Ground Vibrations Induced by Vehicle Loads Moving on an Urban Trunk Road. *Transp. Res. Rec.* **2023**, *2677*, 1500–1519. [CrossRef]

engineering proceedings

MDPI

Proceeding Paper

Data-Driven Analysis for Road Traffic Conditions Using Digital Tachograph Data †

Sung-Bum Yun * and SoonYong Park

Urban Strategy Research Division, Seoul Institute of Technology, Seoul 03909, Republic of Korea; psy@sit.re.kr
* Correspondence: yunsb33@sit.re.kr

† Presented at the Second International Conference on Maintenance and Rehabilitation of Constructed Infrastructure Facilities, Honolulu, HI, USA, 16–19 August 2023.

Abstract: Traffic condition analysis requires various conditions to be met using the conventional method. This has the limitation that it does not reflect the congestion caused by the actual vehicle flow. In this study, we suggest data-based traffic condition analysis. This is a method of determining the dynamic traffic conditions using Digital TachoGraph data, which can reflect the flow rate and the traveling speed of vehicles by the actual time zone. The suggested system could be implemented in both public and private sectors to create new possibilities and insights. For the public sector, police patrol vehicles can implement the system to create a 'dynamic service area' which would enhance efficiency for security patrols. For private sectors, the system could be applied to various call-dispatch systems to minimize waiting time for customers and driving distance for drivers. Also, it could be applied to upcoming autonomous vehicle sharing systems to ensure maximum coverage for autonomous cars.

Keywords: Digital TachoGraph (DTG); traffic analysis; data-driven analysis

Citation: Yun, S.-B.; Park, S. Data-Driven Analysis for Road Traffic Conditions Using Digital Tachograph Data. *Eng. Proc.* **2023**, *36*, 49. https://doi.org/10.3390/engproc2023036049

Academic Editor: Hosin (David) Lee

Published: 17 July 2023

1. Introduction

Conventional methods of traffic condition analysis most frequently rely on traffic flow analysis and traffic system analysis [1] in order to derive the traffic conditions, determining whether the traffic flow will increase or decrease at a certain location at a certain time. The conventional method is effective in terms of city planning and deriving various methods to control the flow of traffic within a city. However, since the conventional method uses systematic analysis of predicted values, it lacks the ability to reflect accidental events or real-world traffic conditions which could vary due to weather conditions or events taking place within a city.

In this study, we suggest a data-driven method to analyze traffic conditions using DTG (Digital TachoGraph) data within Seoul, South Korea. DTG data containing GPS trajectories of 30,000 vehicles within South Korea were used to create a serviceability map for the Seoul area. The map dynamically changed over time showing differences in the serviceable area within certain hours of the day. The serviceability map was drawn using the DTG data with GPS locations by calculating the movement of each vehicle within the given threshold time.

Using the suggested data-driven method could provide various aids for analyzing traffic conditions since it could reflect the actual movement of vehicles, which would react to real-world events such as accidents or the gathering of mass crowds.

2. Data

2.1. DTG Data

For the study, DTG data from January 2020 to December 2020 acquired from the Ministry of Land, Infrastructure and Transport (MOLIT) [2] were used. Information about the data is stated in Table 1. The size of the data is roughly 800 GB per month, since the DTG sensors create 1 row of data every 10 s. The total size of the data for 12 months exceeded 2.5 TB, thus preprocessing steps to remove unnecessary data were conducted beforehand.

Table 1. Data dictionary (DTG data).

Column Name	Description	Example
Date	YYYY/MM/DD	20200104
Time	HH/MM/SS	123002
CARID	Encrypted car ID	A64229AA508BDDCAAE6B9923481C3442
CC	ID for vehicle company	1252361
DTD	Daily travel distance (km)	47
ATD	Accumulated travel distance of vehicle (km)	42,526
TS	km/h (000~255)	69
ACCX	Acceleration for X axis m/s^2	−1.0
ACCY	Acceleration for Y axis m/s^2	5.2
RPM	Rev per minute	1480
BS	0 (off) or 1 (on)	0
X	X GPS coordinate of vehicle (WGS84)	127.188979
Y	Y GPS coordinate of vehicle (WGS84)	36.919575
AZI	GPS azimuth in degrees (0~360)	179

DTG collects data from 'company owned vehicles', such as taxis, buses, freight trucks, etc., which requires continuous management. Originally the data were used for management purposes; however, since the data contain GPS information, for this study, the vehicle data were used as 'agents' depicting the traffic conditions of the Seoul area.

2.2. DTG Data Pre-Processing

The data contained the DTG information of about 30,000 vehicles which travel around South Korea. However, since special vehicles such as freight trucks which usually transport heavy loads throughout the whole of South Korea, using data collected from special vehicles could lead to biased results due to the specific usage of the vehicle. Hence, by using the company ID column, vehicles with specific usages were removed from the study data. Afterwards, the data were processed to only contain information within Seoul, the study area. The preprocessed data only contained vehicles within Seoul which includes in-town buses and private taxis.

3. Method

Service Area Analysis and Data-Driven Method

In order to extract traffic conditions using DTG data containing GPS information, the service area analysis method [3] was used. The conventional method for service area analysis uses the road network as baseline data and implements the Dijkstra algorithm to calculate the reachable distance within a given threshold. However, this method has limitations as it cannot reflect real-world conditions such as rush-hour congestion, resulting in a smaller service area within the same threshold time. Even though the conventional method can use average or maximum speed as inputs for each road segment, it only allows for a fixed value for speed and cannot consider real-world traffic conditions.

On the contrary, the suggested method of data-driven service area analysis uses accumulated GPS location data, which contains actual movement of vehicles in a specified time. By using actual vehicle GPS data, the traffic conditions can be reflected into the service area analysis, allowing for a more accurate representation of the service area size. For example, in late-night hours when traffic flow is low, vehicles tend to travel quicker than usual, while during rush hour with high traffic flow, vehicles will move much slower, providing a smaller service area.

4. Result

The result of the service area analysis within the Seoul area is depicted in Figure 1, where the green areas are areas which could be accessed within a 5 min threshold time. During rush hour (a), the accessibility of Seoul traffic conditions decreased, showing some non-accessible areas. Whereas during non-rush hour times, the green areas tended to cover the whole city (excluding the river and mountain areas).

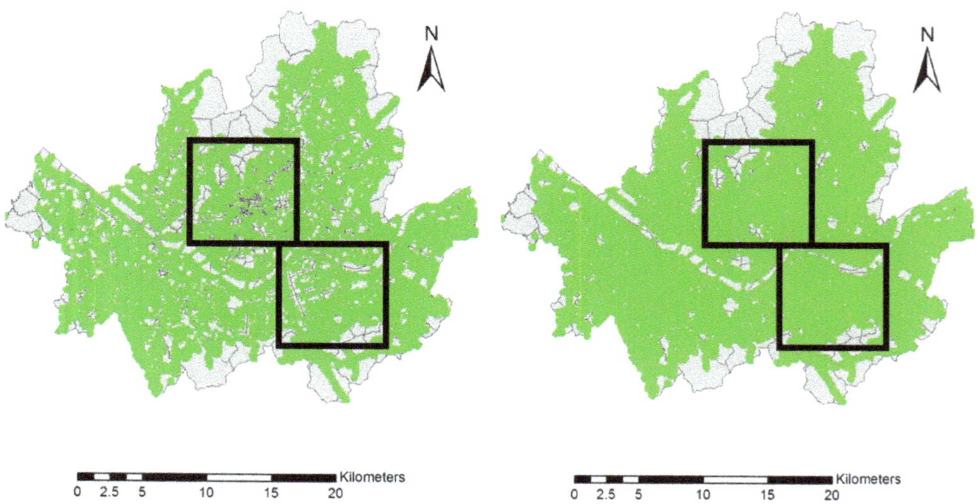

(a) (b)

Figure 1. Seoul traffic accessibility within 5 min during (**a**) rush hour; (**b**) non-rush hour times.

In this study, a proximity analysis method experiment was conducted to derive a data-driven dynamic service area for mobile vehicles. This allowed for the evaluation of the service area for each vehicle within a 5 to 10 min threshold with a time-resolution of 30 min. Additionally, a pre-calculated road segment node dynamic service area was derived to quickly respond to service area queries.

The DTG data used in the study were collected from MOLIT; the data had been accumulating for over 1 year, and due to the large file size (800 GB per month on average), preprocessing was conducted to remove non-significant data. By implementing the data-driven service area analysis method, the dynamic service areas for vehicles within Seoul were derived with a 5 min temporal resolution. The time threshold of the service area was set as 5 min and 10 min, which could be altered by users to derive a user-defined time threshold service area. However, the 5 min threshold dynamic service area still contained some uncovered spaces, especially during rush hours.

Compared to conventional methods, the suggested dynamic service area analysis could consider traffic conditions, as shown in Figure 1a,b, since the method makes use of GPS data, which are also affected by real-world traffic conditions at any given time. The system could be implemented in both public and private sectors to draw new insights. In the emerging field of smart cities, when the system is implemented with police patrol car data, it could result in an optimized route for patrol, leading to a more efficient coverage area and quicker response time. If the system is implemented in private sectors, new insights that can benefit both suppliers and customers can be derived. For example, in the field of taxi call-dispatch systems, customers can be connected to taxis within an accurate 5 min service area. Additionally, when implemented in upcoming autonomous vehicles, the system could present various possibilities for auto-piloting cars to ensure maximum coverage and quick response times.

Author Contributions: Conceptualization, S.-B.Y.; methodology, S.-B.Y.; software, S.-B.Y.; validation, S.-B.Y.; formal analysis, S.-B.Y.; data curation, S.-B.Y.; writing—original draft preparation, S.-B.Y.; writing—review and editing, S.-B.Y., S.P.; visualization, S.-B.Y.; supervision, S.P.; project administration, S.P.; funding acquisition, S.P. All authors have read and agreed to the published version of the manuscript.

Funding: This work was supported by the Korea Institute of Police Technology (KIPoT) grant funded by the Korean government (KNPA) (No. 092021C26S02000, Development of Transportation Safety Infrastructure Technology for Lv.4 Connected Autonomous Driving).

Institutional Review Board Statement: Not applicable.

Informed Consent Statement: Informed consent was obtained from all subjects involved in the study.

Data Availability Statement: DTG data is private data managed by MOLIT (Ministry of Land Infrastructure and Transportation.

Conflicts of Interest: The authors declare no conflict of interest.

References

1. Tanzina, A.; Yodo, N. A survey of road traffic congestion measures towards a sustainable and resilient transportation system. *Sustainability* **2020**, *12*, 4660.
2. MOLIT. Available online: http://www.molit.go.kr/portal.do (accessed on 28 February 2023).
3. Cheng, G.; Zeng, X.; Duan, L.; Lu, X.; Sun, H.; Jiang, T.; Li, Y. Spatial difference analysis for accessibility to high level hospitals based on travel time in Shenzhen, China. *Habitat Int.* **2016**, *53*, 485–494. [CrossRef]

engineering proceedings

MDPI

Proceeding Paper

Real-Time Field Quality Management System for Asphalt Pavement Using Cloud †

Kyu-Dong Jeong [1], Dong-Hyuk Kim [1], Jae-Won Kim [1], Soo-Ahn Kwon [1], Nam-Ho Kim [2] and Sung-Do Hwang [1,*]

1 Korea Institute of Civil Engineering and Building Technology, Goyang-si 10223, Republic of Korea; kdjeong@kict.re.kr (K.-D.J.); my91kim@kict.re.kr (D.-H.K.); jewonkim@kict.re.kr (J.-W.K.); sakwon@kict.re.kr (S.-A.K.)

2 Architectural Engineering, School of Design and Architectural Engineering, Korea University of Technology & Education, Cheonan 31253, Republic of Korea; nhkim@koreatech.ac.kr

* Correspondence: sdhwang@kict.re.kr; Tel.: +(82)-31-910-0180

† Presented at the Second International Conference on Maintenance and Rehabilitation of Constructed Infrastructure Facilities, Honolulu, HI, USA, 16–19 August 2023.

Abstract: If the production and construction information of asphalt mixture are tightly coupled and quality control is performed in real time, it is possible to minimize quality degradation and solve problems early. For these objectives, a cloud-based IoT (Internet of Things) PQMS (Pavement Quality Management System) was developed in this study. As a result, drivers and managers can monitor construction information and identify problems using monitors and apps. In 2023, it will be applied to national road construction sites to verify the effectiveness of the proposed cloud-based IoT PQMS and address potential problems.

Keywords: PQMS; IoT; quality control; real-time quality management; intelligent asphalt pavement

Citation: Jeong, K.-D.; Kim, D.-H.; Kim, J.-W.; Kwon, S.-A.; Kim, N.-H.; Hwang, S.-D. Real-Time Field Quality Management System for Asphalt Pavement Using Cloud. *Eng. Proc.* **2023**, *36*, 50. https://doi.org/10.3390/engproc2023036050

Academic Editor: Hosin (David) Lee

Published: 18 July 2023

1. Introduction

Asphalt pavement has a different performance life depending on quality management during production and construction. However, non-real-time quality management system does not guarantee sufficient performance. Recently, with the rapid development of Internet technology, the era of real-time management of pavement quality has arrived. A lot of research has already been conducted on how to solve the poor quality of pavement construction by monitoring the construction process in real-time [1,2]. Dong et al. applied IoT and 5G technologies to develop a technology to collect and transmit data from mixing plant, transportation vehicles, and paving and compaction processes [3]. And Beainy developed a technology to evaluate the compaction quality in real-time and non-contact by measuring the vibration response of the compaction roller [4]. This study aims to extend the performance life of asphalt pavement through the acquisition, transmission, and processing of quality information at each stage of asphalt pavement construction using IoT technology.

To achieve these objectives, a cloud-based IoT PQMS(Internet of Things Pavement Quality Management System) was developed in this study. A technology has been developed to transmit construction quality information of dump trucks, asphalt pavers, and compaction rollers to the cloud. Through the analysis and processing of transmitted information, drivers and managers can use monitors and apps to monitor construction information and identify problems.

2. Method

Real-time field quality management system for asphalt pavement construction developed in this study is shown in Figure 1. The quality management process in asphalt pavement construction can be largely divided into three categories. First, at the asphalt paver, the quality related to the paving of the mixture can be managed through the paving control IoT module. Second, at the compaction roller, the quality related to compaction can be managed through the compaction control IoT module. Third, after construction, the final compaction quality can be managed with compaction test equipment. Quality management information can be transmitted to the cloud using IoT technology, and through this, the level of quality control can be improved. In this study, the modules to be attached to the equipment of each process was developed, and monitoring software that can check quality control information in real time was developed.

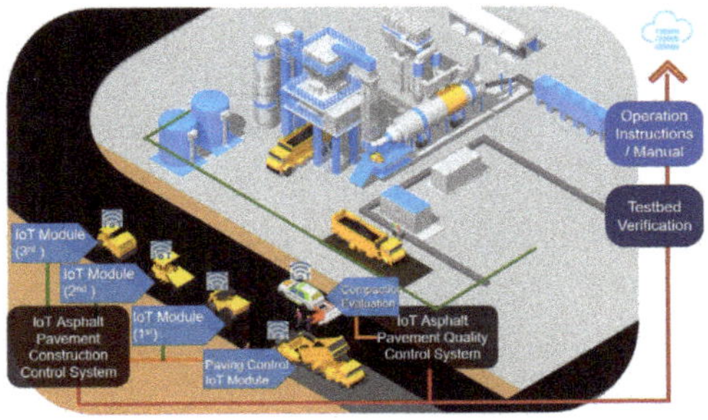

Figure 1. Real-time field quality management system for asphalt pavement construction.

3. Results

The modules including various sensors and devices to be attached to asphalt paver, compaction roller, and compaction test equipment were developed. In addition, quality information and navigation monitoring software that operators of asphalt paver and compaction roller can utilize was developed.

3.1. Asphalt Paver

The climate sensor, RTK GNSS, LTE, infrared camera, monitor, integrated board, and beacon were attached to the asphalt paver as shown in Figure 2a. The climate sensor collected weather information at the time of construction and will be used for big data analysis in the future. The RTK GNSS collected precise location information of asphalt paver during construction. The infrared camera collected the temperature of the mixture and the surface temperature of the pavement. The beacon recognized the truck that has arrived and made it possible to link the production information of the asphalt mixture. The integrated board stored the collected information, and the LTE device transmitted the stored information to the cloud. The information that the operator of the asphalt paver can see was as shown in Figure 2b. The operator can check the collected information such as the location and speed of the paver, the temperature of the transported mixture, and the surface temperature of the pavement through the monitor.

Figure 2. IoT PQMS related to asphalt paver: (**a**) Installed sensor; (**b**) Monitor.

3.2. Compaction Roller

The accelerometer sensor, RTK (Real Time Kinematic) GNSS (Global Navigation Satelite System), LTE (Long Term Evolution), monitor, analysis board, and infrared temperature sensor were attached to the compaction roller as shown in Figure 3a. The accelerometer sensor measured the vibration of the roller and estimated the degree of compaction. The RTK GNSS collected precise location information of compaction roller. The infrared temperature sensor collected the compaction temperature. The analysis board stored the collected information and analyzed the degree of compaction. And the LTE device transmitted the stored information to the cloud. The information that the operator of the compaction roller can see was as shown in the Figure 3b. The operator can check the collected information such as the location and speed of the roller, the compaction temperature, and compaction navigating through the monitor.

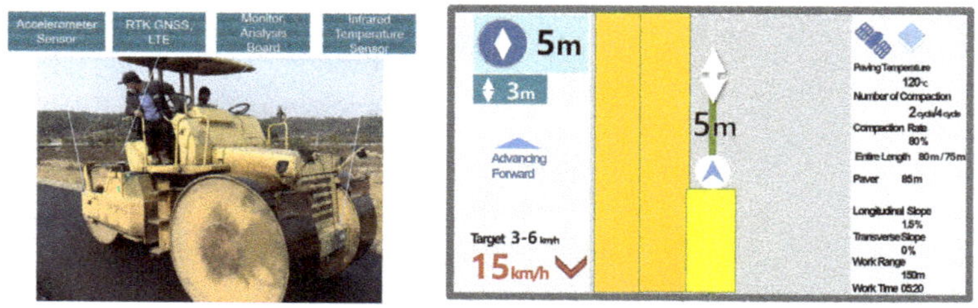

(a) (b)

Figure 3. IoT PQMS related to compaction roller: (**a**) Installed sensor; (**b**) Monitor.

3.3. Compaction Test Equipment

The lift sensor, image capturing sensor, RTK GNSS, road condition detection device, GPR (Ground Penetrating Radar) sensor, and DMI (Directional Movement Index) were attached to the compaction test equipment as shown in Figure 4. The image capturing sensor and road condition detection device collected pavement condition information immediately after construction. The GPR sensor collected information inside the pavement to estimate the degree of compaction. DMI and RTK GNSS collected location information of equipment.

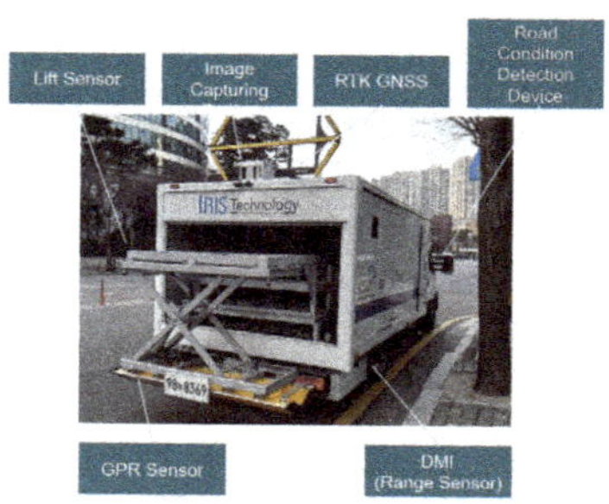

Figure 4. IoT PQMS related to compaction test equipment.

4. Discussion

The real-time quality control system for asphalt pavement construction developed in this study was tested at a test bed in Yeoncheon. The quality control information collected from the test bed is being compared and analyzed with actual information. It is planned to evaluate the applicability by solving the problems found in the analysis and applying it to the actual asphalt pavement section to be constructed by the Ministry of Land, Infrastructure and Transport.

5. Conclusions

In this study, a cloud-based IoT Pavement Quality Management System was developed to minimize quality degradation and address problems early by tightly coupling production and construction information of asphalt mixture and performing real-time quality control. Through the system, drivers and managers can monitor construction information and identify problems using monitors and apps. The proposed cloud-based IoT PQMS will be applied to national road construction sites in 2023 to verify its effectiveness and address potential problems. The results of this study demonstrate the potential of IoT technology in improving pavement quality management and paving the way for further development in this field.

Author Contributions: Conceptualization, K.-D.J. and D.-H.K.; methodology, J.-W.K.; software, D.-H.K.; validation, K.-D.J., D.-H.K. and J.-W.K.; formal analysis, N.-H.K.; investigation, N.-H.K.; resources, S.-A.K.; data curation, J.-W.K.; writing—original draft preparation, K.-D.J.; writing—review and editing, S.-D.H. and S.-A.K.; visualization, D.-H.K.; supervision, S.-D.H.; project administration, S.-D.H.; funding acquisition, S.-D.H. All authors have read and agreed to the published version of the manuscript.

Funding: This research was funded by the Korea Agency for Infrastructure Technology Advancement (KAIA) grant funded by the Ministry of Land, Infrastructure and Transport Grant RS-2019-KA152690.

Institutional Review Board Statement: Not applicable.

Informed Consent Statement: Not applicable.

Data Availability Statement: This paper is not a data-based analysis paper, but a system construction paper, so there is no data that can be shared.

Conflicts of Interest: The authors declare no conflict of interest. The funders had no role in the design of the study; in the collection, analyses, or interpretation of data; in the writing of the manuscript; or in the decision to publish the results.

References

1. Yuan, J.; Li, X.; Ke, Y.; Xu, W.; Skibnewski, M. Developing a Building Information Modeling-Based Performance Management System for Public-Private Partnerships. *Eng. Constr. Archit. Manag.* **2020**, *27*, 1727–1762. [CrossRef]
2. Ma, Z.; Zhang, J.; Philbin, S.P.; Li, H.; Yang, J.; Feng, Y.; Ballesteros-Perez, P.; Skitmore, M. Dynamic Quality Monitoring System to Assess the Quality of Asphalt Concrete Pavement. *Buildings* **2021**, *11*, 577. [CrossRef]
3. Dong, J.; Meng, W.; Liu, Y.; Ti, J. A Framework of Pavement Management System based on IoT and Big Data. *Adv. Eng. Inform.* **2021**, *47*, 101226. [CrossRef]
4. Beainy, F. Non-Contact Sensor for the Real-Time Measurement of Quality of Asphalt Pavements during Compaction. Ph.D. Thesis, University of Oklahoma, Norman, OK, USA, 2011.

engineering proceedings

Proceeding Paper

A Framework for Smart Pavements in Canada †

Pejoohan Tavassoti [1], Hassan Baaj [1,*], Moojan Ghafurian [2], Omran Maadani [3] and Mohammad Shafiee [3]

1 Department of Civil and Environmental Engineering, University of Waterloo, Waterloo, ON N2L 3G1, Canada; ptavasso@uwaterloo.ca
2 Department of Electrical and Computer Engineering, University of Waterloo, Waterloo, ON N2L 3G1, Canada; moojan@uwaterloo.ca
3 Road and Pavement Engineering, National Research Council Canada, Ottawa, ON K1A 0R6, Canada; omran.maadani@nrc-cnrc.gc.ca (O.M.); mohammad.shafiee@nrc-cnrc.gc.ca (M.S.)
* Correspondence: hbaaj@uwaterloo.ca
† Presented at the Second International Conference on Maintenance and Rehabilitation of Constructed Infrastructure Facilities, Honolulu, HI, USA, 16–19 August 2023.

Abstract: Maintaining an acceptable durability and satisfactory in-service condition for pavements is a crucial and relatively complex task, which otherwise can have considerable economic, environmental, and social consequences. Design and management of pavements have traditionally relied mainly on empirical models. However, pavements have been undergoing drastic changes, especially during the new millennium, which can compromise the reliability of the empirical models which were developed based on relatively stagnant historical data. Climate change, traffic loading growth and advancements in pavement materials are some of the main drivers of moving towards more mechanistic-empirical methods which would allow for a better understanding of pavement performance evolution in the future. To this end, this paper discusses the opportunities and challenges of a proposed framework for developing smart pavements in Canada, as well as a summary of the efforts that so far have been made in this regard. The goal of the study is to enable autonomous monitoring and data collection from the instrumented pavement sections in a suitable manner to allow for training Artificial Intelligence models, improving interpretation of the pavement responses and, ultimately, future pavement performance predictions.

Keywords: smart pavements; instrumentation; performance prediction; artificial intelligence; machine learning; mechanistic responses; autonomous monitoring

Citation: Tavassoti, P.; Baaj, H.; Ghafurian, M.; Maadani, O.; Shafiee, M. A Framework for Smart Pavements in Canada. *Eng. Proc.* 2023, *36*, 51. https://doi.org/10.3390/engproc2023036051

Academic Editor: Hosin (David)

Published: 19 July 2023

1. Introduction

Pavement networks play a key role in our modern societies by accommodating the economical, efficient, and safe movement of goods and people [1]. In addition to their significant economic and societal impacts, pavement construction and maintenance activities demand an enormous amount of natural, and mainly non-renewable, resources annually, and, hence, can result in a significant environmental footprint if not managed properly [2]. Therefore, maintaining the long-term performance and durability of pavements is a common goal at different levels including the planning, designing, construction, and management of transportation infrastructure [3,4]. Pavements are typically designed to last for several decades. Traditionally, we have mainly relied on empirical design based on limited experimental studies that were conducted between 1956 and 1961 in the U.S. [4]. Despite the relatively good success rate of these approaches in the past, the appearance of premature distresses and unexpected shortening of pavements' service life during the past two decades have motivated pavement engineers and researchers to further investigate the contributing mechanisms behind these phenomena. To this end, a paradigm shift in terms of pavement design and material characterization has been happening over the course of the past two decades which requires moving away from purely empirical approaches toward linking the empirical and mechanistic characteristics to develop better distress prediction models. Development of the Pavement Mechanistic-Empirical Design (PMED) [5] can be named as one of the well-known examples in this area. Although a robust mechanistic platform for modelling pavement performance is a positive step forward in achieving a better accuracy when designing pavement structures, it is also crucial to calibrate and verify pavement structural responses and performance trends in the eye of the changing climate and introduction of unconventional paving materials [6]. This requires systematic management of the transportation infrastructure, performing regular pavement condition assessments, intervening for maintenance and rehabilitation activities in a timely manner, and improving the practice of design and construction to accommodate for the desired resilience level against climate change effects and the dynamic nature of the exerted traffic loading. The development of autonomous pavement monitoring systems would, therefore, yield several advantages in terms of informing decision making about the timing of maintenance and rehabilitation (M&R) activities based on more realistic performance prediction models [7]. As a result, developing smart pavements has recently become the focus of some research groups across the world.

On the other hand, the application of Machine Learning to predict pavement's performance measures has also been gaining momentum during the past two decades [8,9]. Most of the work conducted to date has been focused on the prediction of the functional indexes such as the International Roughness Index (IRI) [10–12] or general pavement condition metrics such as the Pavement Condition Index (PCI) [13–15]. Predicting the structural response of pavements and distress modes such as rutting and cracking have seen relatively limited attention. Nevertheless, measuring pavement responses in a semi-continuous manner using smart pavement sections also provides an opportunity for utilizing the artificial-intelligence-based methods to improve the pavements' performance prediction as well as locally calibrating the PMED.

This paper provides a summary of the proposed smart pavement framework and the activities undertaken to date in this regard through the instrumented pilot section in Ontario, Canada.

2. Proposed Smart Pavement Framework

The proposed conceptual framework for smart pavements in Canada is composed of five major components. Figure 1 presents a schematic overview of this framework, including: (i) instrumented pavement sections and a Data Acquisition (DAQ) triggering system; (ii) an autonomous and semi-continuous data logging platform with remote data collection/storage capability; (iii) a preliminary (raw data) postprocessing unit; (iv) a secondary data aggregation and metrics computation unit; and (v) a cumulative structured database to store the data pertinent to long-term performance and key structural responses, through the use of both the dynamic and static data types.

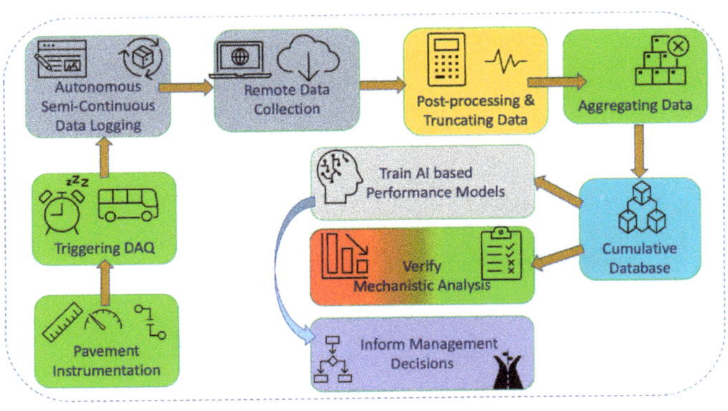

Figure 1. Proposed framework to deploy smart pavements in Canada.

A maximum traffic speed of 70 km/hr would be expected for the pilot section built to implement the proposed framework in Ontario, Canada. This required using dynamic asphalt strain gauges and pressure cells (PC) with a minimum data logging rate of 1000 Hz per channel to capture the full spectrum of the dynamic pulses induced by moving vehicular loads. The selected gauges provided a rate of 1 kHz per channel (not shared). Furthermore, temperature and moisture trees were instrumented to record the temperature and moisture variation at nine different levels within the pavement structure (see Figure 2). The temperature gradient would especially provide valuable information that could help better model the structural response of the pavement at different times of the day and year. Temperature probes (identified by red crosses in Figure 2) recording every 15 min can capture the daily fluctuation and gradient within the pavement structure.

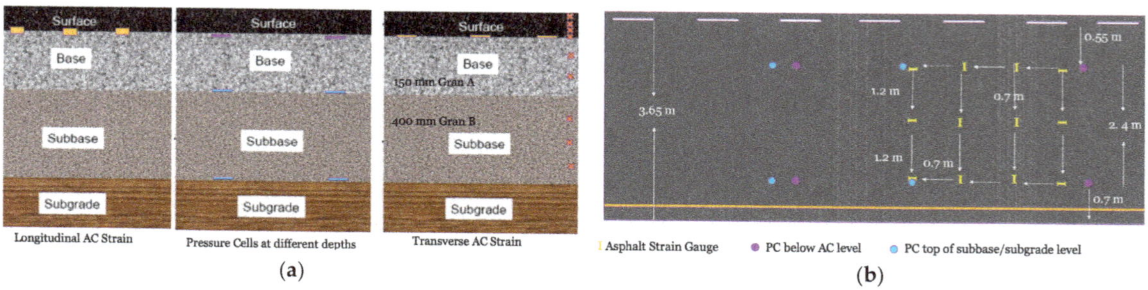

Figure 2. Schematic of the pilot section in this study: (**a**) cross section and (**b**) plane view.

Given the dynamic and semi-continuous nature of the collected data, a triggering system to activate the data collection can help avoid the collection of unnecessary data. This can be achieved through one of the pressure cells embedded underneath the asphalt concrete layer and at a distance from the main cluster of the sensors array, which was used to awaken the DAQ and collect the data until 120 s after the last sensed pulse. Using laser reflectometers or traffic cameras were also identified as viable options, which are not investigated in this study.

The DAQ system is equipped with a wireless modem that allows for remotely accessing the system and periodic downloading of the data. The raw data need to be routinely post-processed for noise removal and to truncate the signals with their corresponding time stamps. These data are then aggregated in terms of engineering metrics to describe the peak magnitudes and frequencies of strain/stress pulses at different depths of the pavement structure, as well as the incremental temperature and moisture changes.

3. Challenges and Opportunities

3.1. Barriers for a Fully Functional System

One of the major hurdles for the dynamic and semi-continuous collection of pavement responses is handling humongous amounts of generated data, which is different than the traditional static data collection. This poses multiple problems, e.g., the need for timely raw data processing and the required data storage logistics. The aggregated processed data in this project will be ultimately transferred to the National Logistics Database in Canada, as a part of the Artificial Intelligence for Logistics (AI4Logistics) program. Another issue preventing the widespread use of such a system remains the high cost for the existing pavement instrumentation. Furthermore, a comprehensive array typically requires a considerable power supply, buried wired connections, and interruptions to the conventional paving operations. Developing a wireless sensor that can eliminate the need for wired connections has been the focus of several research groups, including the authors' research. However, technical barriers such as the need for substantial power and limited operation life on battery power remains a gap in the existing work, along with the short life span of embedded sensors under sever climate conditions, which suggest the need for research on the ruggedness of the sensors. Promising progress has been made during the past few years in terms of wireless temperature measurements. However, this remains a challenge for stress/strain measurements without wired connections and power.

With respect to pavement data analysis and interpretation, most of the existing work on AI applications has focused on the use of supervised learning algorithms to predict functional metrics, such as the International Roughness Index (IRI), or overall pavement condition measures such as PCI, based on the existing databases such as Long-Term Pavement Performance (LTPP). The application of Machine Learning (ML) algorithms to obtain performance-related measures from the structural responses of pavements has seen limited attention. This also requires utilizing a database of a certain size to develop meaningful models.

Finally, unlike the controlled sections and accelerated testing facilities, monitoring live traffic typically becomes more complex due to the wandering of traffic loads, which ultimately affects the measured responses' amplitudes relative to the axles' vertical alignment on top of the sensors array. On the other hand, the axles' load and configuration will be highly variable as compared with the controlled tracks. In the case of live traffic loading, having a Weigh-In-Motion (WIM) station can significantly help with better interpretation of the structural responses to different axles. However, a WIM is not always available, creating one may not be practical for an in-service pavement section, and one-to-one synchronization of detailed traffic data with millions of recorded responses will require a lot of effort. The latter requires a considerable budget and would require further approval from the owner agency.

3.2. *Applications of Smart Pavement Data and Future Steps*

The ability to measure actual pavement responses under live traffic loading and in-service conditions allows for improving the current state of pavement design and management practices. Three major areas that can especially benefit from such a system are: (i) the quantification of climate change implications for pavements, (ii) validating the pavement analysis results with in situ measured data as well as calibration of the MEPDG transfer functions and distress models, and (iii) developing enhanced pavement performance prediction models using suitable ML algorithms. Furthermore, smart pavements can facilitate the implementation of innovative and high-performance paving materials, for which very limited data currently exist in terms of their in-situ responses/performance. This is especially a barrier in reflecting the added value for using such materials in the existing design methods, hence their implementation.

Moving forward, data collection from the proposed instrumented section in Ontario will be continued and once the required data size is available, pertinent ML models will be trained and tested by the research team. In the meantime, the research team has been working on the application of a Random Forest (RF) algorithm using the existing LTPP pavement performance database to predict the IRI and Rutting performance of flexible pavements in North America. In terms of structural response verification, samples of the paving materials have been collected for each layer and mechanical testing in the lab will be performed to correlate the engineering properties of the materials measured at the laboratory scale to their corresponding in-service response under varying temperatures and loading frequencies. In addition to the live-traffic data, verifications will continue to be performed using a truck with a known axle load driving on the instrumented section at different passing speeds, at least once every season, and ideally for several years.

Author Contributions: Conceptualization, P.T., H.B., M.G., O.M. and M.S.; methodology, H.B., M.G., P.T., O.M. and M.S.; investigation, P.T., H.B., M.G., O.M. and M.S.; resources, H.B., O.M. and M.G.; data curation, P.T., O.M. and M.G.; writing—original draft preparation, P.T.; writing—review and editing, P.T., H.B., M.G., O.M. and M.S.; visualization, P.T., M.G. and M.S.; project administration, H.B., O.M. and M.G.; funding acquisition, H.B., O.M. and M.G. All authors have read and agreed to the published version of the manuscript.

Funding: This research was funded by the Artificial Intelligence for Logistics program under the grant number AI4L-102. The authors acknowledge this support.

Institutional Review Board Statement: Not applicable.

Informed Consent Statement: Not applicable.

Data Availability Statement: Not applicable.

Conflicts of Interest: The authors declare no conflict of interest.

References

1. Khisty, C.J.; Lall, B. *Transportation Engineering: An Introduction*; Prentice Hall: Upper Saddle River, NJ, USA, 1990.
2. Wang, F.; Hoff, I.; Yang, F.; Wu, S.; Xie, J.; Li, N.; Zhang, L. Comparative assessments for environmental impacts from three advanced asphalt pavement construction cases. *J. Clean. Prod.* **2021**, *297*, 126659. [CrossRef]
3. Tighe, S. *Pavement Asset Design and Management Guide*; Transportation Association of Canada: Ottawa, ON, Canada, 2013.
4. Haas, R.; Hudson, W.R.; Zaniewski, J.P. *Modern Pavement Management Systems*; Krieger Publishing Co.: Malabar, FL, USA, 1994.
5. Transportation Officials. *Mechanistic-Empirical Pavement Design Guide: A Manual of Practice*; AASHTO: Washington, DC, USA, 2008.
6. Li, Q.; Wang, K.C.; Hall, K.D. Verification of virtual climatic data in MEPDG using the LTPP database. *Int. J. Pavement Res. Technol.* **2010**, *3*, 10.
7. Shon, H.; Cho, C.S.; Byon, Y.J.; Lee, J. Autonomous condition monitoring-based pavement management system. *Autom. Constr.* **2022**, *138*, 104222.
8. Amin, M.S.R. The pavement performance modeling: Deterministic vs. Stochastic approaches. In *Numerical Methods for Reliability and Saf. Assess.: Multiscale and Multiphysics Systems*; Springer International Publishing: New York, NY, USA, 2015; pp. 179–196.
9. Attoh-Okine, N.O. Grouping pavement condition variables for performance modeling using self-organizing maps. *Comput.-Aided Civ. Infrastruct. Eng.* **2001**, *16*, 112–125. [CrossRef]
10. Attoh-Okine, N.O. Analysis of learning rate and momentum term in backpropagation neural network algorithm trained to predict pavement performance. *Adv. Eng. Software* **1999**, *30*, 291–302. [CrossRef]
11. Aultman-Hall, L.; Jackson, E.; Dougan, C.E.; Soon-Nam, C. Models relating pavement quality measures. *Transp. Res. Rec.* **2004**, *1869*, 119–125. [CrossRef]
12. Bashar, M.Z.; Torres-Machi, C. Performance of Machine Learning Algorithms in Predicting the Pavement International Roughness Index. *Transp. Res. Rec. J. Transp. Res. Board* **2021**, *2675*, 226–237. [CrossRef]
13. Amin, S.R.; Amador-Jiménez, L.E. Backpropagation Neural Network to estimate pavement performance: Dealing with measurement errors. *Road Mater. Pavement Des.* **2017**, *18*, 1218–1238. [CrossRef]
14. Amin, M.S.R.; Amador-Jiménez, L.E. Pavement management with dynamic traffic and artificial neural network: A case study of Montreal. *Can. J. Civ. Eng.* **2015**, *43*, 241–251. [CrossRef]
15. Barua, L.; Zou, B.; Noruzoliaee, M.; Derrible, S. A gradient boosting approach to understanding airport runway and taxiway pavement deterioration. *Int. J. Pavement Eng.* **2020**, *22*, 1673–1687. [CrossRef]

engineering proceedings

ng Paper

lementing Public Service Features in Autonomous Vehicles eoul †

m Cho *, SoonYong Park, Junchul Kim and Seolyoung Lee

Seoul Institute of Technology, Seoul 03909, Republic of Korea; psy@sit.re.kr (S.P.); kjc@sit.re.kr (J.K.); lsy0717@sit.re.kr (S.L.)

* Correspondence: hrcho@sit.re.kr; Tel.: +82-2-6912-0942

† Presented at the Second International Conference on Maintenance and Rehabilitation of Constructed Infrastructure Facilities, Honolulu, HI, USA, 16–19 August 2023.

Abstract: Autonomous vehicle technology has been primarily developed by the private sector, with a focus on the automobile industry and fourth industrial revolution technology. However, for self-driving cars to become reliable forms of urban transportation, the private and public sectors must collaborate. The Seoul Metropolitan Government operates three self-driving test beds, namely, Sagam, Gangnam, and Cheonggyecheon, and has implemented various measures to support the use of autonomous vehicles as a means of urban transportation, such as an open platform for traffic signals, roadside safety facilities, dedicated road signs and lanes, and deregulation. This paper explains the technical and institutional support provided for autonomous-driving services in Seoul and suggests ways to build a safer and more effective autonomous driving environment.

Keywords: autonomous vehicle; strategies of transportation in Seoul; urban transportation

1. Introduction

The self-driving car industry has traditionally been led by established vehicle manufacturers like Mercedes-Benz, BMW, and GM. However, in recent years, the focus has shifted towards software platform companies such as Microsoft, Google, and Apple, who prioritize fourth industrial revolution technology. Private companies have primarily driven the industry, but governments are now implementing various support policies to ensure the safe operation of autonomous vehicles. These policies involve reorganizing urban infrastructure, equipping safety facilities where necessary, and sharing traffic signal information with the private sector.

Since 2018, KPMG, a global management-consulting firm, has developed the Autonomous Vehicles Readiness Index (AVRI) to assess the state of the autonomous driving environment worldwide in terms of policy, legislation, consumer acceptance, and technology. As of 2020, Singapore ranked first in terms of policy, legislation, and consumer acceptance, while the Netherlands topped the infrastructure category. Korea is ranked seventh out of thirty countries globally, and its ranking is increasing each year. It has been evaluated as remarkable in terms of 4G communication coverage, mobile connection speed, and ICT, but assessments have determined that its levels of consumer acceptance, government legal regulations, and government readiness for change are low.

In Korea, self-driving car testbeds have been established in local governments across the country, with significant efforts being made to not only develop technology but also increase consumer acceptance and use as a means of urban transportation. This study introduces the public sector's efforts and promotion directions for the introduction of self-driving vehicles as urban transportation in Seoul and suggests future development directions [1].

Citation: Cho, H.; Park, S.; Kim, J.; Lee, S. Implementing Public Service Features in Autonomous Vehicles in Seoul. *Eng. Proc.* **2023**, *36*, 52. https://doi.org/10.3390/engproc2023036052

Academic Editor: Hosin (David) Lee

Published: 24 July 2023

2. Introduction of Testbeds for Autonomous Driving in Seoul

There are three autonomous driving testbeds in Seoul: Sangam, Gangnam, ar Cheonggyecheon (Table 1).

Table 1. Testbeds for Autonomous Driving in Seoul.

Name of Testbed	Area/Length of Road	Test Service
Sangam	6.2 km^2	First/Last Mile
Gangnam	20.4 km^2	Demand response
Cheonggyecheon	8.8 km	Tour bus

The testbed at Sangam, the first of the three testbeds, covers an area of approximately 6.2 km^2 and operates six autonomous vehicles on five routes that connect subway stations and destinations. (Figures 1 and 2) Three of the routes connect subway stations with major commercial and business districts, while two routes operate between nearby tourist attractions and subway stations, including an internal circulation route within the tourist attractions. The fare has been set at around $1.5 per trip—higher than bus or subway fares but lower than taxi fares—to test the functionality and public acceptance of autonomous vehicles in terms of typical urban transportation. By 2025, a total of 25 autonomous vehicles are scheduled to perform test operations in Sangam, with plans to expand the applicability of autonomous vehicles to include not only passenger transportation but also personal logistics [2].

(**a**) (**b**)

Figure 1. Sangam Testbed: (**a**) The location of Sangam autonomous-driving testbed and (**b**) self-driving car in Sangam testbed.

The Gangnam testbed is the most complex of the three testbeds in Seoul, and it is conducting tests on the largest area. The area is a mix of business, commercial, tourism, and residential areas, resulting in high demand for transportation and heavy traffic. The entire area of the Gangnam testbed is about 20.4 km^2, and it is planned to test autonomous vehicles on approximately 85.1 km of roads (Figure 2). The area has a high demand for transportation due to the evenly distributed business, commercial, tourist, and residential areas, resulting in heavy traffic. Additionally, there are many bus and subway routes, and road construction occurs frequently. The main purpose of the Gangnam testbed is to test the safe coexistence of autonomous vehicles in a realistic traffic environment that is closer to reality than the Sangam testbed. The main test service item of the Gangnam testbed is the taxi service, and Hyundai Motor is preparing to test the commercialization of a demand-responsive Robo-Taxi.

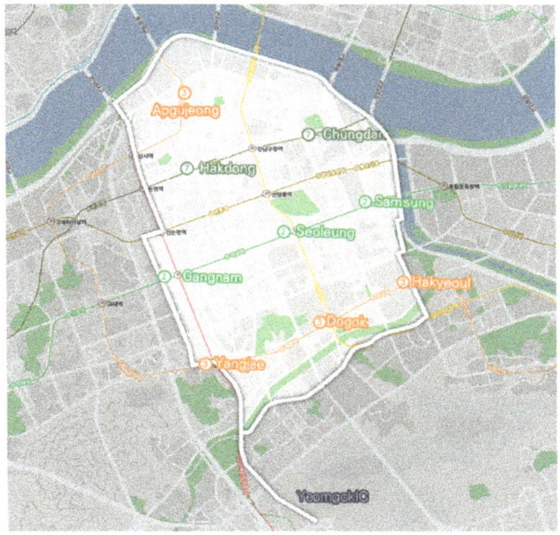

Figure 2. The location of Gangnam autonomous-driving testbed.

The third autonomous driving testbed in Seoul is a self-driving mini-bus for tourism that operates around Cheonggyecheon (Figure 3). The Cheonggyecheon testbed is designed to operate on a total of 8.8 km stretch of the area around the Cheonggyecheon stream, which connects major tourist attractions in Seoul, such as Dongdaemun Market, Deoksugung Palace, Gyeongbokgung Palace, and Jongno. It is also a historically significant location. Currently, the testbed is operating free of charge, with a focus on tourists.

(a) (b)

Figure 3. Cheonggyecheon Testbed: (**a**) The location of Cheonggyecheon autonomous-driving testbed and (**b**) self-driving shuttle.

In particular, the Cheonggyecheon road is a popular tourist destination where many pedestrians gather, and there are also many bikes in the area transporting cargo from the market. Therefore, the main purpose of the Cheonggyecheon testbed is to test the coexistence of autonomous vehicles with pedestrians and bikes compared to the other two testbeds.

All three areas are equipped with infrastructure for safe testing, including basic safety signs and priority lanes for autonomous vehicles. Additionally, they have installed and are operating various RSEs (to cover communication blind spots), CCTV, and a communication system based on 5G to support cooperative driving. Furthermore, the Seoul Metropolitan Government supports the safe operation of autonomous vehicles at intersections by providing private companies with traffic signal information stemming from major intersections. This allows autonomous vehicles on test beds to receive real-time information on the signal operation status ahead and operate safely on the intersections.

3. Operational Results

Since the selection of the Sangam autonomous driving testbed in 2021, the Seoul Metropolitan Government has submitted an annual operational report to the Ministry of Land, Infrastructure, and Transport, which includes the distance traveled by autonomous vehicles and the occurrence of accidents. The city has also been promoting the issuance of temporary driving licenses through operational tests in order to offer the first paid autonomous vehicle service. In addition, a preferred fare survey has been conducted for paid transportation services, and various autonomous driving expos and hackathons are being hosted to increase citizens' acceptance of autonomous vehicles. The Seoul Metropolitan Government is also providing subsidies to autonomous vehicle operators based on their operational performance to promote the autonomous vehicle industry. These efforts culminated in 2022, with Seoul being selected as the most well-operated region among the seven Korean cities and provinces operating autonomous driving testbeds [2,3].

4. Conclusions

The Seoul Metropolitan Government is currently operating three autonomous driving testbeds and plans to expand the area of autonomous vehicle services throughout the city in the future. As part of this plan, the city is introducing a shuttle bus that will operate around the recently renovated park near Cheong-Wa-Dae and is preparing for autonomous vehicle operation tests on late-night bus services. Additionally, the city plans to introduce autonomous vehicles in the public sector, including road cleaning, snow removal, and patrol services, as well as expanding the scope of autonomous vehicle services beyond transportation services. Although the private sector has mainly led the autonomous vehicle industry thus far, various support policies from local governments or the government, such as restructuring urban infrastructure, providing necessary safety facilities, and supplying traffic signal information, will be essential factors for the successful establishment of autonomous vehicles. By creating a well-established governance system through consultations with industry, research, and government in addition to a public-private collaboration system, we can expect a safer and more convenient era of autonomous vehicles to arrive sooner.

Author Contributions: Conceptualization, H.C.; planning, S.L.; methodology, S.P. and J.K. All authors have read and agreed to the published version of the manuscript.

Funding: This work was supported by Korea Institute of Police Technology (KIPoT) grant funded by the Korea government (KNPA) (No. 092021C26S02000, Development of Transportation Safety Infrastructure Technology for Lv.4 Connected Autonomous Driving).

Institutional Review Board Statement: Not applicable.

Informed Consent Statement: Not applicable.

Data Availability Statement: Not applicable.

Conflicts of Interest: The authors declare no conflict of interest.

References

1. Korea Transport Institute. *Introduction of Autonomous;* Mobility Services: Sejong, Republick of Korea, 2021.
2. Lee, S.Y. *Evaluation of the Performance and Establishment of Vision for Autonomous Driving Testbeds;* Seoul Institute of Technology: Seoul, Republic of Korea, 2022.
3. Ministry of Land, Infrastructure and Transport. *Evaluation of the Performance for Autonomous Driving Testbeds 2022;* Ministry of Land, Infrastructure and Transport: Sejong, Republick of Korea, 2022.

engineering proceedings

MDPI

Proceeding Paper

3D Printing Technique for Passive Wireless Strain Sensing †

Joshua Dyogi [1,*], Xi Song [1], Sung-Hwan Jang [2], Sang-Hyeok Nam [3] and Chunhee Cho [1]

1 Department of Civil and Environmental Engineering, University of Hawaii at Manoa, Honolulu, HI 96822, USA; xisong@hawaii.edu (X.S.); chunhee@hawaii.edu (C.C.)
2 Department of Civil and Environmental Engineering, Hanyang University ERICA, Seoul 15588, Republic of Korea; sj2527@hanyang.ac.kr
3 ENGSOFT Co. Ltd., Seoul 07532, Republic of Korea; shnam@engsoft.kr
* Correspondence: dyogi@hawaii.edu
† Presented at the Second International Conference on Maintenance and Rehabilitation of Constructed Infrastructure Facilities, Honolulu, HI, USA, 16–19 August 2023.

Abstract: Passive wireless sensing systems, particularly passive antenna sensors, offer a viable alternative to traditional wired and active sensors for long-term structural health monitoring due to their simplicity, easy installation and maintenance, and ability to measure strain without an external power supply. Customization of antenna shape can also adapt to various structural geometries. However, sensor fabrication using chemical etching is expensive and time-consuming, which is unsuitable for limited-quantity production. To address this, this study explores the potential of extrusion-based additive manufacturing to produce cost-effective passive wireless antenna strain sensors. The study investigates polylactic acid's mechanical and electromagnetic properties for substrate design and uses multi-physics simulation to estimate strain-sensing performance. The obtained results show similar strain-sensing performance to sensors produced through chemical etching, making the manufacturing process a promising alternative.

Keywords: 3D printing; antenna sensors; passive sensing; additive manufacturing; strain sensing; structural health monitoring; battery free

Citation: Dyogi, J.; Song, X.; Jang, S.-H.; Nam, S.-H.; Cho, C. 3D Printing Technique for Passive Wireless Strain Sensing. *Eng. Proc.* **2023**, *36*, 53. https://doi.org/10.3390/engproc2023036053

Academic Editor: Hosin (David) Lee

Published: 25 July 2023

1. Introduction

Structural health monitoring (SHM) is the study of monitoring an engineered structure's safety over time by taking periodic measurements such as strain, displacement, acceleration, humidity, and temperature. The goal is to accurately detect any problems that could impact the structure's reliability and life cycle [1]. Strain is a critical measurement that represents stress concentration and could lead to crack development during service. Wireless sensors are commonly used for SHM, but they require a power source for operating systems. Eventually, the system requires expensive periodic battery replacement, data acquisition, which increases installation time and cost due to cable connectors [2].

To reduce periodic upkeep of the battery system, a passive wireless communication sensor was created. Antenna strain sensors enable long-term measurement without the need for a power source on the sensor side [3]. Previous research attempts and efforts in development of antenna strain sensors that satisfy SHM requirements has been an area of continuing research [4]. However, in these preceding studies, the cost of the antenna sensors relied on the expensive and relatively time-consuming chemical etching process. We propose the use of additive manufacturing (AM) as an alternative to fabricating antenna strain sensors. AM is the method of constructing three-dimensional objects by layering materials according to a computer-aided design (CAD). This technology has proven itself to be an asset to a wide range of fields due to its efficiency, flexibility, and cost-effectiveness compared to traditional subtractive manufacturing techniques [5]. Through Production 3D printing, multiple iterations and functional end-use small-production quantities are possible at a fraction of the cost [6].

2. Method

2.1. Strain Sensing Mechanism of the Antenna Sensor

The passive communication device used in this study was the Ultra High Frequency (UHF) gen 2 standard, which allows for a broad frequency range of 840 to 960 MHz. This standard utilizes the Tagformance Pro© radio frequency identification (RFID) reader (Voyantic, Seattle, WA, USA), which emits an electromagnetic interrogation signal that powers up the receiving antenna sensor, as illustrated in Figure 1a.

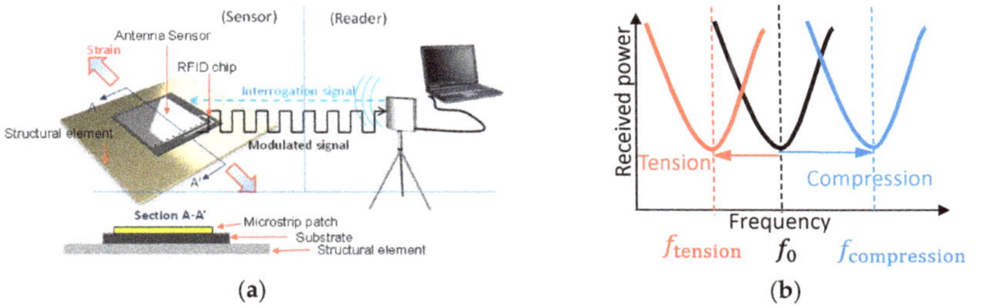

Figure 1. Strain sensing mechanism of antenna sensors: (**a**) Schematic of a passive (battery free) wireless system; (**b**) Strain sensing.

The antenna sensor consists of three layers: (1) microstrip with a metallic pattern aiming for high radiation performance, (2) substrate to separate the microstrip patch and ground plan allowing an electromagnetic field to generate, and (3) a ground plan which is not designed as part of this study. The emitted signal is received by the antenna sensor, which captures a portion of the power and transfers it to the Radio Frequency Identification (RFID) transponder chip. If the transferred power is higher than the activation power threshold of the chip, it is activated. In this state, the chip modulates the return signal in the form of a backscattered signal, which is then received by the reader. This backscattered signal contains information about the strain level of the sensor.

The resonance frequency of a patch antenna with a different strain level is determined as shown in Equation (1):

$$f_R^{Patch} = \frac{c}{2(1+\varepsilon)(L+L')\sqrt{\beta_{reff}}} \tag{1}$$

where c is the speed of the light, L is the physical length of the patch antenna, β_{reff} is the effective dielectric constant of the antenna substrate, L' is the additional electrical length due to the fringing effect, and ε is an applied strain level.

As shown in Figure 1b, the backscattered signal output forms interrogation curves, where the minimum point of the interrogation curve indicates the resonance frequency of the antenna sensor. As shown in Figure 1b, under tensile strain, the resonance frequency of the antenna sensor will decrease and increase when experiencing compressive strain.

2.2. Design of Substrate and Patch Antenna

Material properties are key components when designing antenna sensors. Under the proposed AM fabrication method, the mechanical and electromagnetic properties were tested to gauge the impact of the anisotropic behavior inherent in AM printing technology; this remains a primary concern for end-use printing [6]. The anisotropic behavior is governed by the interlayer bond of the microstructure latices created during the layering process while printing. The A2200 3D Multi-Material Electronics Printer© (nano3Dprint, Burlingame, CA, USA) was used to fabricate the sensor and substrate using a concurrent printing technique. The substrate was printed in polylactic acid (PLA), while the metal pattern of the patch antenna was printed in silver nano ink which had a conductivity of 2 Ω/cm.

A preliminary material study of the extrusion-based 3D printing process was conducted, following ASTM D638-14 on the 810 Material Test System. Two Rectilinear (45°, 135° and 0°, 90°) and Full Honeycomb (FHC) patterns were compared as a function of their infill density. The three features that impacted the final material behavior and strength were: layer height, layer density, and printing direction. The stress–strain constitutive relation, shown in Figure 2, was found to be dependent on the infill density, print orientation, and pattern with relation to the tensile loading condition.

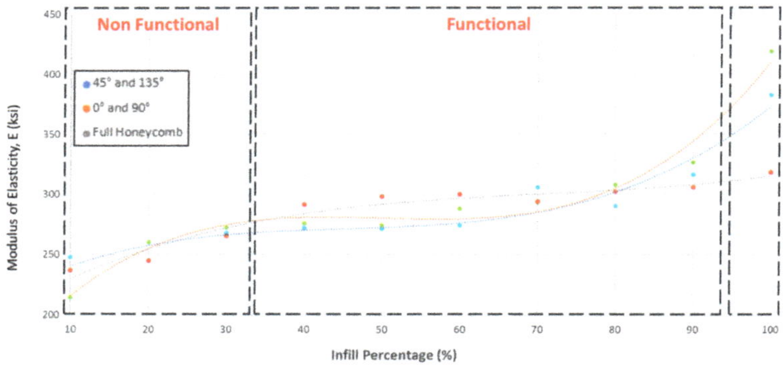

Figure 2. Print pattern's infill comparison for modulus of elasticity (E).

The prototype for the patch antenna was created by building on previous research that utilized a folded-patch topology for the antenna, which can be found in our previous study [7,8]. To construct the antenna, conductive silver nano ink was used to 3D print it onto a PLA substrate. To make the antenna smaller, the antenna and ground plane were connected with vias. In addition, the RFID chip MONZA R6-P (IMPINJ, Seattle, WA, United States of America) was used in the study to prevent data collision. The final dimensions of the printed antenna sensor were determined through an iterative process, where the team experimented with different dimensions and adjusted until they found the optimal size. The resulting dimensions of the antenna sensor were 61 mm in width, 69 mm in length, and 1 mm in thickness, as illustrated in Figure 3.

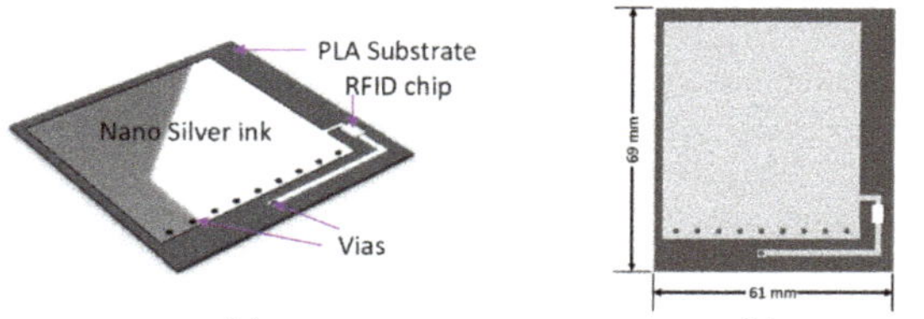

(**a**) (**b**)

Figure 3. Initial design of a patch antenna sensor: (**a**) perspective view (**b**) plan view.

3. Results

To accurately describe the antenna sensor's electromagnetic behavior under strain, we must consider two coupled physical domains, mechanics and electromagnetics. The coupled multi-physics simulation requires mechanical simulations to determine the deformed shape of the antenna, and then estimate electromagnetic response. Detail modeling techniques and procedures are referred to in our previous study [9]. As shown in Figure 4a, scattering parameter (S_{11}) is simulated with different strain levels. Applied strain is up to 2000 $\mu\varepsilon$ with the step size of 500 $\mu\varepsilon$. To estimate the strain sensitivity, we extract the minimum values of S_{11}, and perform a linear regression as shown in Figure 4a. The obtained strain sensitivity is -840 $\mu\varepsilon$/Hz and the coefficient of determination is 0.9978.

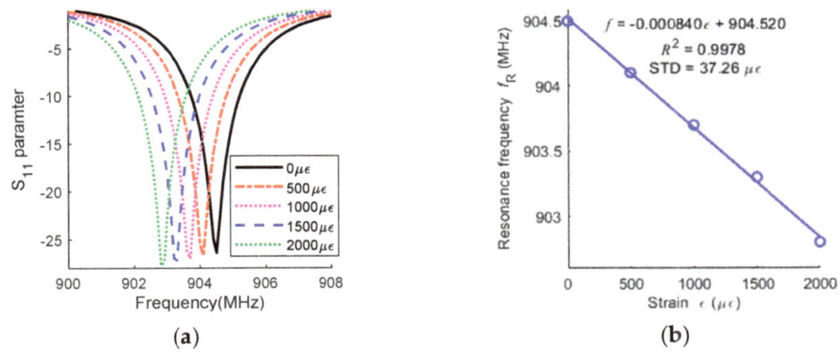

Figure 4. Simulation strain sensitivity of the optimized sensor: (**a**) S_{11} scattering parameter; (**b**) strain vs. frequency plot.

4. Discussion & Summary

In this study, we delved into the various possibilities that 3D printing techniques present in the creation of antenna sensors. We found the details by carrying out extensive mechanical analyses on PLA, with the aim of establishing the elastic modulus in relation to the different infill densities. Additionally, we utilized multi-physics simulation to obtain precise estimations of strain sensing, and from our findings, we were able to conclude that the AM process has immense potential for producing antenna sensors. Our investigation produced a strain sensitivity of -840 με/Hz, and we observed a good linearity in the coefficient of determination of 0.9978. Consequently, we believe that 3D printing technology can be harnessed to fabricate antenna sensors.

As a follow-up study, we intend to carry out nonlinear and fatigue tests to obtain more detailed insights into the mechanical properties of PLA. Furthermore, we plan to develop a method to improve conductivity of a nano ink. Ultimately, a more systematic approach to antenna sensor design and 3D printing will be required to maximize the strain sensing and wireless interrogation distance.

Author Contributions: Conceptualization, S.-H.J., S.-H.N. and C.C.; methodology, J.D., S.-H.J., S.-H.N. and C.C. software, J.D., X.S. and C.C.; validation, J.D. and X.S. formal analysis, J.D., X.S., S.-H.J., S.-H.N. and C.C.; investigation, J.D.; resources, J.D. and C.C.; data curation, J.D. and X.S.; writing—original draft preparation, J.D.; writing—review and editing, C.C.; visualization, J.D. and C.C.; supervision, C.C.; project administration, C.C.; funding acquisition, C.C. All authors have read and agreed to the published version of the manuscript.

Funding: The work was supported by the ATC+ Program (20014127, Development of a smart monitoring system integrating 3D printed battery-free antenna sensor technology with AI optimization) funded by the Ministry of Trade, Industry & Energy (MOTIE, Republic of Korea).

Data Availability Statement: The data presented in this study are available upon request.

Conflicts of Interest: The authors declare no conflict of interest.

References

1. Lynch, J.P.; Loh, K.J. A summary review of wireless sensors and sensor networks for structural health monitoring. *Shock. Vib. Dig.* **2006**, *38*, 91–128. [CrossRef]
2. Abdulkarem, M.; Samsudin, K.; Rokhani, F.Z.; Rasid, M.F.A. Wireless sensor network for structural health monitoring: A contemporary review of technologies, challenges, and future Direction. *Struct. Health Monit.* **2020**, *19*, 693–735. [CrossRef]
3. Yi, X.; Cho, C.; Cook, B.; Wang, Y.; Tentzeris, M.M.; Leon, R.T. A slotted patch antenna for wireless strain sensing. In Proceedings of the ASCE 2014 Structures Congress, Boston, MA, USA, 3–5 April 2014.
4. Deivasigamani, A.; Daliri, A.; Wang, C.; John, S. A review of passive wireless sensors for structural health monitoring. *Mod. Appl. Sci.* **2013**, *7*, 57–76. [CrossRef]
5. Zhang, J.; Tian, G.Y.; Marindra, A.M.; Sunny, A.; Zhao, A.B. A Review of Passive RFID Tag Antenna-Based Sensors and Systems for Structural Health Monitoring Applications. *Sensors* **2017**, *17*, 265. [CrossRef] [PubMed]

6. Somireddy, M.; Czekanski, A. Anisotropic material behavior of 3D printed composite structures—Material extrusion additive manufacturing. *Mater. Des.* **2020**, *195*, 108953. [CrossRef]
7. Yi, X.; Cho, C.; Cooper, J.; Wang, Y.; Tentzeris, M.M.; Leon, R.T. Passive wireless antenna sensor for strain and crack sensing—Electromagnetic modeling, simulation, and testing. *Smart Mater. Struct.* **2013**, *22*, 085009. [CrossRef]
8. Finkenzeller, K. *RFID Handbook*, 2nd ed.; John Wiley & Sons: New York, NY, USA, 2003.
9. Cho, C.; Long, L.; Park, J.; Jang, S.-H. A multi-physics informed antenna sensor model through the deep neural network regression. *Smart Struct. Syst.* **2021**, *28*, 335–362.

engineering proceedings

Proceeding Paper

Is Maintaining a Train Network in New Zealand Worth the Cost? †

Eric Scheepbouwer * and Daniel van der Walt

Department of CNRE, Faculty of Engineering, University of Canterbury, Christchurch 8041, New Zealand; daniel.vanderwalt@canterbury.ac.nz
* Correspondence: eric.scheepbouwer@canterbury.ac.nz
† Presented at the Second International Conference on Maintenance and Rehabilitation of Constructed Infrastructure Facilities, Honolulu, HI, USA, 16–19 August 2023.

Abstract: The IPCC highlighted the potential for rail transport to play a vital role in transitioning to a low-carbon economy. In many countries, rail networks are operated and maintained by private companies. However, in N.Z., the privatization was undone in 2008. The nationalization was necessary to enable public investments to gain long-term benefits for N.Z. as a whole. Literature shows that accurate life cycle cost analysis of networks is very complex, and benefits are financial, environmental, and social. The latter two categories of benefits are often not considered by private companies. The uncertainties in calculations are often so significant that perhaps a more relevant issue is determining the long-term benefits of a network for a country.

Keywords: rail transport network; rail maintenance; rail rehabilitation; sustainable transport; IPCC; state-owned enterprises; sustainability

Citation: Scheepbouwer, E.; van der Walt, D. Is Maintaining a Train Network in New Zealand Worth the Cost? *Eng. Proc.* **2023**, *36*, 54. https://doi.org/10.3390/engproc2023036054

Academic Editor: Hosin (David) Lee

Published: 25 July 2023

1. Introduction

The Intergovernmental Panel on Climate Change (IPCC) noted in its fifth assessment report that rail transport could be a more environmentally friendly option for passenger and freight transport than other modes of transportation [1]. Specifically for New Zealand (N.Z.), the transport fuel emissions factor for intercity passenger transport is two-thirds that of air transport and less than half compared to a passenger car, although twice that of an intercity bus. The same calculation shows that freight transport by truck is more than ten times less efficient than rail transport [2].

New Zealand is an island country located in the Southwest of the Pacific Ocean, with a population of 5.1 million. The primary export commodities are dairy, wool, meat, timber, and wood [3]. The increase in export over recent decades has also increased the need for transportation. KiwiRail Holding Limited manages the New Zealand train network, operates passenger trains, and transports more than 19 million tons of goods annually.

The efficiency of operating a rail network has long been a topic of concern, leading to privatizing the rail industry in several countries. By allowing independent train operators to operate on a network, competition should lead to greater efficiency. However, in some cases, privatization did not lead to increased efficiency; the cost of maintaining and improving rail networks was too high. In 2008, the N.Z. government undid the privatization to invest public funds to address maintenance deficiencies.

A decade later, N.Z. committed to reaching 1990 emission levels for transport by 2050, and the public rail network again received a significant investment boost. While the environmental benefits of rail transport seem straightforward, it is also important to consider the ongoing financial feasibility of maintaining such a significant asset.

A comprehensive literature analysis has been conducted to determine the costs and benefits of rail networks as a mode of transport in New Zealand. This research aims to clarify the long and short-term costs and give insight into the asset management challenges and opportunities of rail networks which could be used in business cases in other small island nations.

2. Internal Costs and Benefits

Performing an estimate based on the first principles is a detailed and precise way of calculating a cost estimate derived from adding the costs of individual components. It needs a clear description of the cost items like maintenance, operations, and incidentals.

The costs can be divided into fixed costs which are not dependent on the number of passengers and quantities of transported freight [4]. Examples are expenses related to constructing components like road crossings, tunnels, culverts, transport nodes like air and seaports, renting buildings, permanent labour, administration, etc. There will also be variable costs associated with existing and new infrastructure maintenance. Maintenance requirements depend on the traffic volume, cost, availability of labour and material, and energy. After doing a comprehensive literature review [5], maintenance was defined as "All maintenance activities, actions, strategies, and processes carried out and adopted such as routine, preventive, predictive, scheduled and unscheduled actions aimed at preventing asset failure or downtime, including all logistics and admin activities which will achieve the goal of increasing efficiency, reliability, and safety of the engineering assets" [6]. Specific studies were found that cover life cycle evaluations of elements of rail networks. For instance, rail bridge construction and transition mitigation, drainage systems for rock tunnels, noise and vibration mitigation measures, and (elements of) the track bed [7–11].

The above shows the enormity of the task of calculating a precise estimate based on the components of a national rail network. Moreover, the described fixed and variable costs are examples of internal costs. The main external costs include air pollution, noise, and environmental effects. These costs are not easy to describe objectively [11,12].

Another way of calculating the cost is the top-down method, related to parametric estimation techniques. An example is developing a decision support system, as described by several authors with varying results. A decision support system for evaluating the infrastructure costs to justify expenses and investments in Europe has been set up [13]. However, the author states that this system also has many uncertainties and requires continuous expert data collection and validation [14].

Finally, some reports compare the cost–benefit analysis of maintaining a rail network to the cost of maintaining a road network. For instance, in N.Z., the maintenance cost saving of the rail network against the comparative static situation of no rail is estimated to be around 64M$ per year [13]. This report, however, has not detailed any external costs and benefits.

An example of external costs is the refurbishment of electric locomotives in N.Z. In 2017, KiwiRail decided to move away from its electric locomotives, but lobbying to reduce carbon emissions led the New Zealand Cabinet to intervene, creating the option to electrify more rail segments in N.Z. to reduce carbon emissions. This is an example where the external benefits outweighed the internal costs. The following civil engineering problems and increases in scope related to the mentioned electrification raised the costs considerably, aggravating uncertainties that need to be considered in future cost–benefit analyses. The latter problems are not confined to New Zealand, as Europe's rail infrastructure costs, on average, 45% more than projected [12].

3. External Costs and Benefits

It is unlikely that the transformation of rail transport to a more sustainable form will occur in any meaningful way without government participation. Such participation has been only modest economically and often short-term [14].

In contrast, the strategy of KiwiRail contains economic, environmental, and social goals. KiwiRail is a leader in low-emissions freight transport and supports New Zealand's transition to net zero carbon by 2050. KiwiRail has 4500 employees spread across 50 often small towns and cities in New Zealand. These goals benefit N.Z. as a whole. These variables have been included in the calculation of the value of rail in New Zealand, estimated to be between USD 1.70 billion and USD 2.14 billion each year [15]. Several external benefits and costs are included in this figure, like reduced congestion by taking cars and trucks off our roads. The rail networks in many countries are monopolies due to their high set-up costs. This is also the case in New Zealand, as large rail infrastructure stretches throughout both islands' geographically challenging landscapes. Table 1 shows the ownership of rail networks in several countries; public ownership often indicates a need for investments, while private ownership could indicate a need for more efficiency.

Table 1. Rail networks, size, and ownership in various countries.

Country	Owners Rail Network	Size in Rail km	Persons/Freight
Fiji	State-owned enterprise	597 km	Freight
New Zealand	State-owned enterprise (bought back in 2008)	3700 km	Both
Australia	Private and state-owned enterprises	33,819 km	Both
Indonesia	(Mostly) state-owned	7032 km	Both
Malaysia	(Mostly) state-owned	2783 km	Both

4. Discussion

Calculating the costs and benefits of a national rail network can quickly become complex. In the literature, decision systems and various other methods are described, but all have inherent uncertainties. These systems still only describe the internal costs and benefits. The economic value of rail to a country or region is not captured in traditional profit and loss calculations. Whether the maintenance of a rail network is worth the cost cannot easily be calculated reliably.

The benefits can be divided into short-term economic benefits and long-term national benefits. National benefits could be sustainability and social issues like the development of a region or industry. This begs the question of whether rail networks should be privatized or be in the hands of private entities. The answer depends on whether the external benefits to a country outweigh the need for efficiency. Private entities can bring more efficiency but will concentrate on short-term profit-seeking. This attitude can interfere with the long-term economic development of (parts of) a country. State-owned enterprises have the advantage that they can have strategic goals which bring long-term profits. State ownership can also be a significant economic enabler. Perhaps deciding whether the rail networks are part of a national strategy to create long-term benefits, such as lower emissions and a more sustainable transport system, is more important because the external costs can outweigh the internal costs.

Author Contributions: Conceptualization, E.S.; methodology, E.S.; validation, E.S. and D.v.d.W.; formal analysis, E.S. and D.v.d.W.; investigation, E.S.; resources, E.S. and D.v.d.W.; data curation, E.S.; writing—original draft preparation, E.S.; writing—review and editing, D.v.d.W.; visualization, E.S.; supervision, E.S. and D.v.d.W.; project administration, E.S.; funding acquisition, E.S. All authors have read and agreed to the published version of the manuscript.

Funding: The paper acknowledges the funding of the New Zealand Ministry of Business, Innovation, and Employment, Endeavour Programme Research Grant. Contract Number: MAUX2005.

Institutional Review Board Statement: Not applicable.

Informed Consent Statement: Not applicable.

Data Availability Statement: Not applicable.

Conflicts of Interest: The authors declare no conflict of interest. The funders had no role in the design of this study; in the collection, analyses, or interpretation of data; in the writing of the manuscript; or in the decision to publish the results.

References

1. IPCC. *IPCC: Climate Change 2021: The Physical Science Basis. Contribution of Working Group I to the Sixth Assessment Report of the Intergovernmental Panel on Climate Change*; Masson-Delmotte, V., Zhai, P., Pirani, A., Connors, S.L., Péan, C., Berger, S., Caud, N., Chen, Y., Goldfarb, L., Gomis, M.I., et al., Eds.; Cambridge University Press: Cambridge, UK; New York, NY, USA, 2021; *in press*. [CrossRef]
2. Callister, P.; O'Callahan, H. How to Decarbonize New Zealand's Transport Sector. Working Paper 21/09. Available online: https://www.wgtn.ac.nz/__data/assets/pdf_file/0009/1926639/WP-21-09-how-to-decarbonise-New-Zealands-transport-sector.pdf (accessed on 11 January 2023).
3. Latief, M.; Rimapradesi, Y.; Riswandha, F. Diplomasi komersial pt inka (persero) indonesia dalam kegiatan ekspor gerbong barang un-tuk kiwirail new zealand tahun. *J. Int. Stud.* **2022**, *7*, 31–53.
4. Tuler, M.V.; Kaewunruen, S. Life cycle analysis of mitigation methodologies for railway rolling noise and ground bourne vibration. *J. Environ. Manag.* **2017**, *191*, 75–82. [CrossRef] [PubMed]
5. Dušan, T.; Milan, J. Chapter 10—Transport Economics. In *Transportation Engineering: Theory, Practice, and Modeling*, 2nd ed.; Butterworth-Heinemann: Oxford, UK, 2017; pp. 663–745, ISBN 9780323908139. [CrossRef]
6. Alawaysheh, I.; Alsyouf, I.; Tahboub, Z.E.A. Selecting maintenance practices based on environmental criteria: A comparative analysis of theory and practice in the public transport sector in UAE/DUBAI. *Int. J. Syst. Assur. Eng. Manag.* **2020**, *11*, 1133–1155. [CrossRef]
7. Bizjak, K.F.; Knez, F.; Lenart, S.; Slanc, K. Life-cycle assessment and repair of the railway transition zones of an existing bridge using geocomposite materials. *Struct. Infrastruct. Eng.* **2017**, *13*, 331–344. [CrossRef]
8. Setsobhonkul, S.; Kaewunruen, S.; Sussman, J.M. Life-cycle assessments of railway bridge transitions exposed to extreme climate events. *Front. Built Environ.* **2017**, *3*, 35. [CrossRef]
9. Rempelos, G.; Preston, J.; Blainey, S. A carbon footprint analysis of railway sleepers in the United Kingdom. *Transp. Res. Part D Transp. Environ.* **2020**, *81*, 102285. [CrossRef]
10. Liljenström, C.; Björklund, A.; Toller, S. Including maintenance in life cycle assessment of road and rail infrastructure—A literature review. *Int. J. Life Cycle Assess.* **2022**, *27*, 316–341. [CrossRef]
11. Nijkamp, P. Book Review: Priemus, H., Flyvbjerg, B. and Van Wee, B. (Eds.) Decision-Making on Mega-Projects–Cost-Benefit Analysis, Planning and Innovation. *Eur. J. Transp. Infrastruct. Res.* **2008**, *8*, 3352. [CrossRef]
12. Zoeteman, A. Life cycle cost analysis for managing rail infrastructure: Concept of a decision support system for railway design and maintenance. *Eur. J. Transp. Infrastruct. Res.* **2001**, *1*, 391–413. [CrossRef]
13. Boshier, J. *Power Surge: How Think Big and Rogernomics Transformed*; New Holland Publishers: Auckland, New Zealand, 2022.
14. Ernst & Young. The Value of Rail in N.Z. 2016. Available online: https://www.kiwirail.co.nz/assets/Uploads/documents/70bd71037f/The-Value-of-the-Rail-in-New-Zealand.pdf (accessed on 2 February 2023).
15. The World Bank Group, World Development Indicators Database. Available online: https://data.worldbank.org/ (accessed on 29 December 2022).

engineering proceedings

Proceeding Paper

Field Application of Hydro-Demolition and Dry-Mix Shotcrete for Repairing the Understructure of Bridge Deck †

Kyongku Yun [1,*], Seunghak Choi [1], Taeho Ha [1], Changseok Song [1], Mohammad Shakhawat Hossain [1], Valerii Panov [1] and Yonggon Kim [2,*]

1 Civil Engineering Department, Kangwon National University, Chuncheon 24341, Republic of Korea; donghaebi@kangwon.ac.kr (S.C.); gkxogh10@kangwon.ac.kr (T.H.); changsuk22c@kangwon.ac.kr (C.S.); shakhawat@kangwon.ac.kr (M.S.H.); panovvalerii97@kangwon.ac.kr (V.P.)
2 Daesang E&C, Wonju 26336, Republic of Korea
* Correspondence: kkyun@kangwon.ac.kr (K.Y.); kyg@idseng.co.kr (Y.K.)
† Presented at the Second International Conference on Maintenance and Rehabilitation of Constructed Infrastructure Facilities, Honolulu, HI, USA, 16–19 August 2023.

Abstract: Since it is impossible to reconstruct the top level that has collapsed, a formwork is constructed to squeegee mortar or spray mortar, and repair kits are being used in Korea to chip away the damaged concrete of the bridge deck structure. In Korea, a technique called hydro-demolition replaces water blasting and water jetting by using high-pressure water to remove not only asphalt but also old and broken concrete. Additionally, dry materials including cement, aggregate, and additives are carried via the inside of a hose to the field using compressed air, where they meet water and are ejected at a high rate of speed. This technique is known as dry mix shotcrete. Using the devised automatic hydraulic dismantling technology and high-performance dry-mix shotcrete, field application results are discussed in this study.

Keywords: hydro-demolition; shotcrete; overhead shooting; repair; field application

Citation: Yun, K.; Choi, S.; Ha, T.; Song, C.; Hossain, M.S.; Panov, V.; Kim, Y. Field Application of Hydro-Demolition and Dry-Mix Shotcrete for Repairing the Understructure of Bridge Deck. *Eng. Proc.* **2023**, *36*, 55. https://doi.org/10.3390/engproc2023036055

Academic Editor: Hosin (David) Lee

Published: 26 July 2023

1. Introduction

A technique for removing concrete called hydro-demolition, which is also known as water blasting and water jetting in Korea, uses high-pressure water to blast away asphalt as well as hazardous and deteriorated concrete [1]. In order for this method to be effective, high-pressure water must be generated by a power pack and applied to concrete surfaces by a robot. This process creates an excellent bonding surface for the repaired material [2].

The dry components, such as cement, aggregates, and admixtures, are transported by compressed air through a delivery hose to the nozzle where water is added under pressure in the dry-mix shotcrete process. As the mixture is sprayed onto the surface from the nozzle at a high rate of speed, the water fully moistens the other ingredients. The water moistens the dry mix ingredients as it shoots. The nozzleman controls how much water is supplied during the dry mix shotcrete process [3].

To remove the cracked concrete from the bridge deck's substructure, an automated hydro-demolition technique was developed. Additionally, a high-performance dry-mix shotcrete material was created in order to increase build-up thickness for overhead firing while decreasing rebound, dust, and dry shrinkage cracking. This paper details the field application outcomes employing the high-performance dry-mix shotcrete and the proposed automatic hydro-demolition process.

2. Methodology and Materials

2.1. Hydro Demolition

Utilizing high-pressure water to blast asphalt hazardous concrete, hydro-demolition, also known as water blasting and water jetting in Korea, is a technique for removing concrete [1]. Specialized machinery to create high-pressure water and specialized rails to apply to the concrete surface are needed for this technology to function. This procedure eliminates the damaged surface without creating micro cracks or shaking the entire structure.

Automated hydraulic demolition technology has been developed to remove cracked concrete from bridge deck substructures. This paper details field application results using high-performance dry mix shotcrete and the proposed automated hydraulic demolition process.

2.2. Dry-Mix Shotcrete

Dry shotcrete is a method in which dry mixed materials such as cement, aggregate, and quick-setting agent are moved inside the hose by compressed air and sprayed by applying pressure water from a nozzle. Dry shotcrete requires small equipment, so there are few restrictions on supply work, and it can be pumped over a relatively long distance. It has the advantage of being easy to clean and repair and can be easily applied to leaky areas [4,5].

2.3. Materials and Mix Design

Considering the mixed aggregate particle size gradation curve, the N1 variable altered the aggregate composition in comparison to the other mix designs in Table 1, and it also increased the aggregate amount. Through literature review, silica fume was also added at a rate of 3% of the total weight, and the amount of polymer was also changed. The shrinkage-reducing substance used was of the powder type. Concrete shrinkage is prevented with shrinkage inhibitors, which also increase cracking resistance. High-performance nylon fibers were substituted for the natural fibers that were originally integrated to minimize plastic shrinkage cracking since they agglomerated when the material was added. In addition to preventing fiber aggregation, nylon fiber is a hydrophilic, high-performance nylon-based fiber reinforcing material with outstanding dispersibility, chemical resistance, and crack resistance. Dust-reducing agents were eliminated since they had little impact and had a detrimental impact on compressive strength. In order to incorporate entrained air into the concrete, an AE agent in powder form was used.

Table 1. Mix design.

	Production Volume	Aggregate			OPC	Early Strength Cement	Expansion Agent	SF	Polymer	Fiber	Dust Reducing Agent	Shrinkage Reducing Agent
		Coarse	Sand No.5	Sand No.6								
T1	700	8.00%	23.00%	23.00%	0.00%	36.00%	2.00%	5.00%	2.50%	0.10%	0.30%	0.10%
T2	700	8.00%	23.00%	23.00%	0.00%	34.30%	2.00%	7.00%	2.50%	0.10%	0.00%	0.10%
T3	700	8.00%	23.00%	23.00%	36.30%	0.00%	2.00%	5.00%	2.50%	0.10%	0.00%	0.10%
T4	700	8.00%	23.00%	23.00%	0.00%	36.30%	2.00%	5.00%	2.50%	0.10%	0.00%	0.10%
N1	700	34.00%	34.00%	10.00%	19.50%	0.00%	0.00%	1.80%	0.40%	0.10%	0.00%	0.30%

3. Results and Discussion

To implement the bridge ceiling test, the test was carried out using the provided materials on an abandoned bridge in Yeongdong-gun, Chungcheongnam-do. With a depth of around 250 mm, the hydro-demolition process completely damaged the ceiling part.

Dry shotcrete equipment was set up for spraying the ceiling, and a ceiling shotcrete experiment was carried out. The experiment was carried out by splitting the ceiling into two layers in order to provide enough adhesive thickness. It was determined that the actual adhesive thickness was roughly 300 mm, which was greater than the maximum depth of 250 mm, as shown in Figure 1a–c.

The initial compressive strength of the modified new blend composition was higher than that of the previous blend composition based on the 3-day compressive strength measurement. Given that the dust reducer tends to diminish compressive strength, the T1 variable with the dust reducer included exhibited the lowest compressive strength.

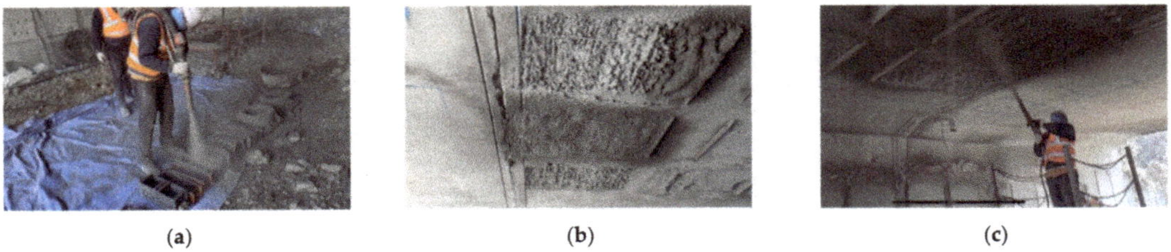

Figure 1. (**a–c**) Hydrodemoltion and Dry shotcrete process in abandoned bridge. (**a**) Field sample preparation. (**b**) Repair section of after shooting. (**c**) Panoramic view of shooting.

The T4 variable, which used early-strength cement and contained 5% silica fume, exhibited the highest compressive strength, while the T3 variable, which used type 1 normal Portland cement, reported a lower compressive strength.

The rightmost column of the Table 2 contains the test results, and it displays the Korea Expressway Corporation's concrete structure repair material quality criteria (2022). The test was conducted after asking a company that specializes in quality testing to conduct it. The test was performed by directly producing samples from the abandoned bridge. It is demonstrable that all test items produced test results that met quality test standards, and that the newly enhanced formulation table secured durability results appropriate for a repair material.

Table 2. On-field sample test results based in Korea Expressway Corporation concrete structure repair material quality standards (2022).

Test for 28 Days		Performance Standard	Test Methods	Test Results
Mechanical Properties	Compressive strength	More than 25.0 MPa	KS F 2405	32.6 MPa (7 days) 42.9 MPa (28 days)
	Flexural strength	More than 4.8 MPa	KS F 2408	4.84 MPa (7 days) 5.89 MPa (28 days)
	Adhesion strength	More than1.7 MPa	KS F 2762	1.74 MPa
Volume Stability	Length change rate	Less than ±0.05%	KS F 2424	−0.023%
	Crack resistance	No cracks for 56 days	AASHTO T334-08	no cracks
	Coefficient of thermal expansion	4.0~25.0 × 10^{-6}/°C	KS F 2608	13.15 × 10^{-6}/°C
	Modulus of elasticity	6.8~38.0 GPa	KS F 2438	2.55 × 10^{-4} MPa
Durability	Salt permeation resistance	less than 1000 C	KS F 2711	524 C
	Freeze-thaw resistance	more than 80 DF	KS F 2456	89.3%

4. Conclusions

In this study, the damaged concrete cross section on the bridge deck ceiling was repaired using hydro demolition and dry-mix shotcrete that was placed directly in the field. Traditionally, manual chipping of the concrete cross section required a person to climb to a great height; however, it can be argued that worker safety was ensured by crushing it using a mechanical device. Dry-mix shotcrete is more suited for cross-section repair using less material than wet-mix shotcrete since it requires less equipment and is simpler to clean. In conclusion, it was discovered that the actual adhesive thickness was close to 300 mm, exceeding the maximum depth of 250 mm. The highest compressive strength measured in this study was T4 variable, which included early-strength cement, and 5% silica fume. Because T1 has the largest amount of dust reducer, it also has the lowest compressive strength. Compressive strength afterwards tends to decline as a result of the dust reducer being increased.

Author Contributions: Conceptualization, K.Y., Y.K. and S.C.; methodology, K.Y., S.C. and M.S.H.; validation, M.S.H., K.Y. and C.S.; formal analysis, T.H. and C.S.; investigation, S.C., C.S., T.H. and V.P.; resources, K.Y. and M.S.H.; writing—original draft preparation, M.S.H. and C.S.; writing—review and editing, K.Y. and M.S.H.; supervision, K.Y. and V.P.; project administration, K.Y., Y.K. and S.C.; funding acquisition, K.Y. All authors have read and agreed to the published version of the manuscript.

Funding: This research was conducted by the Ministry of Land, Infrastructure and Transport/Korea Land, Transport, Science and Technology Promotion Agency (Task number: 21POQW-B152690-03, RS-2021-KA161910) and the National Research Foundation of Korea (assignment number: RS-2023-00208664) funded by the Ministry of Science and ICT.

Institutional Review Board Statement: Not applicable.

Informed Consent Statement: Not applicable.

Data Availability Statement: All data are new and experimented.

Conflicts of Interest: The authors declare no conflict of interest.

References

1. YUN, K.-K.; Lee, K.-R.; HAN, S.-Y.; KIM, Y.-G.; KWON, S.-A. Rehabilitation of Marine Concrete Structure with Under-Water Hydrodemolition and Sprayed Concrete. In Proceedings of the MATEC Web of Conferences, Cape Town, South Africa, 19–21 November 2018; EDP Sciences: Les Ulis, France, 2018; Volume 199, p. 07009.
2. Chynoweth, G.; Stankie, R.R.; Allen, W.L.; Anderson, R.R.; Babcock, W.N.; Barlow, P.; Bartholomew, J.J.; Bergemann, G.O.; Bullock, R.E.; Constantino, F.J. Concrete repair guide. *ACI Comm. Concr. Repair Man.* **1996**, *546*, 287–327.
3. Mukhopadhyay, M. Right Shotcrete Equipment and Skill of Nozzleman Ensure Good Quality Shotcrete in Tunnel Construction. *INCOLD J. A Half Yrly. Tech. J. Indian Comm. Large Dams* **2016**, *5*, 49–51.
4. Yoggy, G.D. The history of shotcrete. *Shotcrete* **2000**, *2*, 28–29.
5. Bernard, S. (Ed.) *Shotcrete: Elements of a System*; CRC Press: Boca Raton, FL, USA, 2010.

engineering proceedings

MDPI

Proceeding Paper

Systematic Evaluation of the Field Constructability and Performance of Asphalt Mixtures Containing High Percentages of Recycled Asphalt †

Logan Cantrell [1] and Haifang Wen [2,*]

[1] Granite Construction, Olympia, WA 98512, USA; logan.cantrell@gcinc.com
[2] Washington Center for Asphalt Technology, Department of Civil and Environmental Engineering, Washington State University, Pullman, WA 99164, USA
* Correspondence: haifang_wen@wsu.edu
† Presented at the Second International Conference on Maintenance and Rehabilitation of Constructed Infrastructure Facilities, Honolulu, HI, USA, 16–19 August 2023.

Abstract: This study evaluates the use of high amounts of recycled asphalt pavement (RAP) and recycled asphalt shingle (RAS) in asphalt mixes for sustainable construction. While past research has focused on asphalt binders and laboratory performance testing, this study assesses the effect of high recycle content on constructability and long-term field performance. A total of 72 mix designs placed from 2016 to 2020 were evaluated for compaction characteristics, while the rutting, cracking, and roughness of 16 projects placed from 2011 to 2015 were assessed based on recycled asphalt levels and mix components. Results showed that high recycled mix projects had equivalent compaction characteristics to low RAP mix projects, except that high RAP mixes had lower variability. High RAP/RAS mixes with rejuvenators had a higher density than those without, and high recycled mix projects had comparable field performance to that of low RAP mix projects, except for lower longitudinal cracking in high RAP projects.

Keywords: high RAP; compaction; density; field performance; cracking

Citation: Cantrell, L.; Wen, H. Systematic Evaluation of the Field Constructability and Performance of Asphalt Mixtures Containing High Percentages of Recycled Asphalt. *Eng. Proc.* **2023**, *36*, 56. https://doi.org/10.3390/engproc2023036056

Academic Editor: Hosin (David) Lee

Published: 26 July 2023

1. Introduction

The constructability characteristics of asphalt mixes, such as compactability and consistency during production and construction, are crucial for the pavement's overall performance. Compaction reduces air voids and provides interlocking of aggregates, affecting the pavement's resistance to rutting, cracking, and moisture. Higher field density leads to better results and extended service life [1]. The compaction of mixes depends on various factors such as aggregate, recycled asphalt pavement (RAP), recycled asphalt shingle (RAS), binder characteristics, and field conditions. Compaction has been identified as the most important construction factor for long-term serviceability [1]. Individual mix properties influence a mixes compaction in different ways [2]. Incorporating high RAP with and without a rejuvenator was reported to produce similar density results to a conventional mix based on test section construction [3]. It has been suggested high RAP mixtures may require more attention due to increased stiffness because of the RAP [4].

After compaction, the pavement's long-term performance is evaluated for rutting and cracking; the main structural distresses of concern. RAP improves rutting resistance, while the aging and stiffening of asphalt increase the cracking potential [5]. Few studies have evaluated the long-term field performance of RAP mixes [4]. Study results varied, with all studies showing similar or decreased rutting and roughness, while some showed increased cracking risk depending on the section [5–7]. Overall, the associated risk of increased cracking with the addition of RAP is concerning.

Superior field performance of asphalt materials is the ultimate goal for which all laboratory testing and pavement design considerations strive. As the use of highly recycled mixes increases, the need to understand its effects on field constructability and performance is imperative. This study compared compaction and performance between high-recycled and low-recycled mixes based on field data.

2. Materials and Methods

Seventy-two mix designs from Washington State Department of Transportation (WSDOT) paving projects performed from the 2016 to 2020 construction seasons were obtained from the WSDOT SAM and analyzed for compaction attributes [8]. These results represent one contractor's mix designs from 13 asphalt plants because only one contractor had been using highly recycled mixes for WSDOT. Three recycle levels were evaluated in this study, including low recycled (LR), high RAP (HR), and super RAP (SR). LR mixes were produced with 0–20% RAP, HR at >20% RAP (maximum 40% binder replacement), and SR used the combination of RAP and RAS (maximum 20% binder replacement from RAS, 40% for the combination). HR and SR mixes analyzed together are referred to as high RAP/RAS (HRR). Overall, 6 SR, 34 HR, and 32 LR mixes were produced from 2016 to 2020, with 48% 1/2″ nominal maximum aggregate size (NMAS) mixes and 52% 3/8″ NMAS mixes.

Sixteen mix designs (6 LR and 10 HR) placed on WSDOT projects from 2011 to 2015 were analyzed for field performance attributes and obtained from the WSPMS [9]. These projects were evaluated for rutting, roughness, and cracking characteristics which are used to determine maintenance and rehabilitation decisions [10]. The selected LR and HR projects have at least four years since placement to potentially have cracking issues, with HR data only existing since 2013. Note that SR mixes were not included, as RAS was not allowed by WSDOT until 2016 [11]. Rutting and IRI metrics had continuous data every year, and non-zero starting points were displayed as an increase per year since construction. Cracking results were not continuous and thus were displayed only as the maximum cracking for that section.

3. Results and Discussion

The constructability characteristics of the 72 mixes were evaluated from WSDOT testing results. The field density, standard deviation, and composite pay factor (CPF) results were first compared between LR and HRR projects. Then, a two-sample t-test was conducted to determine if the mean difference was statistically significant. For field density, the t-test results indicate no statistical difference in the mean density and CPF between LR and HRR mixes, with p-values of 0.94 and 0.36, respectively. However, the t-test results for the standard deviation of field density indicate that HRR mixes had a statistically lower standard deviation by 0.24 than the LR mixes, with a p-value of 0.033. The lower standard deviation of HRR mixes indicates that HRR has better compactability in terms of consistency, contributing to more consistent performance.

The HRR results were then split into HR and SR categories for further evaluation, as shown in Figure 1. SR mixes showed a higher average density (93.9) than that HR (93.1) and LR (93.3) mixes. Analysis of variance analysis (ANOVA) was conducted on the three combinations for density, standard deviation, and CPF returning a p-value of 0.13 for density, 0.004 for standard deviation, and 0.62 for CPF. These results indicate no statistical difference in the density and CPF for LR, HR, and SR mixes. The standard deviation showed a statistical difference between the groups, and based on the Tukey post hoc test, the HR mixes had statistically lower standard deviation than both the SR and LR, with p-values of 0.035 and 0.014, respectively. HR standard deviation was lower than SR by 0.50 and LR by 0.32, respectively. No statistical difference in standard deviation was found between LR and SR mixes. These results contradict the presumption that a larger variability may be expected when a high percentage of recycled asphalt is included.

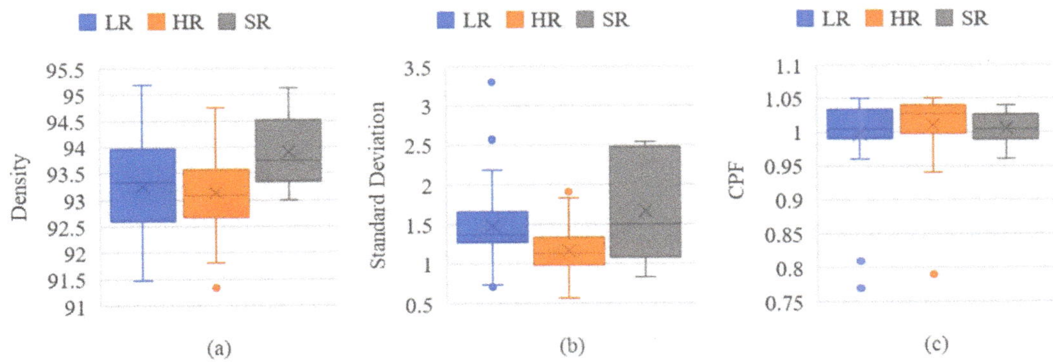

Figure 1. Comparison between LR, HR, and SR for (**a**) Density, (**b**) Standard Deviation, and (**c**) CPF.

Some mix properties, or construction factors, were analyzed to determine the causes of the difference in the compactability of the above mixes. The statistical analysis results show that the factors having significance are rejuvenator use on HRR densities, as well as gradation and the number of density lots on CPF. HRR mixes using a rejuvenator averaged a 0.6% higher density than non-rejuvenated mixes. This indicates that the rejuvenator may be more effective in softening the aged binder in HRR mixes than using a softer virgin binder alone. The rejuvenator could either physically soften the aged binder via mechanical blending and/or chemically dissolve large oxidized molecules. The fact that rejuvenator is more effective than softer virgin binder alone indicates these aged binders were likely chemically dissolved by the rejuvenator. Additionally, since coarse and fine gradations are defined differently for each NMAS based on the percent passing #8 sieve, 3/8″ and 1/2″ NMAS mixes were analyzed individually. The results showed that while 3/8″ NMAS mixes showed no statistical difference in the standard deviation of density, coarse 1/2″ mixes had, on average, 0.4 lower standard deviations than 1/2″ fine mixes with a *p*-value of 0.003.

The field performance of the 16 projects was evaluated in terms of rutting, IRI, transverse cracking, and longitudinal cracking. The mean values of performance results of HR and LR projects, along with the *p*-values of the two-sample t-tests for each performance metric evaluated, are shown in Table 1. Note that no SR mixes were used in these 16 projects. The average first crack year was 3.5 for LR and 4.0 for HR but showed no statistical difference with a *p*-value of 0.17. The only metric showing significant differences were the percent of longitudinal cracking being lower for HR than LR mixes. One possibility is that the HR mixes are oxidizing slower and therefore are less brittle as they age, possibly due to already having an aged binder included or the softer virgin binder aging at a lower rate. Further study is needed to test these theories to determine why cracking performance varies from LR to HR.

Table 1. Significance of RAP on Field Performance.

Performance Metric	LR	HR	*p*-Value
Rutting/year (in)	0.02	0.02	0.42
IRI/year (in/mi)	1.11	1.20	0.43
Transverse Cracking (count/100ft)	3.00	3.96	0.22
Longitudinal Cracking (% of section)	13.02	4.56	*0.001*

4. Conclusions

As recycled material usage increases, the performance of highly recycled mixes in the field becomes more critical. Most studies on performance involved durability in the laboratory. Few studies addressed the field constructability and field performance of highly recycled mixes. In this study, the constructability of HR, LR, and SR mixes was evaluated, and mix properties were evaluated to explain why differences may exist. The field performance of LR and HR projects was compared. It was found:

1. Overall, the compactability of HRR mixes was similar to or better than typical LR mixes. There was no statistical difference between HRR mixes and LR mixes in density means.
2. HR mixes showed a lower standard deviation than LR and SR mixes. For the 1/2″ NMAS mix, fine-graded mixes showed a higher standard deviation than coarse-graded mixes. Second, HR mixes had lower asphalt contents than LR mixes statistically. In combination, having lower asphalt content (statistically insignificant, though) and coarser mixes appear to be giving these HR mixes lower potential mix tenderness, leading to lower standard deviation.
3. HRR mixes that used a rejuvenator were shown to have increased density results significantly over those HRR mixes that used softer virgin binders only without a rejuvenator. It is believed the rejuvenator may facilitate the compaction by chemically dissolving the highly oxidized asphalt molecules.
4. The use of a rejuvenator significantly increases densities in HRR mixes. The air voids were significantly lower for SR mixes, potentially increasing compactability.
5. There is no statistical difference in rutting, IRI, or transverse cracking performance between LR and HR projects, showing similar rates of change. However, the longitudinal cracking of HR mixes was statistically lower than that of LR mixes, with HR having less than half the cracking of the LR section.

Further studies are needed to verify the findings in this study by incorporating larger populations of data that cover different geographies, climates, material sources, etc.

Author Contributions: Conceptualization, L.C. and H.W.; methodology, L.C. and H.W.; formal analysis, L.C. and H.W.; writing—original draft preparation, L.C. and H.W.; writing—review and editing, L.C. and H.W.; All authors have read and agreed to the published version of the manuscript.

Funding: This research received no external funding.

Institutional Review Board Statement: Not applicable.

Informed Consent Statement: Not applicable.

Data Availability Statement: Not available upon request.

Conflicts of Interest: The authors declare no conflict of interest.

References

1. Tran, N.; Turner, P.; Shambley, J. *Enhanced Compaction to Improve Durability and Extend Pavement Service Life: A Literature Review*; NCAT: Greensboro, NC, USA, 2016.
2. Transportation Research Board. *Factors Affecting Compaction of Asphalt Pavements*; Transportation Research Circular E-C105; Transportation Research Board: Washington, DC, USA, 2006.
3. Moon, B.; Lee, H.; Ledtje, P.; Williams, C. Evaluating effect of rejuvenators on high RAP mixtures through laboratory perfor-mance tests and construction of field test sections. *Constr. Build. Mater.* **2022**, *360*, 127698. [CrossRef]
4. Copeland, A. *Reclaimed Asphalt Pavement in Asphalt Mixtures: State of the Practice*; FHWA-HRT-11-021; Federal Highway Administration: Raleigh, NC, USA, 2011.
5. West, R.; Timm, D.; Powell, B.; Tran, N.; Yin, F.; Bowers, B.; Rodezno, C.; Leiva, F.; Vargas, A.; Gu, F.; et al. *Phase VII (2018–2021) NCAT Test Track Findings*; NCAT: Greensboro, NC, USA, 2021.
6. Gong, H.; Huang, B.; Shu, X. Field performance evaluation of asphalt mixtures containing high percentage of RAP using LTPP data. *Constr. Build. Mater.* **2018**, *176*, 118–128. [CrossRef]
7. Azzam, A.; Kim, Y.-R.; Rahmani, M. *Data Analysis of Nebraska Pavements Containing RAP*; Nebraska Department of Transportation: Lincoln, Nebraska, 2020.
8. WSDOT. Materials Lab-Statistical Analysis of Materials (SAM). 2021. Available online: https://wsdot.wa.gov/Business/MaterialsLab/StatisticalAnalysisOfMaterials.htm (accessed on 31 July 2021).
9. Kay, R.K.; Mahoney, J.; Jackson, N. *The WSDOT Pavement Management System-A 1993 Update*; Washington State Transportation Commission: Olympia, WA, USA, 1993; Volume WA-RD 274.1.

10. Baker, M.; Mahoney, J. *Identification and Assessment of Washington State Pavements with Superior and Inferior Performance*; Washington State Transportation Commission: Olympia, WA, USA, 2000; Volume WA-RD 437.1.
11. *Standard Specifications for Road, Bridge, and Municipal Construction*; WSDOT: Olympia, WA, USA, 2016.

engineering proceedings

MDPI

Proceeding Paper

Plastic Recycling in Asphalt Concrete Pavements: Preliminary Observations from Hawaii's Pilot Project †

Seyed Yashar Beheshti Shirazi, Saroj Pathak, Arthur Sickels, Jr. and Adrian R. Archilla *

Department of Civil and Environmental Engineering, University of Hawaii at Manoa, Honolulu, HI 96822, USA; shirazi8@hawaii.edu (S.Y.B.S.); spathak@hawaii.edu (S.P.); asickels@hawaii.edu (A.S.J.)

* Correspondence: archilla@hawaii.edu

† Presented at the Second International Conference on Maintenance and Rehabilitation of Constructed Infrastructure Facilities, Honolulu, HI, USA, 16–19 August 2023.

Abstract: This paper presents findings of testing performed to date from three field asphalt concrete mixes obtained from paving performed in November 2022 for a pilot project in Hawaii. The control mix meets Hawaii State IV mix requirements, with 20% reclaimed asphalt pavement (RAP) and polymer modified asphalt (PMA) binder PG64E-22. The other two mixes, which have the same gradation and RAP content, were prepared with 2 lb. per ton of NewRoad pellets consisting mostly of post-industrial high-density polyethylene (HDPE). One of these was prepared with PMA PG64E-22 and the other with neat binder, PG64-16. Testing results to date show benefits in rutting and expected results in dynamic modulus. They are inconclusive with regard to cracking because of high variability and inconsistencies in IDEAL-CT results without and with moisture sensitivity conditioning.

Keywords: plastics; asphalt concrete; recycling

Citation: Shirazi, S.Y.B.; Pathak, S.; Sickels, A., Jr.; Archilla, A.R. Plastic Recycling in Asphalt Concrete Pavements: Preliminary Observations from Hawaii's Pilot Project. *Eng. Proc.* **2023**, *36*, 57. https://doi.org/10.3390/engproc2023036057

Academic Editor: Hosin (David) Lee

Published: 1 August 2023

1. Introduction

The recycling of plastics in Hot Mix Asphalt (HMA) pavements is gaining attention worldwide [1] as plastic waste (PW) is a primary source of environmental pollution [2] with a current generation that is twice as much as it was two decades ago [3]. According to the OECD, the bulk of PW ends up in landfill, incinerated or leaking into the environment, and only 9% is successfully recycled [3]. As stated in [1], "research is needed to establish a better understanding of the impact of recycled plastics on the performance, especially durability and cracking resistance, of asphalt binders and mixtures".

This paper presents preliminary mechanical testing findings of a study involving both mechanical testing and microplastics and plastic additives testing performed from three field HMA mixes, of which two contained recycled PW. The mixes were obtained from paving performed for the Hawaii Department of Transportation (HDOT) for a pilot project in Honolulu, Hawaii. Testing results to date with a Hamburg Wheel Tracker (HWT) and an Asphalt Mixture Performance Tester (AMPT) have shown benefits in terms of rutting and a priori expected variations in the dynamic modulus. On the other hand, the results of Cracking Tolerance Index (CT_{Index}) (ASTM D8225-19) obtained in IDEAL-CT testing are inconclusive in terms of fatigue cracking performance. The CT_{Index} results show high variability within each mix, unexpected trends between mixes and, for each mix, inconsistent results between tests performed without and with moisture sensitivity conditioning using a Moisture Induced Stress Tester (MIST). The initial findings of this characterization effort have been encouraging, but alternative tests for evaluating cracking performance in Hawaii need to be evaluated.

2. Materials and Methods

The three mixes evaluated were obtained from paving for a pilot project performed in November 2022 on the southern end of Fort Weaver Road, a two-lane bi-directional road in Oahu, Hawaii, that carries 6200 vehicles per day on average. The 1.3-mile-long roadway segment was divided into three sections with the design gradations shown in Table 1, meeting the Hawaii State IV mix requirements (similar to those for a Superpave 12.5 mm mix) and incorporating 20% Reclaimed Asphalt Pavement (RAP).

Table 1. Design gradations.

Mix	Sieve Size (mm)									
	3/4″ 19 mm	1/2″ 12.5 mm	3/8″ 9.5 mm	#4 4.75 mm	#8 2.36 mm	#16 1.18 mm	#30 0.60 mm	#50 0.30 mm	#100 0.15 mm	#200 0.075 mm
Control	100.0	93.7	86.8	60.4	38.4	23.7	16.2	11.4	8.8	6.98
PMA NR	100.0	92.6	85.8	59.0	37.7	23.7	16.2	11.5	8.8	7.05
HMA NR	100.0	93.1	86.1	59.8	38.8	23.9	16.3	11.5	8.8	7.07

1. Control section: 5.3% asphalt content by total weight of mix (TWM) of Polymer Modified Asphalt or PMA (PG64E-22) (0.95% by TWM contributed by RAP binder);
2. PMA NR Plastic section: 5.2% asphalt content by TWM of PMA PG64E-22 binder (0.96% by TWM contributed by RAP), incorporating 2 lb. per ton of mix of plastic pellets (0.1% by mass of mix) as recommended by the supplier;
3. HMA NR Plastic section: 5.1% asphalt content by TWM of neat PG64-16 binder (0.96% by TWM contributed by RAP) with 2 lb. per ton of mix of plastic pellets.

Pacific GeoSourse provided NewRoad plastic pellets consisting mostly of post-industrial high-density polyethylene (HDPE), which were added to the mix at the plant. No pellets were visible in the field samples, indicating that they had mostly melted into the mix.

Trial specimens were compacted to find the appropriate amount of mix to compact specimens with 7 ± 0.5% air voids for the different tests: HWT, AMPT dynamic modulus and permanent deformation, and IDEAL-CT. The specimens were re-heated to the compaction temperature and tested without any additional aging, as specified in ASTM D8225 [4].

3. Results

Figure 1 shows the HWT results at 50 °C for the three mixes. None of the mixes exhibited a stripping point, and they had low final rutting values, ranging from slightly above 1 mm to 2.25 mm. Thus, all mixes were expected to be rutting resistant. Despite the small values, it can be seen that the addition of NR to the control mix reduced the rutting. The HMA NR mix did not perform as well as the control mix, but it still exhibited a good performance.

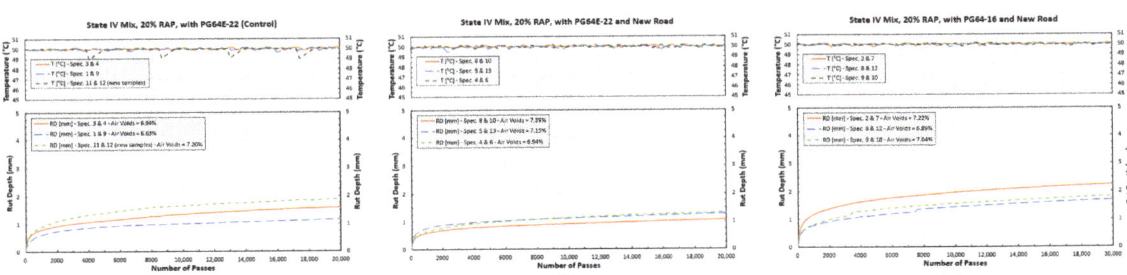

(a) (b) (c)

Figure 1. HWT test results: (**a**) control, (**b**) PMA NR, and (**c**) HMA NR.

As shown in Figure 2, cylindrical samples of the control and PMA NR mixes tested for permanent deformation (PD) at 54 °C without and with MiST conditioning (20 h at 50 °C followed by 3500 pressure cycles at 276 kPa) provided basically the same results. These PD tests were carried out after dynamic modulus (|E*|) testing on the same samples. Both without and with conditioning, the PMA NR (PG64-22NR) sample exhibited less rutting than the control mix (PG64E-22) sample with the same type of conditioning.

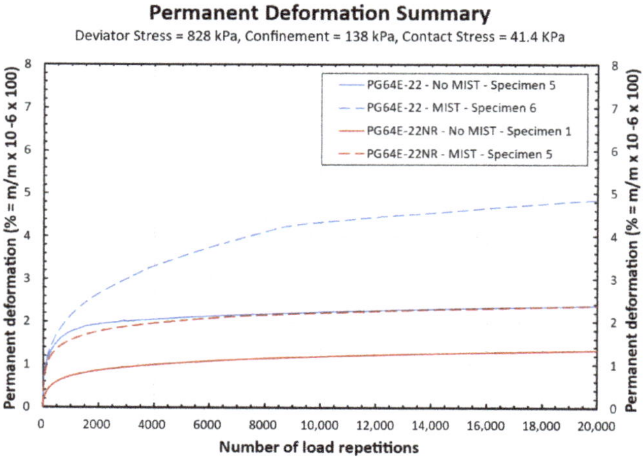

Figure 2. AMPT permanent deformation results for the control and PMS NR mixes with and without MiST conditioning.

Furthermore, for each mix, conditioning in the MiST resulted in larger PD values, which was the a priori expected result.

The dynamic modulus (|E*|) of the same four samples was also consistent in terms of MiST conditioning. As shown in Figure 3, for each mix, the master curves consistently dropped after MiST conditioning (compare (a) to (b) and (c) to (d)). The comparison between mixes indicated that regardless of conditioning, the master curve of the PMA NR mix was lower than the control mix without confinement, and higher with 138 kPa and 207 kPa confinement (compare (a) to (c) and (b) to (d)).

The |E*| and PD results presented above are based on a single specimen under each condition. Consequently, one should be cautious when drawing a strong conclusion from these. Nevertheless, the results are consistent with previous testing by the last author, with similar PMA mixes in Hawaii, and with the a priori expectations of the effect of moisture damage on these properties.

Figure 4 shows the CT_{Index} values obtained to date. The Figure presents a stem and leaf plot, but since there are only three samples for most combinations of mix and conditioning, it is mostly helpful to visualize the mean value (identified with a cross symbol), the range of values, and the standard deviation (SD). Note that, in general, there is very high variability. It can also be seen that without MiST conditioning, the values for the PMA NR mix were generally lower than for the control mix. Although there is an overlap for the PMA NR and HMA NR, the average tended to increase with the HMA NR, which is counterintuitive. Similar differences can be seen after mix conditioning, but the variability for the control mix was substantially higher. Even more concerning is the fact that better (higher) values were obtained on average after MiST conditioning for each mix, which is again contrary to expectations.

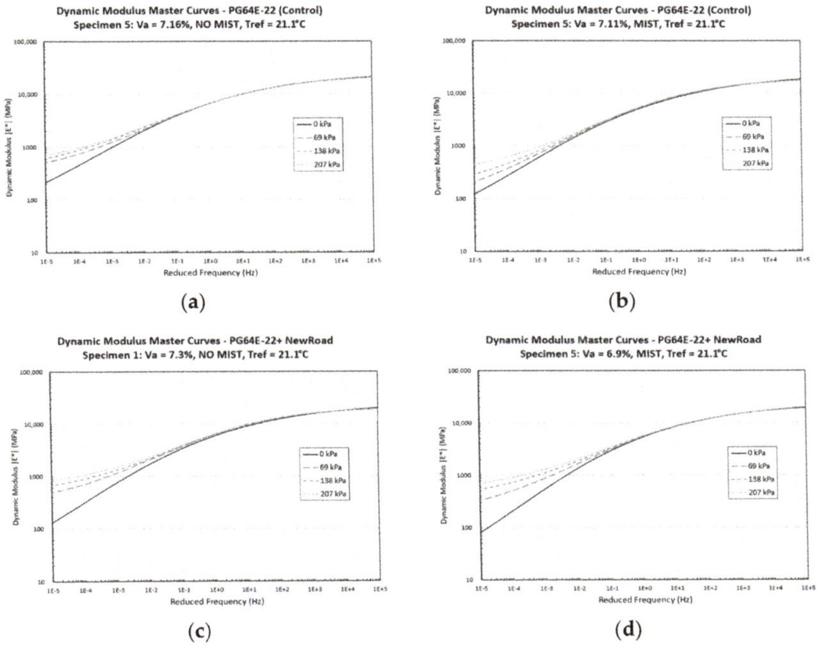

Figure 3. Dynamic modulus of the control and PMS NR mixes with and without MiST conditioning.

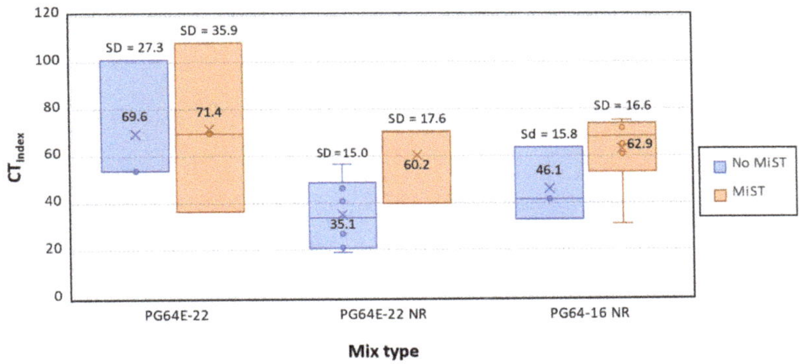

Figure 4. CT_{Index} results of the control mix (PG64E-22), PMA NR (PG64E-22 NR) mix, and HMA NR (PG64-16 NR) mix without and with MiST conditioning.

4. Conclusions

The testing of mixes from the field pilot study with sections containing recycled PW showed benefits in rutting and expected variations in dynamic modulus with moisture sensitivity conditioning. The CT_{Index} results were inconclusive, since they were counterintuitive: without MiST conditioning, the results for the control, PMA NR, and HMA NR mixes implied not only that the addition of PW is detrimental to cracking, but that it also negates all the well-known benefits of polymer modification relative to the mix with the neat binder. Also, the increases of the means for the three mixes after MiST conditioning were counterintuitive and the opposite of what the dynamic modulus and permanent deformation results with the AMPT indicated, albeit not for the same performance measure. Consequently, it appears that the CT_{Index} does not capture the benefits of polymer modification and the effects of moisture damage for these mixes. These factors also make the comparison between the mixes less reliable. Consequently, it is recommended that in future HDOT efforts IDEAL-CT testing be complemented or replaced with a test more capable of capturing the effects of these generally well-known factors for these Hawaiian mixes on long-term-aged samples.

Author Contributions: Conceptualization, S.Y.B.S. 10%, S.P. 10%, A.S.J. 10% and A.R.A. 70%; methodology, S.Y.B.S. 10%, S.P. 10%, A.S.J. 10% and A.R.A. 70%; formal analysis, S.Y.B.S. 10%, S.P. 10%, A.S.J. 10% and A.R.A. 70%; data collection, S.Y.B.S. 25%, S.P. 25%, A.S.J. 25% and A.R.A. 25%; writing—original draft preparation, S.Y.B.S. 5%, S.P. 5%, A.S.J. 5% and A.R.A. 85%; writing—review and editing, A.R.A. 100%. All authors have read and agreed to the published version of the manuscript.

Funding: This research was funded by Hawaii Department of Transportation—Project Number DOT-10-030. The contents of this paper reflect the view of the writers, who are responsible for the facts and accuracy of the data presented herein. The contents do not necessarily reflect the official views or policies of the State of Hawaii Department of Transportation or the Federal Highway Administration.

Institutional Review Board Statement: Not applicable.

Informed Consent Statement: Not applicable.

Data Availability Statement: The data used in this study was collected for the Hawaii Department of Transportation (HDOT). The data are not publicly available due to restrictions imposed by the DOT, and access may be subject to approval and compliance with their data sharing policies.

Conflicts of Interest: The authors declare no conflict of interest.

References

1. Yin, F.; Moraes, R.; Fortunatus, M.; Tran, N.; Elwardany, M.; Planche, J. *Performance Evaluation and Chemical Characterization of Asphalt Binders and Mixtures Containing Recycled Polyethylene*; Final Report for Plastics Industry Association: Washington, DC, USA, 2019.
2. Gourmelon, G. *Global Plastic Production Rises, Recycling Lags*; World Watch Institute: Washington, DC, USA, 2016.
3. OECD. *Global Plastics Outlook: Economic Drivers, Environmental Impacts and Policy Options*; OECD Publishing: Paris, France, 2022. [CrossRef]
4. *ASTM D8225-19*; Standard Test Method for Determination of Cracking Tolerance Index of Asphalt Mixture Using the Indirect Tensile Cracking Test at Intermediate Temperature. ASTM International: West Conshohocken, PA, USA, 2019.

engineering proceedings

MDPI

Proceeding Paper

Automated Distress Detection, Classification and Measurement for Asphalt Urban Pavements Using YOLO †

Paulina Gómez-Conti [1,*], Alelí Osorio-Lird [1] and Héctor Allende-Cid [2]

1 School of Civil Works, Federico Santa María Technical University, San Joaquín 8940000, Chile; aleli.osorio@usm.cl
2 School of Informatics Engineering, Pontificia Universidad Católica de Valparaíso, Valparaíso 2340000, Chile; hector.allende@pucv.cl
* Correspondence: paulina.gomezc@usm.cl

† Presented at the Second International Conference on Maintenance and Rehabilitation of Constructed Infrastructure Facilities, Honolulu, HI, USA, 16–19 August 2023.

Abstract: In pavement management, it is essential to have a good database with information on the condition of the roads that compose the corresponding network. In Chile, such a database does not currently exist, and there is no technology that can evaluate urban pavement condition in an efficient way. On this research, more than 50,000 images of 13.2 × 2.6 m of asphalt pavement from different zones of Santiago, Chile, were obtained. These images were processed, and the following distresses were labeled with two different levels of severities: patches; potholes; and transversal, longitudinal, and fatigue cracking. These data were used to train and evaluate the following object detection convolutional neural network models: YOLOv5 and YOLOv7.

Keywords: asphalt pavement; urban pavements; distress detection; distress classification; deep learning; convolutional neural network; object detection

Citation: Gómez-Conti, P.; Osorio-Lird, A.; Allende-Cid, H. Automated Distress Detection, Classification and Measurement for Asphalt Urban Pavements Using YOLO. *Eng. Proc.* **2023**, *36*, 58. https://doi.org/10.3390/engproc2023036058

Academic Editor: Hosin (David) Lee

Published: 7 August 2023

1. Introduction

This study was carried out in Santiago, Chile, with the purpose of creating a database for pavement management, specifically to automate and improve the efficiency of urban pavement monitoring. Using low-cost technology, pavement images are taken and used to train a YOLO neural network to automate the detection, classification, and measurement of deterioration in urban pavements.

2. Materials and Methods

Asphalt pavement recordings were obtained by using a GoPro Hero 8 black camera mounted to a car by using a bicycle rack, as shown in Figure 1 [1].

Figure 1. System used to obtain pavement images. Reprinted with permission from Ref. [1]. Copyright 2022 Venegas, J.

By using the camera's telemetry, frames were extracted to obtain an image of every section of the pavement.

Due to the position of the camera, the images obtained are not in plan view, and since the objective is to be able to measure distresses, a perspective transformation is applied [2], as shown in Figure 2. The red square was used to calibrate the transformation, since it had previously known dimensions (400 × 400 mm).

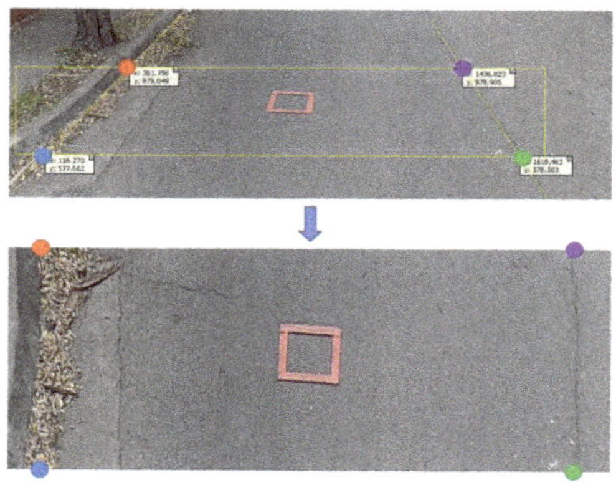

Figure 2. Perspective transformation applied to pavement images.

For a better understanding of the context of the distresses shown in the image, five of these images are joined together, working with a 13.55 × 2.66 m pavement section image. These images are normalized for better performance of the artificial neural network. Some examples of pavement sections are shown in Figure 3.

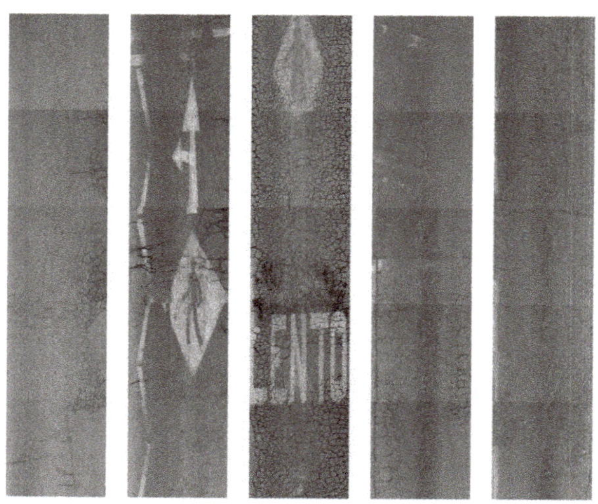

Figure 3. Example of images of asphalt urban pavements from Chile.

Asphalt pavement images from six different municipalities were obtained, with a total length of 104.0 km and a total surface of 276.7 km^2.

The total number of distresses labeled are shown in Table 1. This was performed manually with the help of some engineering students by using the rectangle labels in VGG Image Annotator [3].

Table 1. Number of distresses found in the images obtained in Santiago, Chile.

	Severity	
Distress	**Medium**	**High**
Fatigue	2128	4240
Transversal Cracking	6783	4701
Longitudinal Cracking	2176	1603
Patch	296	193
Potholes	129	313

It should be noted that the most common singularities found in pavement sections were also labeled in order to avoid confusion with distresses, such as manhole covers, drains, and core drilling.

The images are randomly split into 80% training, 10% validation, and 10% testing sets, while maintaining the same percentages for each type of distress. For both YOLOv5 and YOLOv7, training with 300 epochs each is carried out using the training and validation set, while for performance evaluation, the test set is used, i.e., images that have not been previously seen by the network.

The training was performed using a Lenovo Legion T5i Tower 6ta Gen with a NVIDIA GeForce® RTX™ 3060 12 GB GDDR6 graphic card.

3. Results

Table 2 shows the performance using both YOLOv5 and YOLOv7 with the test set. YOLOv5 and YOLOv7 took 144 and 75 h to run, respectively.

Table 2. Results obtained using YOLO.

YOLO	Precision (%)	Recall (%)	mAP 0.05 (%)
v5	41.4	43.3	37.4
v7	42.2	39.9	36.8

The confusion matrices are shown in Figure 4.

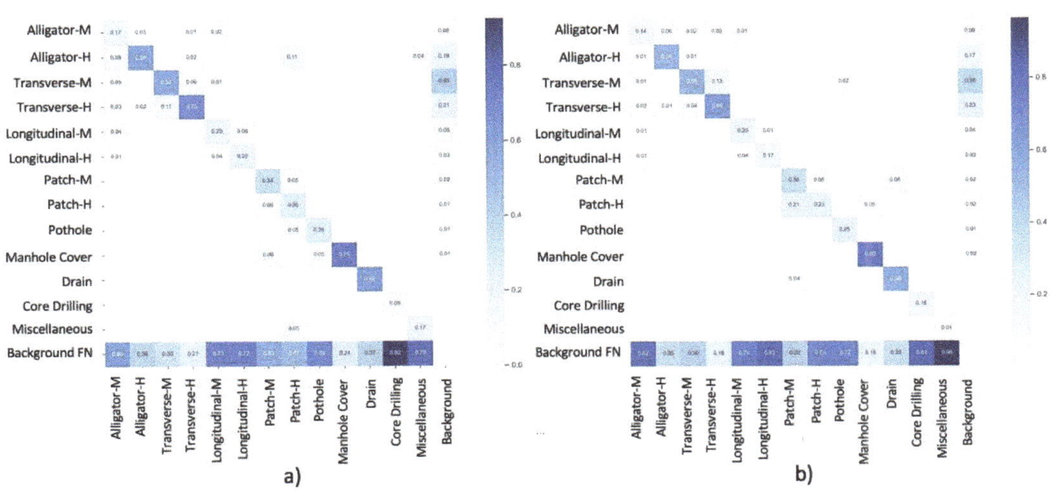

Figure 4. Confusion matrices obtained by evaluating the test set for (**a**) YOLOv5 and (**b**) YOLOv7.

An example of the results obtained by evaluating the test set in YOLOv5 is shown in Figure 5.

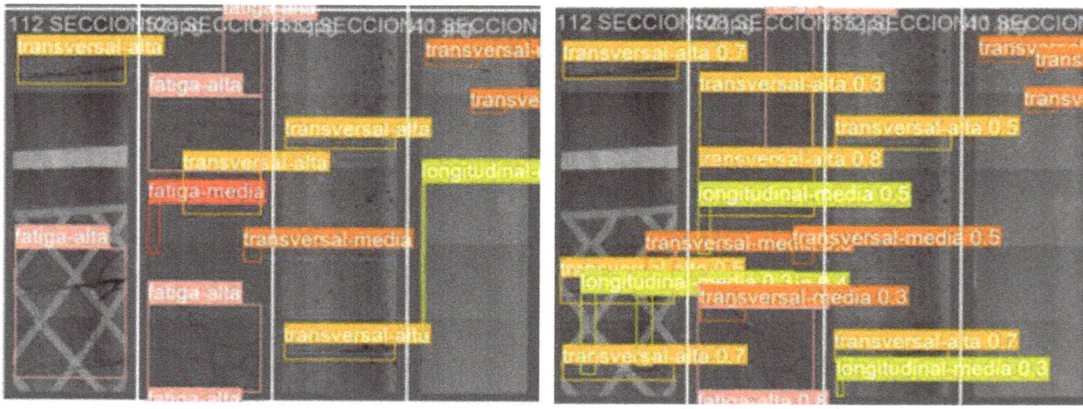

a) b)

Figure 5. Example of labels obtained by evaluating the test set for (**a**) the labels assigned and (**b**) those obtained by YOLOv5.

4. Discussion

As shown in Table 2, both versions of YOLO achieved a similar performance.

As for the distresses, alligator and transverse cracking demonstrated better performance (over 50%). However, longitudinal cracking, patches, and potholes are mostly undetectable.

Although the network can identify and classify the distresses, it is unreliable in terms of severity classification, which can be seen in the diagonal of the confusion matrices in Figure 4.

Finally, YOLO was originally trained with the COCO dataset, in which objects are clearly defined, unlike distresses, where different observers might classify cracks differently.

5. Conclusions

In conclusion, there is no significant difference between the performances of the different YOLO versions. However, YOLOv7 took about half the time it took to train YOLOv5, which is a significant difference.

The pavement distresses that can be found with these results are alligator and transverse cracking; the network developed in this investigation may be useful for the detection of manhole covers and drains.

Author Contributions: Conceptualization, A.O.-L.; methodology, A.O.-L., P.G.-C. and H.A.-C.; algorithm application, P.G.-C. and H.A.-C.; validation, P.G.-C. and H.A.-C.; formal analysis, P.G.-C., A.O.-L. and H.A.-C.; resources, A.O.-L.; data curation, P.G.-C.; writing—original draft preparation, P.G.-C.; writing—review and editing, A.O.-L.; supervision, A.O.-L. and H.A.-C.; project administration, A.O.-L.; funding acquisition, A.O.-L. All authors have read and agreed to the published version of the manuscript.

Funding: This research was funded by ANID Fondecyt Project 11201150. Chilean government.

Institutional Review Board Statement: Not applicable.

Informed Consent Statement: Not applicable.

Data Availability Statement: Data are available on request from the corresponding author.

Conflicts of Interest: The authors declare no conflict of interest.

References

1. Venegas, J. *Desarrollo de Una Metodología de Utilización de Cámaras de Bajo Costo Para la Evaluación de Pavimentos Urbanos*; Technical report Anid Fondecyt 11201150; Chilean Government: Santiago, Chile, 2022.
2. Rozas, D. *Validación de Metodología Experimental Para la Medición de Deterioros Superficiales en Pavimentos Urbanos a Partir de Imágenes Recopiladas por Instrumentos de Bajo Costo*; Technical report Anid Fondecyt 11201150; Chilean Government: Santiago, Chile, 2022.
3. Dutta, A.; Zisserman, A. The VIA Annotation Software for Images, Audio and Video. In Proceedings of the 27th ACM International Conference on Multimedia, Nice, France, 21–25 October 2019; pp. 2276–2279. [CrossRef]

engineering proceedings

MDPI

Proceeding Paper

Design and Evaluation of Ultra-Thin Overlay with High Viscosity and High Elasticity †

Yijia Chen [1], Zhi Liao [1], Lide Chen [1], Tao Ma [2,*], Susan Tighe [3] and Ningyuan Li [3]

1 Road Intellitech Co., Ltd., 9 Shengyun Rd., Nanjing 210008, China; jackchenyj@gmail.com (Y.C.); liaoz0919@gmail.com (Z.L.); chenld3536@gmail.com (L.C.)
2 School of Transportation, Southeast University, 2 Southeast University Rd., Nanjing 211189, China
3 Department of Civil Engineering, McMaster University, 1280 Main St W, Hamilton, ON L8S 4L8, Canada; tighes1@mcmaster.ca (S.T.); nli@uwaterloo.ca (N.L.)
* Correspondence: matao@seu.edu.cn
† Presented at the Second International Conference on Maintenance and Rehabilitation of Constructed Infrastructure Facilities, Honolulu, HI, USA, 16–19 August 2023.

Abstract: Ultra-thin asphalt overlay, which is considered one of the main pavement maintenance strategies, has been widely used to maintain and restore pavements. However, the structural properties of traditional ultra-thin overlay materials, such as anti-friction and anti-cracking pavement surfaces, do not last longer under the climate change and traffic loading conditions. This paper introduces an innovative design of ultra-thin asphalt overlays with high viscosity and high elasticity, which provide not only a long service life of anti-resistance and anti-cracking performance, but also lower traffic noise and smoother riding quality. The process of designing such ultra-thin lift overlays involves multi-objective optimization of the overlay's structural and functional performances, including the quality and quantity of asphalt additives, gradation of coarse aggregates and materials' engineering, and cohesive and adhesive properties of asphalt overlays. During the lab tests prepared for this study, the compound-modified asphalt was prepared by modifying base asphalt with the high viscosity and high elasticity modifier. The gradation design was performed to improve coarse aggregate voids' filling and the density of the mixture, and the trackless tack coat emulsified asphalt was used as an adhesive layer material. Laboratory tests were conducted to evaluate the performance of the asphalt mixture and bonding effect of trackless tack coat emulsified asphalt. Results showed that the high viscosity and elasticity ultra-thin overlay exhibited excellent performance in terms of skid resistance and noise reduction. The interlocking effect of the coarse aggregate skeleton and the optimal asphalt film contribute to the resilient and durable properties of an ultra-thin asphalt overlay.

Keywords: asphalt pavement; asphalt mixture; ultra-thin overlay; high viscosity; high elasticity; skid resistance; molecular structure-activity; gradation design

Citation: Chen, Y.; Liao, Z.; Chen, L.; Ma, T.; Tighe, S.; Li, N. Design and Evaluation of Ultra-Thin Overlay with High Viscosity and High Elasticity. *Eng. Proc.* **2023**, *36*, 59. https://doi.org/10.3390/engproc2023036059

Academic Editor: Hosin (David) Lee

Published: 14 August 2023

1. Introduction

Among the various pavement maintenance technologies, ultra-thin overlay has been increasingly used because of its engineering economy and construction convenience compared to traditional maintenance technologies such as micro-surfacing, slurry seal, and hot in-place recycling. Currently, the main ultra-thin overlay technologies include Nova Chip, UTAC, BBTM, etc. Ultra-thin asphalt concrete and super-thin asphalt concrete were firstly proposed in France. Subsequently, it has been widely used in many countries, including the United States, China, Australia, and South Africa [1]. Huang et al. carried out the performance evaluation of UTA-10 asphalt mixture and applied it to the maintenance engineering of cement bridge deck pavement [2]. Various types of ultra-thin overlay were constantly proposed, including warm mix ultra-thin overlay, cold mix ultra-thin overlay, and anti-icing ultra-thin overlay. Mixture design methods and asphalt modification techniques were studied [3–8].

Recently, more research projects have focused on the development of functional ultra-thin overlay. Li et al. proposed a colored ultrathin overlay which had good skid-resistance, water permeability, and, most importantly, a cooling effect [9]. Budiarto et al. proposed an ultra-thin surfacing hot mix asphalt (UTSHMA) with strong water permeability resistance [10]. Other functional ultra-thin overlay technologies such as porous ultra-thin overlay (PUAO) and self-healing ultra-thin overlay were also tested.

In summary, substantial studies have been conducted to develop ultra-thin overlay technologies and achieved many positive results. However, research on forming a comprehensive performance evaluation method and customized design method has rarely been reported. The purpose of this study is to verify the application of high-viscosity and high-elasticity asphalt in ultra-thin overlay and propose a design and evaluation method considering the durability of skid resistance and crack resistance.

2. Materials and Methods

2.1. Materials

The HVHE (high-viscosity and high-elasticity) asphalt was prepared by modifying base asphalt with HVHE modifier. The technical indexes and test result of the HVHE asphalt were listed in Table 1. The trackless tack coat emulsified asphalt was used as adhesive layer material to resist damage during construction and improve interlayer bonding. Basalt aggregates are used as coarse aggregates to ensure slip resistance. Limestone aggregates are used as fine aggregates to enhance the adhesion of asphalt mortar. The aggregates met the technical specifications of JTG E20-2011 [11].

Table 1. Technical indexes and test result of the HVHE asphalt.

Technical Indexes		Unit	Results	Requirements	Test Protocols
Penetration (25 °C)		0.1 mm	61.9	40–80	T 0604-2011
Soft point		°C	92	≥90	T 0606-2011
Ductility (5 °C)		cm	45.4	≥40	T 0605-2011
Brinell rotational viscosity (60 °C)		Pa·s	>580,000	≥300,000	T 0620-2000
Elastic recovery		%	99	≥95	T 0662-2000
Evaporated residue	Quality loss	%	0.44	±1.0	T 0610-2011
	Penetration (25 °C)	0.1 mm	91	≥80	T 0604-2011
	Ductility (5 °C)	cm	33	≥25	T 0605-2011

2.2. Methods

Gradation Design of UT-10

The thickness of the ultra-thin overlay ranges from 1.5 to 2.5 cm. A gradation with a maximum nominal particle size of 10 mm was applied. Based on both OGFC-10 and SMA-10 mixtures, nine gradations, as shown in Figure 1, were developed to determine the range of the UT-10 mixture gradation. Five asphalt aggregate ratios were set from 5.2% to 7.2% with an interval of 0.5%. The immersion rutting test was conducted to determine the optimum air void ratio. The texture depth decay test and the TSRST test were carried out to obtain the optimum asphalt content.

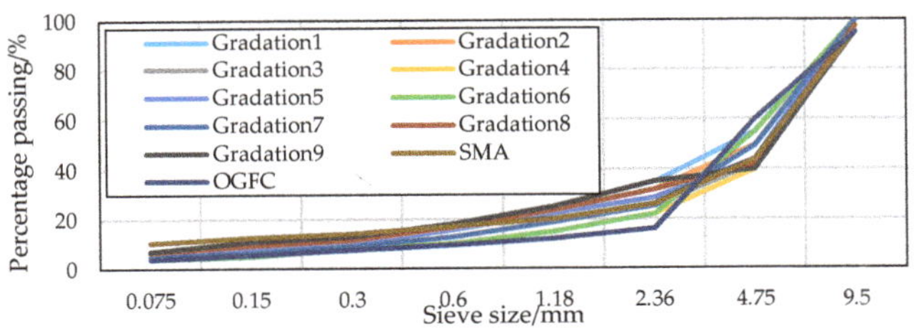

Figure 1. Nine preset gradations of UT-10.

3. Results and Discussion

3.1. Determination of Air Void Ratio

The results of the immersion rutting test are shown in Figure 2. They showed that the rutting deformation of Gradation 1, 2, 8, and 9 was around 2 mm, which is much larger than other gradations. The air void ratio of these four gradations ranged from 8% to 12%, indicating that it was difficult to drain water after it entered the semi-connected voids, and with the application of vehicle loads, water damage occurred, resulting in a larger rutting deformation. Therefore, the air void ratio of UT-10 mixture was determined to range from 12% to 15%.

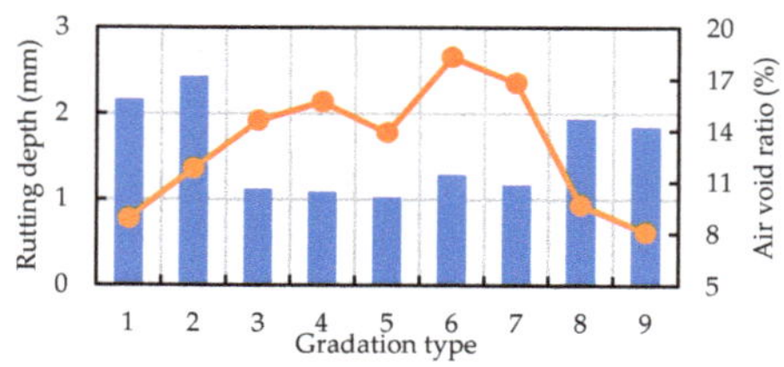

Figure 2. Results of the immersion rutting test.

3.2. Determination of Asphalt Aggregate Ratio

The results of the texture depth decay test are shown in Figure 3. They indicated that the texture depth attenuation increased with the rise in asphalt aggregate ratio, showing a slow growth trend at the three asphalt aggregate ratios of 5.2%, 5.7%, and 6.2%, and the anti-slip performance showed significant decay when the asphalt aggregate ratio exceeded 6.7%.

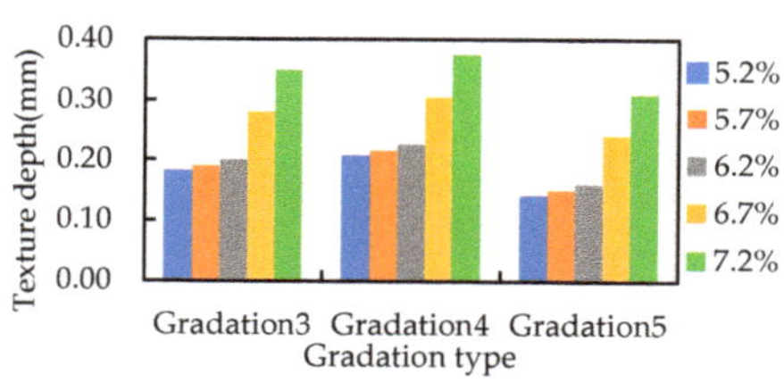

Figure 3. Results of the texture depth decay test.

The results of the TSRST test are shown in Figure 4. They indicated that the fracture temperature increased with the rise in asphalt aggregate ratio, and the anti-slip performance showed significant decay when the asphalt aggregate ratio exceeded 6.7%. For Gradation 3 and 5, when the asphalt aggregate ratio reaches 6.7%, the fracture temperature instead showed a slight increase.

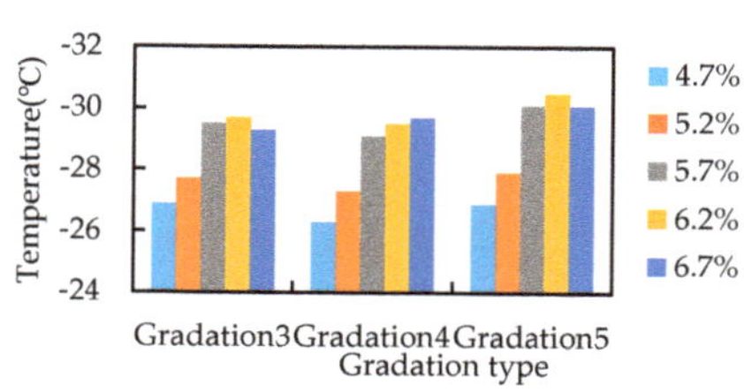

Figure 4. Results of TSRST test.

Based on the results of the texture depth decay test and the TSRST test, the final determined asphalt aggregate ratio of UT-10 mixture ranged from 5.2% to 6.2%.

3.3. Evaluation of Mixture Performance

The results of the evaluation tests are shown in Table 2. It can be seen that the performance of the HVHE mixture with gradation 5, 5.7% asphalt aggregate ratios, and 13.9% air void ratio fully meets the technical requirements, and has excellent high- and low-temperature performance.

Table 2. Results of the evaluation tests.

Technical Indexes	Unit	Results	Requirements
Leakage loss	%	0.1	≤0.3
Cantabro loss	%	1.71	≤8
Residential stability	%	87.0	≥85
TSR	%	81.6	≥80
Dynamic stability (60 °C)	loads/mm	6631	≥3200
Bending strain (−10 °C)	με	3864	≥2500

4. Conclusions

This study proposed an innovative design of ultra-thin asphalt overlays with high viscosity and high elasticity. The air void ratio and asphalt aggregate ratio were determined by designed test programs. The performance of HVHE asphalt mixture was verified. Several conclusions can be drawn as follows:

1. The performance of HVHE asphalt with gradation 5, 5.7% asphalt aggregate ratios, and 13.9% air void ratio fully meets the technical requirements (JTG E20-2011); the Brinell rotational viscosity of HVHE asphalt reaches 580,000 Pa·s, exhibiting superior bonding properties;
2. When the air void ratio of the HVHE mixture is between 12 and 15%, based on the immersion rutting test, it can drain the rainwater in time and reduce the water damage under the vehicle loads;
3. Based on the texture depth decay test, When the asphalt aggregate ratio is between 5.2 and 6.2%, it can guarantee the simultaneous durability of skid resistance and crack resistance under low temperature of the HVHE;
4. Based on the TSRST test, The HVHE mixture exhibits excellent high- and low-temperature performance, showing promising applications in North America.

Author Contributions: Conceptualization, T.M. and S.T.; methodology, N.L.; validation, Y.C. and L.C.; formal analysis, Z.L.; writing—original draft preparation, Y.C.; writing—review and editing, T.M. All authors have read and agreed to the published version of the manuscript.

Funding: This research received no external funding.

Institutional Review Board Statement: Not applicable.

Informed Consent Statement: Informed consent was obtained from all subjects involved in the study.

Data Availability Statement: Data used in supporting and analyzing the results reported in this paper is available upon request.

Conflicts of Interest: The authors declare no conflict of interest.

References

1. Kim, Y.K.; Lee, S.W. Performance evaluation of bonded concrete overlay. *Constr Build Mater.* **2013**, *49*, 464–470. [CrossRef]
2. Huang, L.; Fang, G.; Ni, F.; Fan, H. Study on the application of ultra-thin cover in cement concrete bridge deck pavement. *Highw. Traffic Sci. Technol.* **2008**, *4*, 10–12. (In Chinese)
3. Wan, J.; Wu, S.; Xiao, Y.; Fang, M.; Song, W.; Pan, P.; Zhang, D. Enhanced ice and snow melting efficiency of steel slag based ultra-thin friction courses with steel fiber. *J. Clean. Prod.* **2019**, *236*, 117613.
4. Ding, L.; Wang, X.; Cui, X.; Zhang, M.; Chen, B. Development and performance research of new sensitive materials for microwave deicing pavement at different frequencies. *Cold Reg. Sci. Technol.* **2021**, *181*, 103176.

5. Chen, S.; Gong, F.; Ge, D.; You, Z.; Sousa, J.B. Use of reacted and activated rubber in ultra-thin hot mixture asphalt overlay for wet-freeze climates. *J. Clean. Prod.* **2019**, *232*, 369–378. [CrossRef]
6. Geng, L.; Ma, T.; Zhang, J.; Huang, X.; Hu, P. Research on performance of a dense graded ultra-thin wearing course mixture. *Appl. Sci.* **2017**, *7*, 800. [CrossRef]
7. Ding, L.; Wang, X.; Zhang, M.; Chen, Z.; Meng, J.; Shao, X. Morphology and properties changes of virgin and aged asphalt after fusion. *Constr. Build. Mater.* **2021**, *291*, 123284. [CrossRef]
8. Ahmed, S.; Dave, E.V.; Buttlar, W.G.; Exline, M.K. Cracking resistance of thin-bonded overlays using fracture test, numerical simulations and early field performance. *Int. J. Pavement Eng.* **2013**, *14*, 540–552. [CrossRef]
9. Li, X.; Ye, J.; Badjona, Y.; Chen, Y.; Luo, S.; Song, X.; Zhang, H.; Yao, H.; Yang, L.E.; You, L.; et al. Preparation and performance of colored Ultra-Thin overlay for preventive maintenance. *Constr. Build. Mater.* **2020**, *249*, 118619. [CrossRef]
10. Budiarto, A.; Purnomo, H.R.; Setyawan, A. The structural properties evaluation on ultra-thin surfacing hot mixture asphalt concrete. *Appl. Mech. Mater.* **2015**, *754*, 828–832. [CrossRef]
11. *JTG E20-2011*; Standard Test Methods of Bitumen and Bituminous Mixture for Highway Engineering. Ministry of Transportation: Beijing, China, 2011.

engineering proceedings

Proceeding Paper

Innovative Design of Paving Cold Mix and Cohesive Overlays for Sustainable Pavement Maintenance †

Xiang Chen [1], Xiaohu Wang [1], Tao Ma [2,*], Susan Tighe [3] and Ningyuan Li [3]

1 Road Intellitech Co., Ltd., 9 Shengyun Rd., Nanjing 210008, China; 230218396@seu.edu.cn (X.C.); tootlex@163.com (X.W.)
2 School of Transportation, Southeast University, 2 Southeast University Rd., Nanjing 211189, China
3 Department of Civil Engineering, McMaster University, 1280 Main St W, Hamilton, ON L8S 4L8, Canada; tighes1@mcmaster.ca (S.T.); nli@uwaterloo.ca (N.L.)
* Correspondence: matao@seu.edu.cn

† Presented at the Second International Conference on Maintenance and Rehabilitation of Constructed Infrastructure Facilities, Honolulu, HI, USA, 16–19 August 2023.

Abstract: A cold mix overlay is a typical preventive maintenance treatment that is applied to an existing pavement surface. However, the service life of cold mix overlay is limited because of its poor skid resistance and high tendency to crack, especially in cold regions. This study presents a new technology of high-performance cold mix overlay materials that slows skid resistance reduction, increases the resistance to thermal cracking, and shows long-lasting anti−icing performance. The sustainable performance of paving cold mix overlays can be assured by adding high-performance anti−icing agents, fiber, and emulsified asphalt to the cold mix. A series of laboratory tests were conducted to evaluate the performance and anti−icing effect of the cold mix. The results showed that the freezing temperature of the cold mix dropped to more than −10 °C. The open-to-traffic time can be shortened to 3.5 h after construction. The anti-wearing ability and cracking resistance were evidently increased in comparison with traditional micro-surfacing techniques by conducting indoor wet-wheel wearing tests and low-temperature bending beam tests. Based on the study, the new-tech cold mix overlay has shown promising applications in North America.

Keywords: cold mix overlay; preventive maintenance; overlay; anti−icing; freezing point; high performance; wet-wheel wearing test

Citation: Chen, X.; Wang, X.; Ma, T.; Tighe, S.; Li, N. Innovative Design of Paving Cold Mix and Cohesive Overlays for Sustainable Pavement Maintenance. *Eng. Proc.* **2023**, *36*, 60. https://doi.org/10.3390/engproc2023036060

Academic Editor: Hosin (David) Lee

Published: 14 August 2023

1. Introduction

Cold mix overlay is a pavement maintenance overlay technology that has the advantages of a short construction period and rapid opening of traffic. To meet the higher performance requirements for heavy traffic volume and complex climate conditions, researchers have carried out investigations on how to improve the performance of cold mix overlays regarding cracking resistance, water damage resistance, etc.

According to literature, the performance of cold mix overlays could be achieved by optimizing the performance of modified asphalt emulsions or by adding additional special fibers. Ji et al. [1] conducted research on using waterborne epoxy resins to improve the performance of micro-surfacing. The waterborne epoxy resin and SBR composite modified emulsified asphalt were used to enhance the performance of micro-surfacing by Zheng et al. [2]. Yao et al. [3] reported that polypropylene fiber micro-surfacing has the best cracking resistance. Luo et al. [4] investigated the influence of fiber type and fiber content on micro-surfacing cracking and rutting resistance. The recommended optimal fiber content is 0.1–0.2% by weight of the asphalt mixture. Yu et al. [5] used high-viscosity and high elasticity modified asphalt emulsions to pre-pared HCUP-8 cold mix overlay.

In summary, the performance of micro-surfacing and cold mix overlays, such as cracking resistance and water damage resistance, could be effectively enhanced by optimizing the performance of modified asphalt emulsions and adding fibers or polymer resins [6]. However, there was a lack of research on the long-lasting anti−icing cold mix overlay. The effect of anti−icing agents on the performance of the cold mix overlay needs to be further investigated. The purpose of this study is to evaluate the effect of an anti−icing agent on the performance of a cold mix overlay and to optimize the performance of the anti−icing cold mix overlay.

2. Materials and Methods

2.1. Materials

The extended-release anti−icing agent obtained by Road Intellitech Co., Ltd., (Nanjing, China) was consistent with Table 1. The emulsified asphalt, trackless tack coat emulsifier asphalt, and polypropylene fiber were also purchased by Road Intellitech Co., Ltd. The coarse aggregates and fine aggregates were obtained from Zhenjiang Maodi Industrial Co., LTD., (Zhenjiang, China) which met the technical specifications for preventive maintenance of highway asphalt pavement (JTGT 5142-01-2021).

Table 1. Physical and mechanical properties of the extended-release anti−icing agent (ERA).

Test Item	Units	Technical Requirement
Freezing Point (20% aqueous solution)	°C	≤−14
Quality Loss (170 °C)	%	≤0.5
Corrosion Rate	mm/a	≤0.15
Water Content	%	≤3

2.2. Gradation

The basalt aggregates were used in the cold mix overlay. Which grading range and com-posite gradation were shown in Table 2.

Table 2. The gradation of cold mix overlay mixture.

Sieve	Percentage Passing (%)								ERA	PPF	Cement	PREA
Diameter	9.5	4.75	2.36	1.18	0.6	0.3	0.15	0.075 *				
Limits	100	70–90	45–70	28–50	19–34	12–25	7–18	5–15	5%	0.3%	1.5%	11%
Gradation	100	79.0	56.6	35.6	25.2	13.9	8.7	1.8				

* The first gear of 0.075 is replaced by ERA, and the substitution ratio is 9:10.

3. Results and Discussion

The bond property of the ice interface and the cold mix overlay mixture was evaluated by a drawing test of the ice interface, which was used to evaluate the freezing point reduction ability of the anti−icing agent. The test result is shown in Figure 1a. The pulling strength of the binder increased as the test temperature decreased. The pulling strength of ERA increased smoothly as the temperature decreased from −5 °C to −10 °C. However, the pulling strength of ERA significantly increased from −10 °C to −12 °C. This phenomenon occurred because the ice had formed when the temperature reached −12 °C. Thereby, the freezing point of ERA was −10 °C.

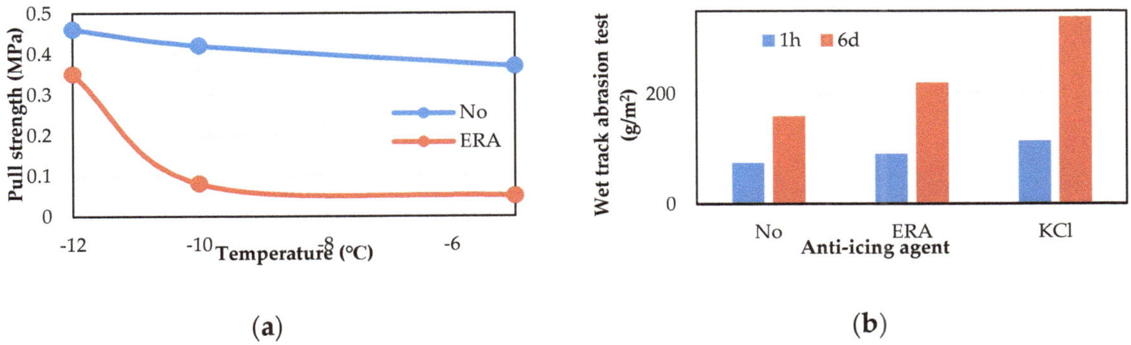

Figure 1. The ice point and wet track abrasion were influenced by anti−icing agents. (**a**) Ice drawing force; (**b**) The wet track abrasion test.

The wet track abrasion test was applied to evaluate the water damage resistance of cold asphalt mixtures. The wet track abrasion times of 1 h and 6 d are shown in Figure 1b. The wet track abrasion of KCl cold mixtures was higher than the other two cold mixtures. The cold mixture of No was smaller than the other two cold mixtures. This was because the release rates of the three cold mixtures were as follows: KCl > ERA > No. The voids formed when the KCl or ERA were released, which could affect the water damage resistance of the mixtures.

The cohesion torque reflects the forming speed and opening traffic time of the cold mix overlay. As shown in Figure 2, the cohesion torque of ERA cold mixtures increased as the anti-emulsification time got longer. When the anti-emulsification time reached 3.5 h, the cohesion torque of ERA cold mixtures was 2.4 N·m.

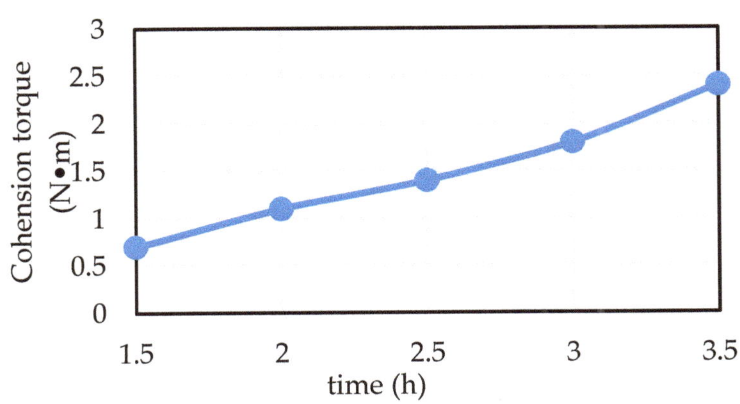

Figure 2. The cohesion torque of the ERA cold mix overlay.

With excellent conditions in hand, the optimal properties of the ERA cold mix overlay mixture were tested, and the results are shown in Table 3 The mixture performance of the ERA cold mix overlay was worse than micro-surfacing.

Table 3. The performance of different asphalt mixtures.

Test Item	ERA	Micro-Surfacing	Units	Technical Requirement
Mixing time/s	180	138	/	≥120
Breaking time/min	20	5	/	/
Cohesion torque /(N·m)	2.40 (3.5 h)	2.15	/	≥2.0
Wet track abrasion test 1 d/(g/m^2)	90.4	74.1	/	≥538
Wet track abrasion test 1 d/(g/m^2)	219.6	158.8	/	≥807
Load wheel test (g/m^2)	198.6	167.3	/	≥538

Flexural tensile strains were used to investigate the low-temperature performance of asphalt mixtures. The results were shown in Figure 3. The ultimate flexural tensile strains of the ERA cold mix overlay could reach 3848.6 με, which was more excellent than the others. This phenomenon was due to the PREA's excellent performance. So, the cracking resistance of the ERA cold mix overlay was the best of all.

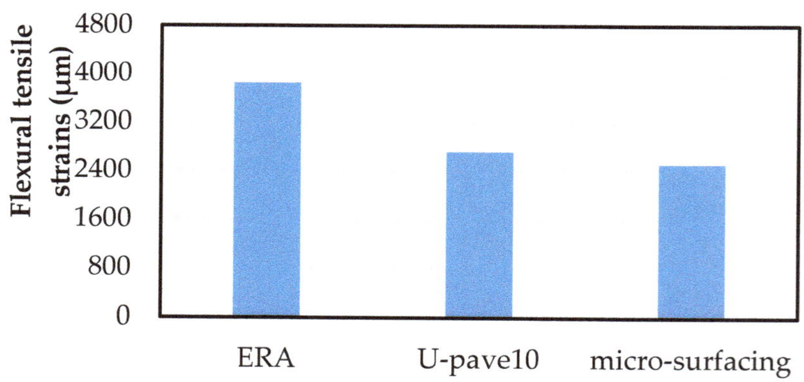

Figure 3. The flexural tensile strains of different asphalt mixtures.

The cumulative release was often used to evaluate the durability of anti−icing. The concentration of chloride ions was detected by a rapid determination instrument of chloride ion content. The results were shown in Figure 4. The chloride release rate of KCl was faster than ERA. The anti−icing function of KCl was finished when the soaking days reached 20 d. However, the anti−icing function of ERA was finished as the soaking days reached 95 d and the single release of chloride ions was 0.00158948 mol/L, which was the critical ice inhibition concentration.

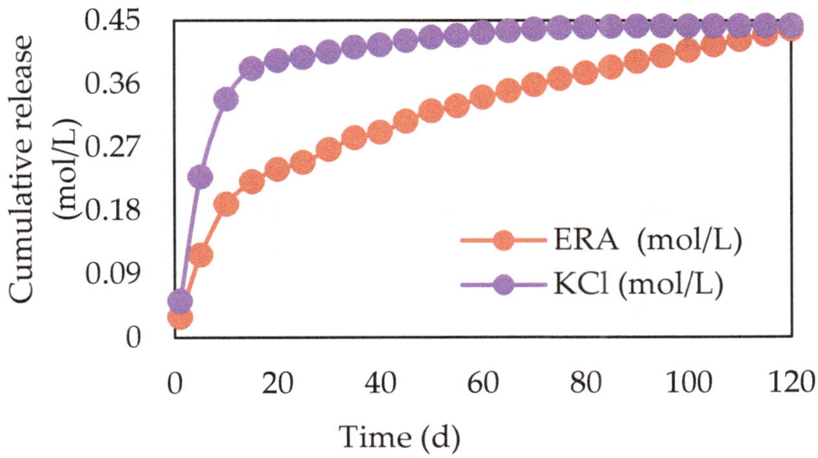

Figure 4. The cumulative release of different anti−icing agents.

4. Conclusions

To obtain an excellent anti−icing cold mix overlay, this paper studied the effects of the ERA anti−icing agent on the freezing point, opening traffic time, wet track abrasion, flexural tensile strains, and mixture performance. The rapid determination instrument for chloride ion content was used to evaluate the durability of anti−icing.

(1) The freezing temperature of cold asphalt mixtures dropped to much lower than −10 °C. This technology shortened the open-to-traffic time to 3.5 h after construction;
(2) The ERA cold mix overlay had excellent water damage and cracking resistance performance. The flexural tensile strains could reach 3848.6 με, which was higher than that of micro-surfacing and U-pave 10 hot asphalt mixtures;
(3) The performance of ERA cold mixtures met the ISSA standards. The durability of anti−icing cold mixtures could reach 95 days when soaked in water. Thereby, the new-tech cold mix overlay had shown promising applications in North America.

Author Contributions: Conceptualization, T.M. and S.T.; methodology, N.L.; validation, X.C. and X.W.; formal analysis, X.C.; writing—original draft preparation, X.C.; writing—review and editing, T.M. All authors have read and agreed to the published version of the manuscript.

Funding: This research received no external funding.

Institutional Review Board Statement: Not applicable.

Informed Consent Statement: Informed consent was obtained from all subjects involved in the study.

Data Availability Statement: Data used in supporting and analyzing the results reported in this paper is available upon request.

Conflicts of Interest: The authors declare no conflict of interest.

References

1. Ji, J.; Liu, L.H.; Suo, Z.; Xu, Y.; Yang, S.; Xu, S.F. Performances of micro-surfacing with waterborne epoxy resin modifified emulsified asphalt. *J. Chang'an Univ. (Nat. Sci. Ed.)* **2017**, *37*, 23–30. (In Chinese)
2. Zheng, M.L.; Fan, X.P.; Liu, F.Q.; Li, H.Y.; Lin, X.Y. Durability of compound waterborne epoxy emulsifified as phalt microsurfacing. *J. Chang'an Univ. (Nat. Sci. Ed.)* **2020**, *40*, 68–76. (In Chinese)
3. Wang, L.; Lu, P.; Hao, P.W. Factors affecting pavement performance of micro-surfacing mixture. *J. Chang'an Univ. (Nat. Sci. Ed.)* **2014**, *34*, 29–33. (In Chinese)

4. Luo, Y.; Zhang, K.; Xie, X.; Yao, X. Performance evaluation and material optimization of Micro-surfacing based on cracking and rutting resistance. *Constr. Build. Mater.* **2019**, *206*, 193–200. [CrossRef]
5. Yu, J.M.; Yang, N.K.; Chen, F.D.; Chen, Y.L.; Lin, Z.H.; Yu, H.Y. Design of Cold-Mixed High-Toughness Ultra-Thin Asphalt Layer towards Sustainable Pavement Construction. *Buildings* **2021**, *11*, 619. [CrossRef]
6. Research Institute of Highway Ministry of Transport. *Standard Test Methods of Bitumen and Bituminous Mixtures for Highway Engineering;* Ministry of Transport of the People's Republic of China: Beijing, China, 2011. (In Chinese)

engineering proceedings

Proceeding Paper

Data-Driven Approach to Decision-Making for Pavement Preservation †

Sara Arezoumand [1,2], Alireza Sassani [1,2] and Omar Smadi [1,2,*]

1 Institute for Transportation, Iowa State University, Ames, IA 50011, USA; sarzmnd@iastate.edu (S.A.); asassani@iastate.edu (A.S.)
2 Department of Civil, Construction, and Environmental Engineering, Iowa State University, Ames, IA 50011, USA
* Correspondence: smadi@iastate.edu; Tel.: +1-515-294-8103
† Presented at the Second International Conference on Maintenance and Rehabilitation of Constructed Infrastructure Facilities, Honolulu, HI, USA, 16–19 August 2023.

Abstract: Selecting the right treatment for the right pavement at the right time is fundamental to pavement preservation success. This research aimed to develop a data-driven decision-making framework for selecting pavement preservation strategies and thereby provide an analytical foundation to enhance and update "Iowa's Pavement Preservation Guide". The research utilized the pavement performance, road characteristics, and preservation project data from Iowa DOT's databases to develop a pavement performance profile before and after different preservation treatments, and this evaluates the effectiveness of the treatment methods in enhancing pavement performance. The results were used in a simple economic analysis framework to assess the economic viability of preservation methods in light of their effectiveness levels.

Keywords: pavement management system; preservation; data-driven decision-making

1. Introduction

Pavement preservation involves proactive efforts to maintain existing pavements in a state of good repair. Unlike rehabilitation and reconstruction projects, pavement preservation methods aim to restore and prolong pavement serviceability rather than add structural capacity. Numerous benefits can be attributed to pavement preservation, including optimized use of scarce resources to improve safety, mobility, and user satisfaction. Engineering judgment is extensively involved in the current practices of pavement preservation decision-making; nevertheless, as the state transportation agencies move away from subjective decision-making toward more objective, data-driven approaches, the pavement preservation programs shall also move along.

The state DOTs can implement various preservation treatments based on pavement surface type, functional class, traffic loading, and the extent and severity of distresses. Today, many transportation agencies have developed decision-making tools, such as decision trees and application matrices, to assist in the timing and treatment selection of pavement preservation programs. These decision-making tools typically link preservation treatments to pavement distress types and severity levels based on decision factors such as service conditions, past repair and maintenance activities, expected treatment effectiveness, and the relative costs of the candidate treatments [1,2].

This research aims to develop an analytical foundation for data-driven pavement preservation strategies by evaluating the effectiveness of different preservation techniques relative to the road type and underlying conditions. The study leverages Iowa DOT's pavement management information system (PMIS) data to model pavement performance and determine the influence of preservation treatments.

Citation: Arezoumand, S.; Sassani, A.; Smadi, O. Data-Driven Approach to Decision-Making for Pavement Preservation. *Eng. Proc.* **2023**, *36*, 61. https://doi.org/10.3390/engproc2023036061

Academic Editor: Hosin (David) Lee

Published: 14 August 2023

2. Data Sources and Description

This study's primary data source was Iowa DOT's pavement management information system (PMIS). Iowa DOT's electronic records management system was also utilized to derive project location information when necessary. Project details and cost information were derived from Iowa DOT's historical bid tabulations (2015 through 2023). The PMIS contains comprehensive road characteristics data for all state roads in Iowa. The collected data include detailed distress measurements, traffic information, and pavement history. Condition data in the PMIS include indexed metrics, such as the pavement condition index (PCI), the rutting index, the international roughness index (IRI), and the cracking index, and the measures of individual distress types, including alligator cracking, transverse cracking, longitudinal cracking, durability cracks, and joint spalling. The study used the performance data of 7255 PMIS road segments from 1998 through 2020.

3. Methodology and Analysis

The overall research methodology adopted in this study is described in the sections below.

3.1. Performance Modeling

The effectiveness of a pavement preservation treatment was evaluated by analyzing pavement performance before and after the particular treatment was implemented. Pavement performance was represented by the absolute values of condition indices and the index degradation rate. Obtaining these quantities enabled the quantification of the amount of pavement life extension by a given preservation project. Then, this life extension calculation was applied to all segments undergoing a given type of preservation treatment. This allowed for the determination of the expected (mean) life extension capability for that treatment method. Furthermore, analyzing life extension data against pavement and road characteristics, particularly the underlying pavement condition and traffic volume, provides insights into when and where the treatments are best implemented. The condition index used for the analysis was the PCI because it encompasses all of the important distress indices for flexible pavements. However, individual distress indices, i.e., the rutting index, riding index, and cracking index, will be considered in the next steps to establish a relationship between treatment effectiveness and individual distress. The treatment methods considered were microsurfacing, slurry seal, patching, crack sealing/filling, and hot mix asphalt (HMA) thin overlays. The Iowa DOT calculates the PCI based on Equation (1) [3]:

$$PCI = 0.4 \times Cracking\ Index + 0.4 \times Riding\ Index + 0.2 \times Rutting\ Index \quad (1)$$

where the cracking index is obtained from Equation (2):

$$Cracking\ Index = 0.2 \times TCI + 0.1 \times LCI + 0.3 \times LWPCI + 0.4 \times ACI \quad (2)$$

where TCI is the transverse cracking index, LCI is the longitudinal cracking index, $LWPCI$ is the longitudinal wheel path cracking index, and ACI is the alligator cracking index.

The service life extensions with respect to the PCI were then determined by solving the post-preservation deterioration curve function for the latest pre-preservation index value. Figure 1 illustrates the improvement in the PCI and the pavement service life extension caused by pavement preservation treatment. The index benefits were defined as the area between the pre-project performance curve (i.e., the do-nothing scenario) and the post-project deterioration curve over the range of extended life; this value was used as an additional metric to evaluate the treatment effect with respect to a specific condition index. Note that the pre-project performance curve must be extrapolated onto the extended life range for this purpose.

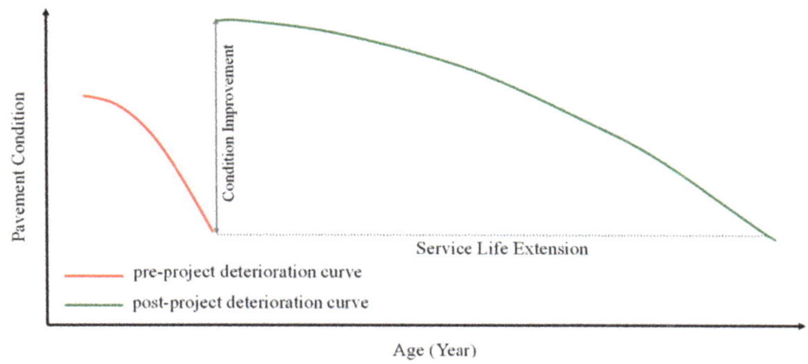

Figure 1. Immediate condition improvement and life extension for an example segment.

In order to determine the average life extension for each treatment type, the results of each project segment were categorized according to the project type and averaged. Table 1 indicates that slurry sealing could extend pavement life by about 2.5 years, which is the lowest of all possible treatments. The highest service life extension was provided by hot mix asphalt (HMA) thin overlays (less than 3 inches) with an average of almost nine years of extension.

Table 1. Average service life extension by each treatment method.

Treatment	Service Life Extension (Years)
Slurry Seal	2.54
Patch	4.24
Microsurfacing	4.52
Crack Seal/Fill	5.62
HMA Overlay	8.93

3.2. *Benefit–Cost Analysis*

Using the service life extension and index benefits, the next step was a benefit-cost analysis. The life cycle cost analysis (LCCA) took two primary inputs for each distinct preservation treatment to determine their cost over the life cycle, namely, implementation cost and life extension. Implementation costs were calculated as shown in Table 2 based on the bidding tabulation data. Life–cycle cost was represented by the equivalent uniform annual cost ($EUAC$) which was obtained from Equation (3):

$$EUAC = NPV \times \frac{discount\ rate(1 + discount\ rate)^{analysis\ period}}{(1 + discount\ rate)^{analysis\ period} - 1} \tag{3}$$

where the net present value (NPV) can be calculated from Equation (4).

$$NPV = \sum\nolimits_0^{n_k} Preservation\ Cost\left[\frac{1}{(1+i)^{n_k}}\right] + \sum\nolimits_0^{n_k} Maintanance\ Cost\left[\frac{1}{(1+i)^{n_k}}\right] - Salvage\ value \times \left[\frac{1}{(1+i)^{n_k}}\right] \tag{4}$$

where n_k is the year of expenditure and i is the real discount rate according to the OMB Circular by the U.S. Government, (1.2 to 2%), and the analysis period was 10 and 20 years. Maintenance costs were capped at USD 2500 annually and calculated using Equation (5) based on a 5-year sliding scale.

$$Maintanance\ cost = \begin{cases} \frac{year}{5} \times 2500 & if\ year \leq 5 \\ 2500 & if\ year > 5 \end{cases} \tag{5}$$

Table 2. Average implementation cost of the preservation treatments.

Treatment	Number of Projects	Total Lane Mile	Cost Per Lane Mile USD *
Microsurfacing	72	2836	20,995
Crack Sealing/filling	200	4800	4412
Patching	1401	1742	66,000
Slurry Sealing	66	1741	7831
HMA Overlay	412	556	98,556
Total	519	-	-

* USD: United States dollar.

Equation (6) can be used to calculate the salvage value, which represents the present value of any remaining service life:

$$Salvage\ value = \frac{remaining\ service\ life\ after\ analysis\ period}{Total\ amount\ of\ service\ life} \times Perservation\ Cost \quad (6)$$

The life cycle cost analysis (LCCA) results, for a 10-year analysis period and 1.5% discount rate, are summarized in Figure 2. As the figure shows, crack/joint sealing demonstrated the lowest EUAC at USD 3253.160. Slurry sealing produced the next lowest value followed by microsurfacing, HMA thin overlay, and patching, with average EUAC values of USD 5920.129, USD 7861.798, USD 17,985.873, and USD 20,896.108, respectively.

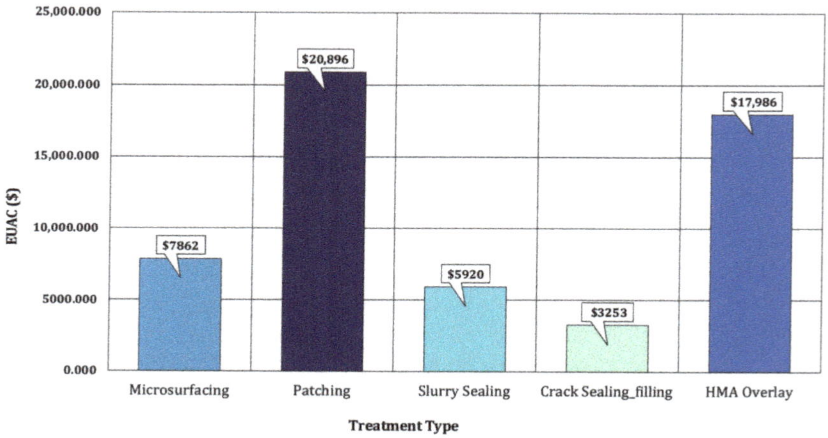

Figure 2. EUAC for the preservation methods based on the PCI (the analysis period is 10 years).

4. Summary and Conclusions

This research pursued the development of a data-driven framework for objective decision-making about pavement preservation strategies. Through statistical performance modeling, the effectiveness of five pavement preservation methods, namely, microsurfacing, slurry sealing, patching, crack sealing/filling, and HMA thin overlay were determined in terms of pavement life extension and PCI index benefits. Then, the results were used in a simple economic analysis framework to evaluate the economic impact of each preservation method in terms of equivalent uniform annual cost (EUAC). The study found that HMA thin overlay resulted in the largest service life extension for the PCI, 8.93 years, while slurry sealing had the lowest service life extension, 2.54 years.

Author Contributions: Conceptualization, O.S. and A.S.; methodology, A.S., S.A. and O.S.; validation, A.S. and S.A.; formal analysis, A.S. and S.A.; investigation, A.S. and S.A.; resources, O.S.; data curation, A.S.; writing—original draft preparation, A.S. and S.A.; writing—review and editing, A.S., S.A. and O.S.; visualization, S.A. and A.S.; supervision, O.S. and A.S.; project administration, O.S.; funding acquisition, O.S. All authors have read and agreed to the published version of the manuscript.

Funding: This research was funded by Iowa Highway Research Board, Iowa Department of Transportation grant number 20-728, TR-784.

Institutional Review Board Statement: Not applicable.

Informed Consent Statement: Not applicable.

Data Availability Statement: The pavement data presented in this study are not publicly available. The data derived from Iowa DOT's pavement management information system (PMIS) and Iowa DOT contracts portal and Bid Express® subscription service.

Conflicts of Interest: The authors declare no conflict of interest.

References

1. Buss, A.; Claypool, B.; Bektas, F. Effectiveness of Pavement Preservation Techniques. 2019. Available online: https://intrans.iastate.edu/research/completed/effectiveness-of-pavement-preservation-techniques/ (accessed on 10 August 2023).
2. Abdelaty, A.; Jeong, H.D.; Smadi, O.; Gransberg, D.D. Iowa Pavement Asset Management Decision-Making Framework. 2015. Available online: https://rosap.ntl.bts.gov/view/dot/29597 (accessed on 10 August 2023).
3. Bektas, F.; Smadi, O.G.; Al-Zoubi, M. Pavement Management Performance Modeling: Evaluating the Existing PCI Equations. 2014. Available online: https://intrans.iastate.edu/research/completed/pavement-management-performance-modeling-evaluating-the-existing-pci-equations/ (accessed on 10 August 2023).

engineering proceedings

MDPI

Proceeding Paper

Consistent Foamed Asphalt Contents Needed for Cold In-Place Recycled Pavement Layers in Practice †

Hosin Lee [1,*], Byungkyu Moon [2], Ashley Buss [3] and Charles T. Jahren [4]

1 Civil and Environmental Engineering, University of Iowa, Iowa City, IA 52242, USA
2 ARA Associates, Albuquerque, NM 87110, USA; civilhawkeye@gmail.com
3 Iowa Department of Transportation, Ames, IA 50010, USA; ashley.buss@iowadot.us
4 Department of Transportation, Civil, Construction and Environmental Engineering, Iowa State University, Ames, IA 50011, USA; cjahren@iastate.edu
* Correspondence: hosin-lee@uiowa.edu

† Presented at the Second International Conference on Maintenance and Rehabilitation of Constructed Infrastructure Facilities, Honolulu, HI, USA, 16–19 August 2023.

Abstract: Cores were recovered from the right wheel path and between wheel paths at 8 locations US 34 in Mills and Wapello Counties. The cores were cut into discs to isolate the pavement layers and the discs were fabricated into semicircular bending test specimens. Asphalt contents among CIR cores varied significantly and that CIR specimens with higher asphalt binder contents exhibited higher flexibility index values. From the output of these tests, the CIR pavement layers were found to be more flexible in comparison to the HMA layers, that CIR layers have higher flexibility that allow them to serve as a stress relieving layer and mitigate reflective cracking.

Keywords: cold in-place recycling; forensic analysis; performance; cores; foamed asphalt

Citation: Lee, H.; Moon, B.; Buss, A.; Jahren, C.T. Consistent Foamed Asphalt Contents Needed for Cold In-Place Recycled Pavement Layers in Practice. *Eng. Proc.* **2023**, *36*, 62. https://doi.org/10.3390/engproc2023036062

Academic Editor: Haifang Wen

Published: 16 August 2023

1. Introduction

Cold in-place recycling (CIR) involves cold milling 3 to 4 in of a deteriorated asphalt pavement; processing and stabilizing the millings; and relaying, compacting, curing, and covering with a hot-mix asphalt (HMA) overlay or a bituminous surface treatment. A mix design procedure was previously developed and validated [1,2]. The long-term field performance of the CIR pavements constructed in Iowa was evaluated [3–6], and the impacts of moisture content in the CIR layer prior to an overlay were identified [7,8]. This paper discusses an analysis of cores collected from CIR/HMA overlay sections in Iowa. It was confirmed that the higher binder contents in the cores produced higher Flexibility Index values based on the Semi-Circular Bending (SCB) test. HMA samples seemed to have aged more than CIR samples, resulting in lower flexibility index values with a brittle behavior.

2. Analysis of HMA/CIR Cores

To investigate the underlying characteristics of CIR, field cores were collected from CIR/HMA overlay sections on US 34 in Mills and Wapello Counties, Iowa, on 16 November 2021. The cores were evaluated to determine the cracking resistance, binder contents, and gradations of both the CIR and HMA layers. A total of 16 cores with 6 in diameters were collected from eight locations at one-mile intervals along US 34. Two cores were collected at each location, one in the wheel path and one between wheel paths.

The HMA and CIR layers were distinguished by a difference in color, with the CIR layer being darker than the HMA layer. The thicknesses of both the HMA and CIR layers were measured from each core and are summarized in Table 1. Cores were labeled with the milepost number followed by either "RWP", which indicates that the core was collected from the pavement's right wheel path, or "¼ pt", which indicates that the core was collected from between the pavement's wheel paths.

Table 1. Thicknesses of HMA and CIR cores from US 34.

Core (MP)	HMA (in/cm)	CIR (in/cm)	Total (in/cm)
12.3 RWP	5.9/15	3.5/9	9.4/24
12.3 ¼ pt	5.9/15	3.5/9	9.4/24
13.2 RWP	2.4/6	2.4/6	4.7/12
13.2 ¼ pt	2.8/7	3.1/8	5.9/15
14 RWP	2.4/6	3.1/8	5.5/14
14 ¼ pt	1.6/4	5.4/13	6.7/17
15.3 RWP	1.4/3.5	5.3/13.5	6.7/17
15.3 ¼ pt	1.6/4	5.1/13	6.7/17
18.4 RWP	1.6/4	3.9/10	5.5/14
18.4 ¼ pt	1.6/4	3.5/9	5.1/13
19.3 RWP	1.4/3.5	3.5/9	4.9/12.5
19.3 ¼ pt	1.4/3.5	5.3/13.5	6.7/17
20.3 RWP	1.6/4	3.5/9	5.1/13
20.3 ¼ pt	1.6/4	3.5/9	5.1/13
181.4 RWP	2.6/6.5	3.7/9.5	6.3/16
181.4 ¼ pt	2.4/6	3.9/10	6.3/16

3. Semicircular Bending Test Results

Semicircular bending (SCB) tests have been used to evaluate both fatigue cracking and low-temperature cracking. The flexibility index (FI) resulting from this test has been reported to have the capability to capture some of the critical changes in mixture variables. Lower *FI* values indicate that the asphalt mixture is more brittle and has a higher crack growth rate.

SCB tests were performed according to AASHTO TP 124 on the cores taken from US 34. As shown in Figure 1, samples for SCB testing were prepared by cutting the cores into semicircular specimens, each with a diameter of 150 mm and a maximum thickness of 50 mm (with some samples being thinner because field core samples have various thicknesses). A notch 15 mm long and 1.5 mm wide was made on the flat side of each semicircular specimen to induce cracking. All specimens were conditioned at 25 °C for 2 h before SCB testing was performed.

Figure 1. Cuts made along the red lines to prepare a semicircular specimen with a thickness of 50 mm.

The SCB test setup and the dimensions of the test specimens are illustrated in Figure 2a. To induce cracking in the middle of the specimens, a 15 mm notch was created in each.

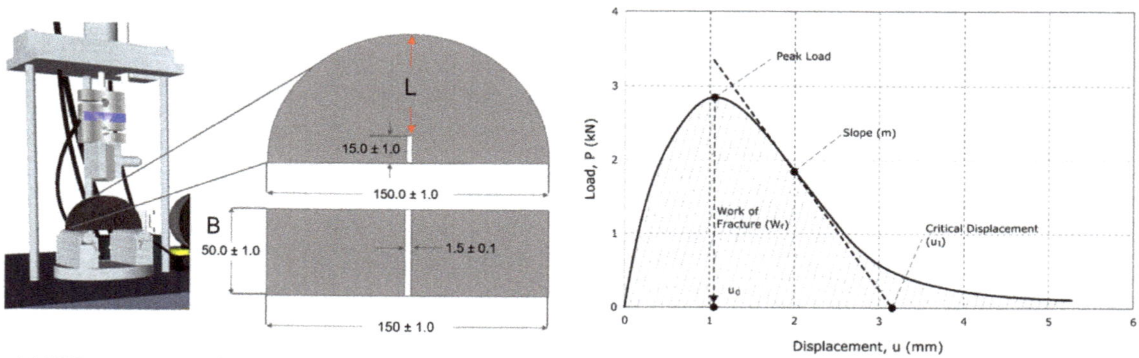

(**a**) SCB test setup and dimensions of the test specimen (**b**) force-displacement curve from SCB test

Figure 2. SCB test setup, dimensions, and force–displacement curve.

A typical force–displacement curve from an SCB test is illustrated in Figure 2b, which shows the work of fracture (W_F) as the area under the curve and the post-peak slope (m) at the inflection point after the peak point. These parameters were used to calculate the fracture energy (G_F) and Flexibility Index (FI).

Fracture energy (G_F) can be calculated using the equation below:

$$G_F = \frac{W_F}{B \times L}$$

where W_F is the work of fracture, B is the specimen thickness, and L is the ligament length.

FI can be calculated using the equation below:

$$FI = \frac{G_F}{|m|} \times A$$

where G_F is the fracture energy calculated by dividing the work of fracture by the ligament area, m is the post-peak slope at the inflection point after the peak point, and A is a unit conversion from field to laboratory (0.01).

4. Test Results

Based on the burn-off and sieve test results for the cores from all eight MP locations, the asphalt contents in the CIR layers were higher than those in the HMA layers and the aggregates of the CIR layers were finer than those of the HMA layers, as expected. However, there was no statistically significant difference in aggregate gradations or AC contents between the cores obtained from the right wheel path (RWP) and the cores obtained from between the wheel paths (¼ pt).

The *FI* values of both the CIR and HMA samples are plotted against binder contents in Figure 3. Based on the SCB test results of the CIR and HMA samples, increased binder contents produced higher *FI* values for both the CIR and HMA samples. However, the *FI* values of the CIR samples were significantly higher than those of the HMA samples. The average *FI* value of the CIR samples was 4.808, whereas that of the HMA samples was 1.422. It can be postulated that the HMA samples were aged for longer than the CIR samples, resulting in lower *FI* values, and that, therefore, relative brittleness was observed in the HMA samples.

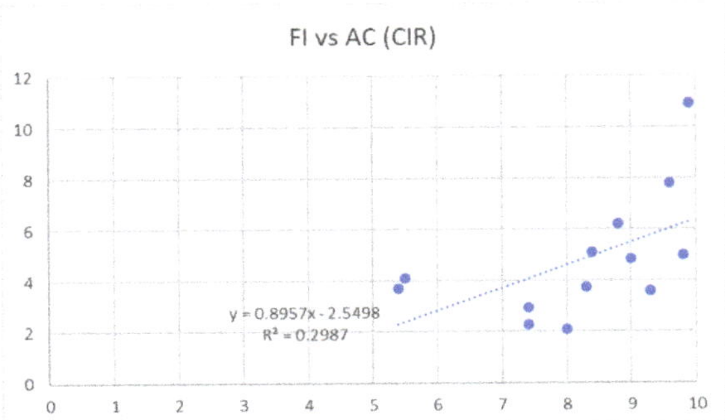

(**a**) *FI* values versus asphalt contents for CIR samples

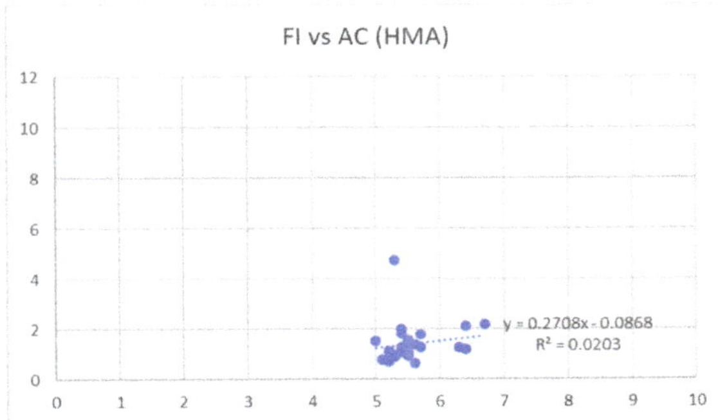

(**b**) *FI* values versus binder contents for HMA samples

Figure 3. Plots of *FI* values versus binder contents for (**a**) CIR and (**b**) HMA samples.

5. Summary and Conclusions

Laboratory testing was conducted on cores extracted from US 34 in Mills and Wapello Counties, Iowa. These cores were cut to isolate the HMA and CIR layers, and the resulting samples were fabricated into specimens for the semicircular bending test. The output of the test was a load versus displacement curve that could be processed to calculate the work of fracture; further calculations were conducted to provide the flexibility index for each specimen. A higher flexibility index was expected to be associated with specimens that are more flexible and less likely to crack. When the HMA and CIR layers of the cores were compared, the HMA layers were found to have lower flexibility index values versus the CIR layers. This would suggest that, for CIR projects with an HMA overlay, a more brittle HMA overlay layer would cover a more flexible CIR layer. It was found that the asphalt contents among CIR cores varied noticeably and that CIR specimens with higher asphalt binder contents exhibited higher flexibility index values.

Asphalt binder contents may vary significantly across a pavement's width due to a limited number of spraying nozzles and subsequent mixing operations for CIR with foamed asphalt. Therefore, adding the correct amount of asphalt is important. Too much asphalt will the reduce air void content in the CIR layer, which will facilitate shearing of the pavement layer under heavy traffic loading. Too much asphalt will also reduce the aggregate-to-aggregate interlock, allowing aggregate particles to move past each other as they are separated by asphalt that shears under load. Too little foamed asphalt will increase the air void content in the CIR layer, which will cause wheel path cracking in the HMA overlay as compaction rutting of the CIR layer develops under heavy traffic loading. The CIR layers are more flexible than the HMA overlay layers. For CIR using foamed asphalt, it is recommended that asphalt binder be more evenly distributed and mixed across the width of the pavement to ensure a more consistent asphalt binder content.

Author Contributions: Conceptualization and methodology, H.L.; validation, A.B. and C.T.J.; investigation, B.M.; writing—original draft preparation, H.L. and B.M.; writing—review and editing, A.B. and C.T.J. All authors have read and agreed to the published version of the manuscript.

Funding: This research was funded by Iowa Department of Transportation (DOT) and the Iowa Highway Research Board (IHRB), grant number IHRB Project TR-774.

Institutional Review Board Statement: Not applicable.

Informed Consent Statement: Not applicable.

Data Availability Statement: The data presented in this study are available on request from the corresponding author.

Acknowledgments: The authors would like to thank the Iowa Department of Transportation (DOT) and the Iowa Highway Research Board (IHRB) for sponsoring this research, as well as the Technical Advisory Committee members.

Conflicts of Interest: The authors declare no conflict of interest. The funders had no role in the design of the study; in the collection, analyses, or interpretation of data; in the writing of the manuscript; or in the decision to publish the results.

References

1. Kim, Y.; Lee, H.D. Development of mix design procedure for cold in-place recycling with foamed asphalt. *J. Mater. Civ. Eng.* **2006**, *18*, 116–124. [CrossRef]
2. Kim, Y.; Lee, H.D.; Heitzman, M. Validation of new mix design procedure for cold in-place recycling with foamed asphalt. *J. Mater. Civ. Eng.* **2007**, *19*, 1000–1010. [CrossRef]
3. Diefenderfer, B.K.; Bowers, B.F.; Apeagyei, A.K. Initial Performance of Virginia's Interstate 81 In-Place Pavement Recycling Project. *Transp. Res. Rec. J. Transp. Res. Board* **2015**, *2524*, 152–159. [CrossRef]
4. Kim, J.J.; Lee, H.D.; Jahren, C.T.; Heitzman, M.; Chen, D. Long-term field performance of cold in-place recycled roads in Iowa. *J. Perform. Constr. Facil.* **2010**, *24*, 265–274. [CrossRef]
5. In-Place and Central-Plant Recycling of Asphalt Pavements in Virginia. Available online: https://www.fhwa.dot.gov/pavement/sustainability/case_studies/hif19078.pdf (accessed on 4 August 2023).
6. Buss, A.; Mercado, M.G.; Schram, S. Long-term evaluation of cold-in-place recycling and factors influencing performance. *J. Perform. Constr. Facil.* **2017**, *31*, 04016111. [CrossRef]
7. Kim, Y.; Im, S.; Lee, H.D. Impacts of curing time and moisture content on engineering properties of cold in-place recycling mixtures using foamed or emulsified asphalt. *J. Mater. Civ. Eng.* **2011**, *23*, 542–553. [CrossRef]
8. Al-Qadi, I.; Cao, Q.; Abufares, L.; Wang, S.; Ali, U.M.; Renshaw, G. Moisture Content and In-Place Density of Cold-Recycling Treatments. Available online: https://apps.ict.illinois.edu/projects/getfile.asp?id=10152 (accessed on 4 August 2023).

engineering proceedings

MDPI

Proceeding Paper

Important Sustainability Determinants Meeting Sustainability Goals of California Infrastructure Construction Projects †

Joseph J. Kim [1,*] and Patricia McCarthy [2]

1 Department of Civil Engineering and Construction Engineering Management, California State University Long Beach, Long Beach, CA 90840, USA
2 Department of Public Works, City of Los Angeles, Los Angeles, CA 90012, USA; patricia.mccarthy@lacity.org
* Correspondence: joseph.kim@csulb.edu; Tel.: +1-562-985-1679

† Presented at the Second International Conference on Maintenance and Rehabilitation of Constructed Infrastructure Facilities, Honolulu, HI, USA, 16–19 August 2023.

Abstract: The United States has developed and is developing multiple rating systems for infrastructure and transportation projects. Although these systems share some commonalities in terms of methods and criteria, decision makers need to deal with which one best fits their project's evaluation and meet their organization goals because the systems are different from one another in certain ways. This paper aims to examine the importance of sustainability determinants and how they affect the success of meeting sustainability goals of infrastructure construction projects. This paper, therefore, presents the statistical results on five major sustainability determinants such as site, water/wastewater, energy, materials/resources, and environmental determinants.

Keywords: sustainability; infrastructure; assessment; determinants; quantitative analysis

Citation: Kim, J.J.; McCarthy, P. Important Sustainability Determinants Meeting Sustainability Goals of California Infrastructure Construction Projects. *Eng. Proc.* **2023**, *36*, 63. https://doi.org/10.3390/engproc2023036063

Academic Editor: Hosin (David) Lee

Published: 24 August 2023

1. Introduction

Building assessment systems have been developed to grade performances of building projects based on a specific set of green or sustainable criteria. Several rating systems for infrastructure and transportation projects have been developed or are under development using a point-based system, such as the United States Green Building Council (USGBC)'s LEED system for building construction [1]. However, these sustainability rating systems are not as commonly used in infrastructure construction projects as they are in building construction. While there is limited industry guidance on sustainable transportation construction practices, several states have developed their own transportation rating systems. For example, Simpson [2] and Simpson et al. [3] used 10 existing rating systems to create a framework for several departments of transportation (DOTs), including Colorado, South Dakota, Utah, and Wyoming. Cormack et al. [4] presented a case study that utilized sustainable concepts based on FHWA's INVEST rating system. Bueno et al. reviewed existing tools and methods for sustainability assessments of transport infrastructure projects. Lineburg [5] conducted a comprehensive analysis on transportation rating systems and social sustainability. Chisholm et al. [6] presented sustainable project rating systems including Envision.

Public agencies face a major problem in determining the best rating system for their infrastructure projects in terms of meeting their organization priorities and improving performance. The difficulty lies in adopting an appropriate sustainability rating system for infrastructure construction projects that is also cost-effective method. The paper is motivated by the necessity of developing a framework for infrastructure construction projects. The research hypotheses examine if five sustainability determinants and their associated factors chosen in this paper show any significant differences in terms of meeting the sustainability goals of infrastructure construction projects.

2. Research Objectives and Methods

The aim of this paper is to examine five major sustainability determinants affecting the success of meeting organizations' sustainability goals for infrastructure construction projects. The authors chose five determinants and related factors from literature reviews on existing sustainability systems available in the U.S.A. The determinants include site, water/wastewater, energy, materials/resources, and environmental determinants. Industry professionals involved in infrastructure construction projects in California were asked to rate five determinants and related factors on a 7-point Likert scale, which is used to rate the survey response, to effectively express their opinions on how important determinants and factors are when successfully meeting sustainability needs in infrastructure construction projects. Statistical methods such as the Kruskal–Wallis test, analysis of variance (ANOVA), and Z-test are used to analyze the responses of survey data.

3. Results

Of the 59 responses received, 25 surveys (42.4%) were used in the data analysis. Participants for the valid surveys had an average experience of 18 years and consisted of engineers/designer, construction managers, and employees of government agencies in proportions of 55.6%, 13.9%, and 30.6%, respectively. The respondents answered nine projects on average and used a sustainability rating system with a median of 3 years.

3.1. Comparison of Medians among Five Major Determinants

The statistical results are shown in Table 1 regarding multiple comparisons for the median values. Kruskal–Wallis tests were used for five major sustainability determinants. For the factors under each determinant, the hypotheses are H0: the medians for five major sustainability determinants are equal and Ha: the medians for five major sustainability determinants are not equal. We cannot reject the null hypothesis because test statistics of H-value are high and because the observed significance values are greater than $\alpha = 0.05$. In other words, no evidence is found to conclude that there is any difference in the median values among five major sustainability determinants.

Table 1. Results for Multiple Comparison of Medians among Five Major Determinants.

Test	Test Statistic (H-Value)		*p* Value		
Determinant	Not Adjusted for Ties	Adjusted for Ties	Not Adjusted for Ties	Adjusted for Ties	Any Difference among Medians?
Site	6.240	6.750	0.182	0.150	No
Water/Wastewater	4.520	5.180	0.345	0.269	No
Energy	2.290	2.620	0.682	0.623	No
Materials/Resources	2.050	2.240	0.727	0.692	No
Environmental	1.900	2.090	0.755	0.719	No

3.2. Comparison of Means among Five Major Determinants

We compared the mean response rates among five major sustainability determinants using an ANOVA test. The purpose of the test is to see if there is any difference in the extent of how respondents rate the importance level. The null and research hypotheses are H_0: $\mu Ci = 0$ for all i, where i is the determinant, and H_a: at least two mean values among five sustainability determinants differ. The ANOVA results are tabulated in Table 2. Since the *p* value of 0.244 is not less than $\alpha = 0.05$, the authors cannot find any significant evidence, which means that we cannot reject the null hypothesis. The results show that the mean value of one determinant is statistically equal to those of other determinants. The test results means that the respondents valued all five determinants closely and did not undervalue any one of the determinants.

Table 2. Results of ANOVA for Multiple Comparison of Five Major Determinants.

Source	Degree of Freedom	Adjusted Sum of Squares	Adjusted Mean Squares	F-Value	*p* Value
Factor	4	7.686	1.922	1.37	0.244
Error	620	871.072	1.405		
Total	624	878.758			

3.3. Importance of Factors Associated with Each Determinant

We examined the importance of factors associated with each determinant using Kruskal–Wallis test. The purpose of this test is to see if respondents prefer a specific factor. The Kruskal–Wallis test is useful because its outcomes generate the mean rank of factors. The mean rank is computed as the average of the all-responses' ranks within each factor and shows its corresponding z-value. If a higher mean rank is shown, this indicates that the observation values in the group are higher than those in the other groups. Additionally, its corresponding z-values show how the average rank for each group compares to the average rank of all the observations. If a factor's average rank is less than the overall average rank, then the corresponding z-value is negative. If a factor's average rank is greater than the overall average rank, then the corresponding z-value is positive. The greater the absolute value, the more distant a factor's average rank is from the overall average rank [7]. As an example, Table 3 tabulates the descriptive statistical results for site-related factors.

Table 3. Results of Responses for Site Factors.

Factor	n	Mode	Median	Mean	Std. Dev.	Mean Rank	Z-Value
(S1) Location of sustainable sites	25	6	6	5.48	1.33	55.6	−1.14
(S2) Effective and efficient design	25	6	6	5.68	1.22	60.5	−0.39
(S3) Mobility enhancement and sustainable transportation	25	7	7	6.28	0.98	78.7	2.42
(S4) Environmental impact reduction on ecology and biodiversity	25	7	6	5.68	1.31	61.6	−0.21
(S5) Noise, vibration, and light pollution minimization	25	5	6	5.64	1.15	58.6	−0.68

The mean ranks having positive z-values indicates their relative importance among the factors under each of the five major sustainability determinants. For the site determinant, the respondents ranked mobility enhancement as well as sustainable transportation as the most important. For the water and wastewater determinant, they put more importance on water quality protection, stormwater treatment and management, and water consumption reduction than others. For the energy determinant, they ranked greenhouse gas emission reduction as the highest factor, followed by pollution reduction. For material and resource determinants, the most important factor was hazardous waste elimination, followed by effective and efficient material utilization. For the environmental determinant, respondents ranked the people's life quality improvement as the most important factor, followed by air quality improvement, factoring in traffic flow, bicycle and pedestrian facilities improvement, and wildlife protection, enhancement, and restoration. The results from respondents show that a few factors under each of five major sustainability determinants received higher ranks than other factors.

4. Concluding Remarks

The authors presented statistical results on sustainability determinants and their associated factors. The results showed that there was no significant difference in the median response values for five major sustainability determinants. The mean response values for the five major sustainability determinants considered showed no statistically significant difference. The pairwise comparison results showed that there was no difference among the five major sustainability determinants. The findings indicated that the determinants considered are equally important for the successful implementation of sustainability in infrastructure construction projects. However, the mean ranks values, determined based on the positive z-values, indicated the relative importance of the factors under each of the five major sustainability determinants. The quantitative analysis results presented in this paper will aid in evaluating the awareness of infrastructure industry professionals regarding key high-performance sustainability requirements that are being incorporated into the design of infrastructure construction projects. The determinants can be used to develop a framework that provides a sustainability rating system for infrastructure construction projects.

Author Contributions: Conceptualization, J.J.K.; methodology, J.J.K.; software, J.J.K. and P.M.; validation, J.J.K. and P.M.; formal analysis, J.J.K. and P.M.; investigation, J.J.K. and P.M.; resources, J.J.K. and P.M.; data curation, J.J.K. and P.M.; writing—original draft preparation, J.J.K.; writing—review and editing, J.J.K. and P.M.; visualization, J.J.K. and P.M.; supervision, J.J.K.; project administration, J.J.K.; funding acquisition, J.J.K. All authors have read and agreed to the published version of the manuscript.

Funding: This work was supported by TRANSPORT-2021 under SB1 grant (ZSB12017-SJAUX).

Institutional Review Board Statement: The study was conducted in accordance with 45 CFR 46 104 (d)(2) approved by Institutional Review Board for the Protection of Human Subjects of California State University, Long Beach on 18 October 2021.

Informed Consent Statement: Not applicable.

Data Availability Statement: The data presented in this study are available on request from the corresponding author.

Conflicts of Interest: The authors declare no conflict of interest.

References

1. U.S. Green Building Council. LEED Rating System. 2018. Available online: https://www.usgbc.org/leed (accessed on 31 December 2018).
2. Simpson, S.P. A Framework for Assessing Transportation Sustainability Rating Systems for Implementation in U.S. State Departments of Transportation. Mater's Thesis, Colorado State University, Fort Collins, CO, USA, 2013.
3. Simpson, S.; Ozbek, M.; Clevenger, C.; Atadero, R. *A Framework for Assessing Transportation Sustainability Rating Systems for Implementation in U.S. State Departments of Transportation*; MPC-14-268; North Dakota State University-Upper Great Plains Transportation Institute, Mountain-Plains Consortium: Fargo, ND, USA, 2014.
4. Cormack, S.M.; Sturgill, R.; Howell, B.; Dyke, C.V.; Kreis, D. Green Infrastructure. Kentucky Transportation Center, 2014. KTC-14-10/SPR447-12-1F. Available online: https://uknowledge.uky.edu/cgi/viewcontent.cgi?article=2461&context=ktc_researchreports (accessed on 15 July 2018).
5. Lineburg, K.S. Transportation Rating Systems and Social Sustainability: A Comprehensive Analysis. Honors Program Thesis, James Madison University, Harrisonburg, VA, USA, 2016.
6. Chisholm, D.; Reddy, K.; Beiler, M.R.O. Sustainable project rating systems including envision. In *Engineering for Sustainable Communities*; Chapter 20; ASCE: Reston, VA, USA, 2017. [CrossRef]
7. Minitab. 2022. Available online: http://www.minitab.com/en-US/products/minitab/ (accessed on 15 December 2022).

engineering proceedings

MDPI

Proceeding Paper

Evaluation of Longitudinal Irregularity in Airport Pavements and Unpaved Runway †

Livia Merighi *, Claudia Pereira and Jose Schiavon

Aeronautics Institute of Technology—ITA, Sao Jose dos Campos 12228-900, Brazil; claudia.azevedo@ita.br (C.P.); schiavon@ita.br (J.S.)

* Correspondence: livia.merighi@gmail.com

† Presented at the Second International Conference on Maintenance and Rehabilitation of Constructed Infrastructure Facilities, Honolulu, HI, USA, 16–19 August 2023.

Abstract: This paper presents a study about the effect of unevenness on conventional airport pavements and unpaved runways. During landing and take-off operations, aircraft tires are at high levels of tension and possible surface roughness can contribute to aircraft damage, landing gear fatigue, as well as the loss of aircraft directional stability, thus increasing the chances of accidents or incidents. The National Civil Aviation Agency (ANAC), responsible for regulating and supervising civil aviation activities in Brazil, regulates the need to evaluate longitudinal irregularity through the International Roughness Index (IRI) parameter on paved runways. In addition to the IRI, the Boeing Bump Index (BBI) and Runway Roughness Index (RRI) are indices also recommended by the Federal Aviation Administration (FAA). Moreover, this study understands the concept of unpaved runways and how these indexes can be evaluated on unpaved runways, however, with minimum requirements. Therefore, the present study addresses these bearing quality indices whose purpose is to guarantee the safety of operations.

Keywords: airport pavement; Roughness Index; paved runway; unpaved runway

Citation: Merighi, L.; Pereira, C.; Schiavon, J. Evaluation of Longitudinal Irregularity in Airport Pavements and Unpaved Runway. *Eng. Proc.* **2023**, *36*, 64. https://doi.org/10.3390/engproc2023036064

Academic Editor: Hosin (David) Lee

Published: 28 August 2023

1. Introduction

Surface condition indexes are important to airport operations where the lack of a minimum surface condition on runways can contribute to accidents or incidents.

During the process when the aircraft is in contact with the pavement, the existence of possible unevenness in the surface may increase structural fatigue on the landing gear.

So, this paper presents a study of the anomalies represented by surface deviations, characteristic of the International Roughness Index (IRI), Boeing Bump Index (BBI), and Runway Roughness Index (RRI) on paved and unpaved runways, parameters that aim to guarantee the adequate safety and riding quality.

2. Roughness

Roughness can be measured by the International Roughness Index (IRI) and is an important parameter linked to riding and the comfort and quality experienced by a passenger. As defined by ASTM E 867 it is formed by surface deviations that can affect vehicle dynamics as well as surface drainage of the road, influencing the comfort and safety of the vehicle [1,2].

Some indicators of pavement condition through the assessment of unevenness are listed in Figure 1 [3]:

Figure 1 presents some anomalies that may contribute to the presence of unevenness on pavement, such as corrugation, defect usually of structural origin, depression, disintegration, slippage, swelling, as well as longitudinal and transversal cracking and patches usually have a functional origin [4].

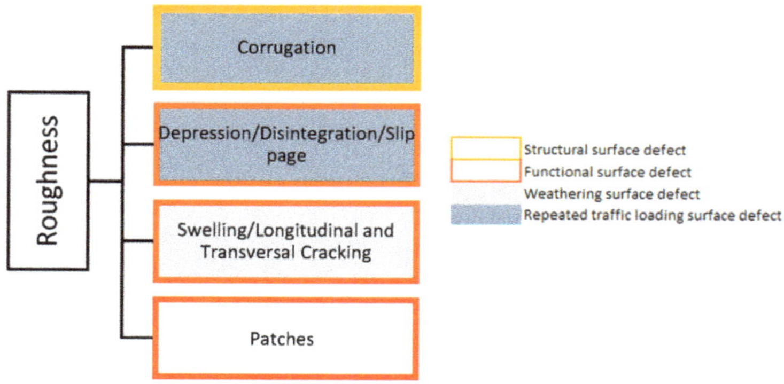

Figure 1. Condition indicators of airport pavements [3].

It is worth noting that pathologies such as corrugation, depression, disintegration, and slippage are usually derived from the cyclic loading of traffic while swelling and longitudinal/transversal cracking pathologies can occur due to climate change [4].

Some studies mention that the use of longitudinal irregularity (IRI), an index that is based on the response of the vibrational model of a quarter of a car, would not be suitable for the airport modal, due to the speeds during landing and takeoff operations, where the aircraft works with an approximate speed of 185 km/h (115 mph), diverging from the highway model, in addition to this parameter not considering the critical wavelengths and their amplitudes that can directly impact operations [5,6].

While the IRI, although has been adopted by many agencies and countries including Brazil, China, Italy, Canada, Mexico, and South Africa, would not be suitable for runways not only because of the speed but also due to the vehicle configuration. The Boeing method, Boeing Bump Index (BBI), the method used in the United States and recommended by the International Civil Aviation Organization (ICAO) and Federal Aviation Administration (FAA) through document AC 150/5380-9, is highly used to verify the functional parameters of co1mfort and rolling quality on takeoff and landing [7,8].

It is important to mention that the Boeing method was developed and is used on paved runways, either with asphalt concrete or Portland cement concrete, as well as for unpaved runways, an item that will be discussed later in this study [7].

Boeing's method consists of drawing a virtual line between two points, and checking the deviation from the surface, called "bump height". This index considers wavelengths from 0.5 m (1.6 ft) to 120m (393 ft), where wavelengths greater than 120 m (393 ft) do not contribute to the dynamic response of the aircraft [9].

It was observed that for the operating speeds for the landing and takeoff operations (approximately 185 km/h or 115 mph), wavelengths equivalent to 73 m (240 ft), would be critical, while for the taxiways (approximate speed of 37 km/h or 23 mph), even though this method has been developed for runways, 15 m (50 ft) wavelength can significantly impact the operation on taxiways [5].

As a limitation of the BBI parameter, it is worth mentioning that the method only identifies isolated events. However, the successive events of longitudinal irregularity are the most harmful to the aircraft [5].

Moreover, the Runway Roughness Index (RRI), an American index recently developed by the FAA's research and technology area, using the BBI, Pilot Subjective Rating (PSR) as a basis, evaluates the subjective rating by the pilots, as well as the resulting of the weighted root mean square of the vertical acceleration—WtRMS [10].

The RRI index is like the BBI when considering the pavement surface deviations; however, the RRI does not allow us to obtain the exact location of pavement anomalies as the BBI index, it presents the locations of acceleration events that would be experienced by the pilot in an aircraft, with a short wavelength [10].

Furthermore, it was identified that the RRI is a method only valid for runways, not applicable for taxiways or aprons where typical operating speeds are less than 185 km/h (115 mph) [10].

Therefore, the FAA suggests using more than one index to assess surface deviations on airport pavements, as well as allocating the necessary resources to correct the pathology [10].

3. Unpaved Runway

Unpaved runways are characterized by runways that do not have the structure of a paved runway but are free of debris (debris) and/or obstructions. They have been used by military aviation since World War II, where the aircraft must undergo fuselage adjustments, among others, to operate on this type of runway avoiding damage to the aircraft [11].

It is important to point out that short waves on pavement can cause fatigue on the aircraft's landing gear. If the unevenness is located on the wheel track, approximately 7.6 cm (3 inches) is enough to cause damage in some aircraft models [12].

For this type of runway, the document regarding the Certification Authorities for Large Transport Aircraft (CATA), represented by ANAC, European Union Aviation Safety Agency (EASA), FAA, and Transport Canada (TCCA), recommends that surface deviations be evaluated through wavelengths and peaks or by Power Spectral Density (PSD) to verify the assessment of fatigue in the landing gear caused by irregularity [13].

One of the challenges when working with a semi-prepared runway is the shear stress. Rutting is also a pathology that can be found on semi-prepared runways, as well as dust, foreign object damage (FOD), and uncertainties in takeoff performance [11].

Therefore, the analysis of surface deviations becomes important both on the paved and unpaved runway.

4. Conclusions

Based on the identified studies, it is possible to verify the importance of the proper evaluation of surface deviations aiming to guarantee the safety of operations, as well as the allocation of financial resources involved in correcting the anomaly when identified.

As previously mentioned, the IRI has a gap in the airport modal due to the critical wavelengths, as well as their amplitudes that can impact mechanically on aircraft operations, while the BBI results in a parameter to assist in the maintenance strategy through existing deviations in the pavement.

It is worth noting that the RRI is an index recently developed by the FAA for runways and is not suitable for taxiways and aprons due to speed operation. In addition to that, the FAA mentions that this parameter should not be used for acceptance of construction quality.

It is important to consider that the BBI is like the RRI in terms of deviations from the surface; however, the RRI does not allow the identification of the location of the anomaly. Therefore, this index should be considered an acceleration index and not a pavement index.

Thus, the replacement of the indices is not recommended by the FAA, but their use in a complementary way, where the BBI contributes to pavement defects, while the RRI helps with the interference effects due to the condition of the pavement.

So, it is suggested for paved runways use the three indices in a complementary way to define the most appropriate maintenance strategy due to the limitation of each index, as well as monitoring through IRI and BBI in taxiways.

Regarding surface deviations on semi-prepared runways, it is worth mentioning that this information is relevant to guarantee the safety of operations since it can interfere with the performance of the aircraft.

Therefore, it is recommended to carry out studies to define the most appropriate parameter for this type of survey.

Author Contributions: Conceptualization, L.M.; methodology, L.M., J.S. and C.P.; formal analysis, L.M.; writing—original draft preparation, L.M.; supervision, J.S. and C.P.; writing—review editing, J.S. and C.P.; project administration, J.S. and C.P. All authors have read and agreed to the published version of the manuscript.

Funding: This research received no external funding.

Institutional Review Board Statement: Not applicable.

Informed Consent Statement: Not applicable.

Data Availability Statement: Not applicable.

Conflicts of Interest: The authors declare no conflict of interest.

References

1. *E867-06*; Standard Terminology Relating to Vehicle-Pavement Systems. ASTM International: West Conshohocken, PA, USA, 2020.
2. Merighi, L.F. Functional evaluation of Asphaltic Pavements at Airports with the Purpose of Establishing Goals for their Maintenance. Master's Thesis, Polytechnic School of the University of Sao Paulo, Sao Paulo, Brazil, 2017.
3. Fernandes Junior, J.L.; de Oliveira, F.H.L.; Pereira, C.A. Studies on airport pavements at Brazilian Universities. In Proceedings of the XIV ALACPA Seminar, Quito, Ecuador, 28 May–1 June 2018. Available online: https://www.alacpa.org/ALACPA2018_FernandesJR_Oliveira_Azevedo.pdf (accessed on 12 February 2023).
4. Roginski, M. Runway roughness evaluation—Boeing Bump Methodology. In Proceedings of the XI ALACPA Seminar, Santiago, Chile, 1–5 September 2014. Available online: https://www.icao.int/nacc/documents/meetings/2013/alacpa10/alacpa10-p24.pdf (accessed on 10 February 2023).
5. Sousa, E.S. Analysis of International Irregularity Index Limits on Airport Runways Based on Computer Simulations. Master's Thesis, Department of Transport Engineering at the Federal University of Ceara, Fortaleza, Brazil, 2021.
6. FAA. *DOT/FAA/TC-21/32—New Index Testing and Verification—Runway Roughness Index*; Federal Aviation Administration: Washington, DC, USA, 2021.
7. FAA. *AC No. 150/5380-9: Advisory Circular—Guidelines and Procedures for Measuring Airfield Pavement Roughness*; Federal Aviation Administration: Washington, DC, USA, 2009.
8. Tian, Y.; Liu, S.; Liu, L.; Xiang, P. Optimization of International Roughness Index Model Parameters for Sustainable Runway. *Sustainability* **2021**, *13*, 2184. [CrossRef]
9. Merighi, L.F.; Pereira, C.A.; Schiavon, J.A. Evaluation of Longitudinal Irregularity in Airport Pavements and semi-prepared runways. In Proceedings of the Sitraer Air Transportation Symposium 2022, São José, Brazil, 25–27 October 2022.
10. FAA. *Runway Roughness Index*; Federal Aviation Administration: Washington, DC, USA, 2021. Available online: https://www.faa.gov/sites/faa.gov/files/2022-05/508.FAA_RunwayRoughnessIndex_Brynick_200302.pdf (accessed on 7 February 2023).
11. Hansen, E. Evaluating the C-17 Unpaved Runway Capability—An off Road Map. Master's Thesis, Wright-Patterson Air Force Base, Air Force Institute of Technology, Air Force Air University, Dayton, OH, USA, 2002.
12. ICAO. Pavement Maintenance and Aircraft/Pavement Interaction. In Proceedings of the Short Course on The Aircraft/Pavement Interaction, Santa Cruz de la Sierra, Bolivia, 22–23 July 2002; International Civil Aviation Organization: Montreal, QC, Canada, 2002.
13. CATA. Certification Authorities for Large Transport Aircraft. Unusual Landing Operations. 25.301, 25.235, 25.491, 25.571, Appendix Q. 2021. Available online: https://www.easa.europa.eu/sites/default/files/dfu/proposed_cwi_easa-005_unusual_landing_operations_v5_final_industry_sign_.pdf (accessed on 1 February 2023).

engineering proceedings

MDPI

Proceeding Paper

Improving Fatigue and Rutting Resistance of Road Pavements Using Aramid Fibers †

Jorge Pais [1,*], Grigório Neto [1], Johnny Coelho [2] and Paulo Pereira [1]

1 Department of Civil Engineering, University of Minho, 4800-048 Guimarães, Portugal; grigor.io@hotmail.com (G.N.); ppereira@civil.uminho.pt (P.P.)
2 Departamento de Engenharia Civil, Instituto Federal de Santa Catarina, São Carlos, SC 89885-000, Brazil; johnny.coelho@ifsc.edu.br
* Correspondence: jpais@civil.uminho.pt
† Presented at the Second International Conference on Maintenance and Rehabilitation of Constructed Infrastructure Facilities, Honolulu, HI, USA, 16–19 August 2023.

Abstract: Fibers in asphalt mixtures have been adopted to improve road pavement performance for constructing new pavements and rehabilitating existing pavements. Fibers in asphalt mixtures improve flexural strength and toughness, reducing cracking and rutting in asphalt pavements. Because there is a continuous need to enhance pavement performance, and using fibers is a good strategy for this improvement, this paper aims to discuss how aramid fibers can improve asphalt pavements' fatigue and rutting performance by studying the behavior of asphalt binders modified with fibers.

Keywords: road pavements; asphalt mixtures; fibers; aramid; cracking; rutting

1. Introduction

Asphalt mixtures reinforced with fibers can be adopted to improve pavement performance. Fibers have been used to strengthen paving materials for decades [1,2]. They are commonly used in Stone Matrix Asphalt (SMA), porous asphalt mixtures, and open gradation asphalt mixtures to mitigate binder drain-down [3,4]. Incorporating fibers can improve fatigue cracking and the permanent deformation resistance of asphalt mixtures [5].

Cracking is the primary concern for road pavements [6]; so, identifying a potential material to reduce cracking can be very beneficial. Due to their inherent compatibility with the asphalt binder and better mechanical properties, fibers offer an excellent resource for reinforcing asphalt mixtures [7]. Many types of fibers are available for incorporation into asphalt mixtures. Cellulose and mineral fibers are commonly used in gap-graded SMA and open-graded or porous mixtures [8]. Polypropylene and polyester fibers have been used in dense-graded asphalt mixtures and are still somewhat used in SMA [9]. Various polymers and steel wool fibers have been added to asphalt mixtures [10]. Recently, aramid fibers have been used to reinforce asphalt mixtures due to their strength [11].

Research has been conducted to study the impact of aramid fibers on asphalt mixtures and road pavements, mainly to reduce pavement cracking and rutting. An extensive testing program was carried out by [7], where they performed binder and mixture characterization tests. They performed triaxial strength, permanent deformation, dynamic modulus, fatigue cracking flexural strength indirect tensile and cracking propagation tests for asphalt mixture. Despite the conclusions obtained where it is visible that the asphalt mixture improves performance using aramid fibers, some aspects should be studied to clarify the effect of these fibers that are characterized by a diameter less than 0.1 mm, and high strength. Their length varies from around 20 mm to 50 mm, influencing how fibers support the distress modes. So, this paper contributes to understanding how fibers work on fatigue and rutting by looking at the behavior of asphalt binders with aramid fibers.

Citation: Pais, J.; Neto, G.; Coelho, J.; Pereira, P. Improving Fatigue and Rutting Resistance of Road Pavements Using Aramid Fibers. *Eng. Proc.* **2023**, *36*, 65. https://doi.org/10.3390/engproc2023036065

Academic Editor: Hosin (David) Lee

Published: 5 September 2023

2. Materials

This work was carried out using 0.012 mm diameter aramid fibers with three different lengths, namely 25, 38, and 50 mm. These lengths suit asphalt mixtures with 12.5, 19, and 25 mm maximum aggregate sizes. These fibers were mixed into asphalt binder in 0.12%, 0.14%. 0.17% and 0.20% by weight of binder, meaning 12 asphalt binders were produced. They were tested for rheological properties through the MSCR (Multiple Stress Creep and Recovery), LAS (Linear Amplitude Sweep), and shear complex modulus tests, as represented in Figure 1. Test specimen names were defined based on the fiber content and fiber length, such as C12L25, meaning that fiber content is 0.12% and the fiber length is 25 mm. The SF specimen refers to asphalt binder without fibers, used as reference. The 35/50 pen asphalt binder was used.

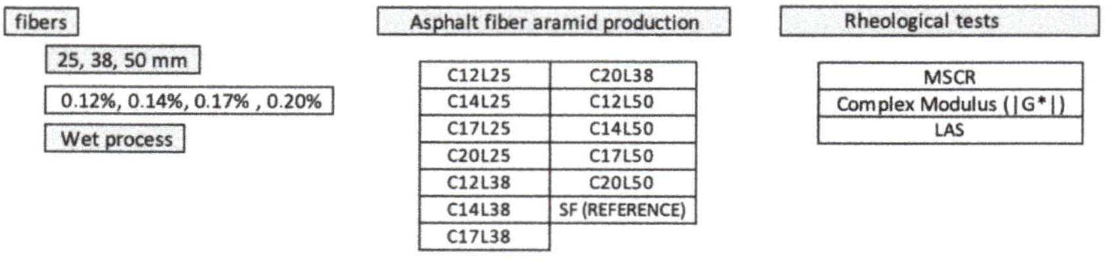

Figure 1. Testing program.

Aramid fibers with a melt temperature greater than 500 °C were added to the asphalt binder by the wet process, mixing them on the hot asphalt (150 °C), as observed in Figure 2 (left), obtaining test specimens as observed on the right-hand part of the figure. Aramid fibers do not melt in the asphalt. The 8 and 25 mm diameter specimens were cut from the hot-prepared samples to be used in a dynamic shear rheometer, where tests were performed. A 1 mm gap was set in the rheologic tests.

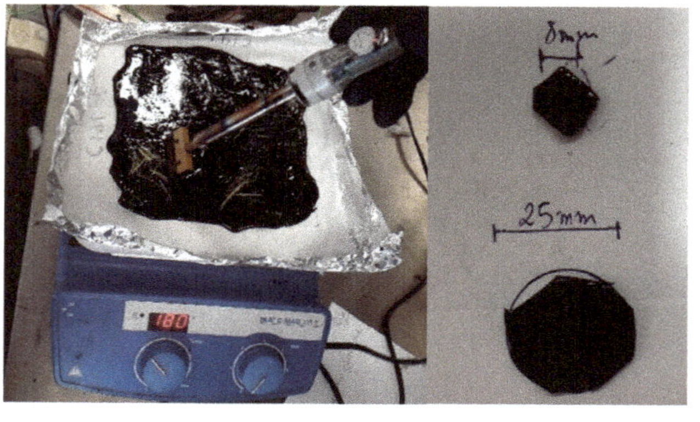

Figure 2. Specimen preparation.

3. Results

MSCR tests were carried out on all specimens to obtain the fiber-modified binder recovery. The evolution of the shear strain during the test for an applied 3.2 kPa shear strain is presented in Figure 3, where it is possible to observe a minimal recovery obtained in this experiment. It was expected for the base bitumen. The presence of fiber reduced the accumulation of rutting more than that obtained for the case without fibers. It can also be observed that longer fibers and their content minimize the accumulation of rutting.

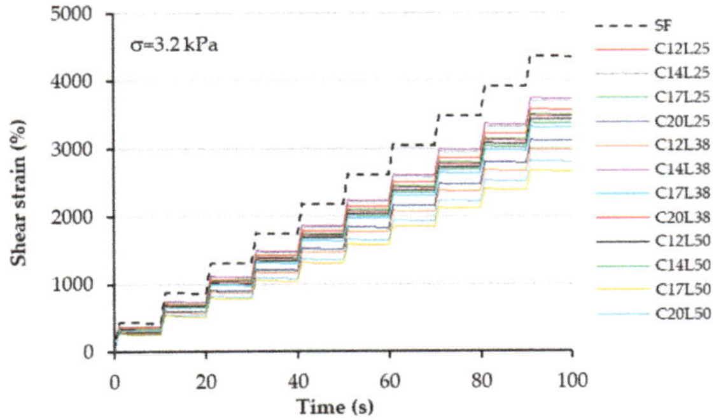

Figure 3. MSCR test results.

LAS tests on all specimens produced the results presented in Figures 4 and 5, where a significant difference is visible between the cases studied in terms of curve slopes and energy (Figure 4). In terms of C-S characteristic curves (Figure 5), different behaviors are visible, meaning that the presence of fiber influences the asphalt-binder behavior. It is also observed that increasing the fiber's content and length has a positive effect on the cracking resistance of the binder, meaning that there is an expected improvement of the fatigue resistance of the asphalt mixtures.

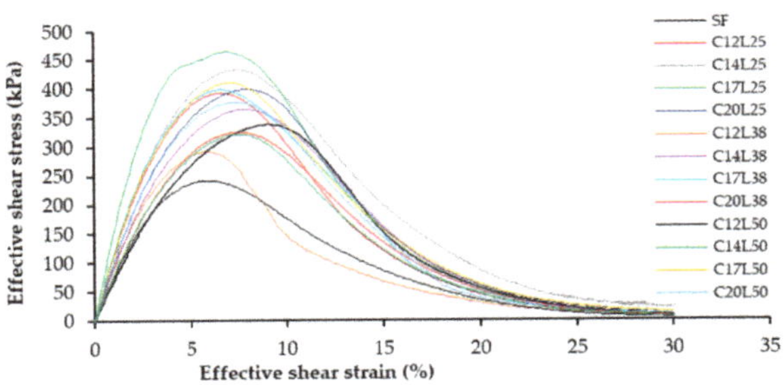

Figure 4. LAS stress–strain curves.

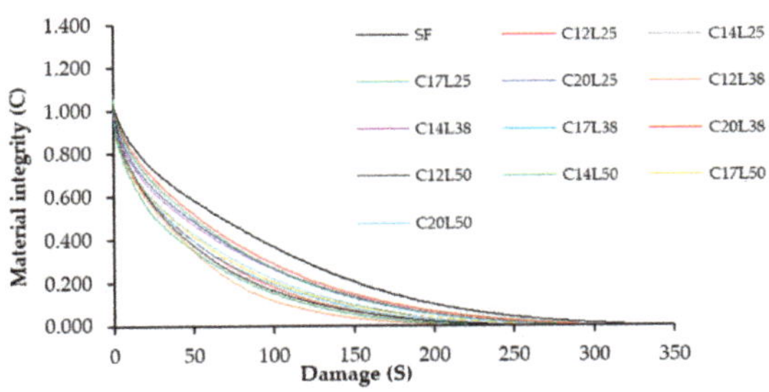

Figure 5. LAS C-S characteristic curves.

Tests to obtain the complex shear modulus (Figure 6) show a slight variation of the modulus amount in the different case studies (fiber content and length). These master curves were obtained for 20 °C using the WLF model. All asphalt binders with fiber seem to exhibit the same complex shear modulus as the reference asphalt binder.

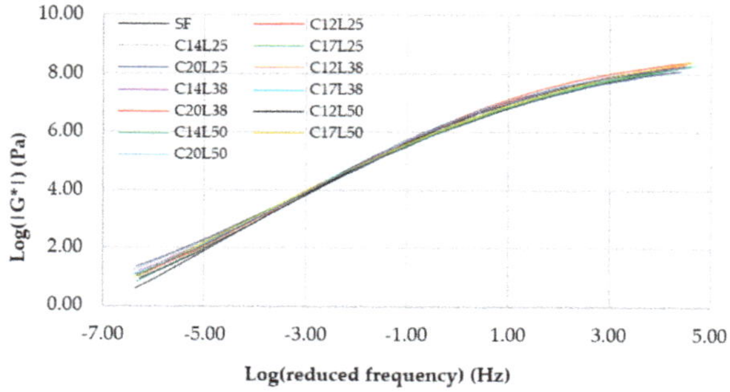

Figure 6. Complex shear modulus master curves for 20 °C.

Despite the minor differences observed in the master curves, they are represented in a log–log scale which means that these minor differences represent significant variations in the complex shear modulus. There is a crossover frequency where all master curves cross, around log(−3.0) Hz.

4. Conclusions

This paper presented a procedure to assess the behavior of asphalt binders with aramid fibers where those fibers were added to the asphalt while wet and proceeded to produce specimens for dynamic shear tests, namely, to evaluate the creep and recovery response, linear analysis, and complex shear modulus.

The developed process seems useful for adding aramid fibers to the asphalt binder. Still, the dispersion of the fiber into the asphalt binder is problematic due to the microscopic fibers' diameter. This problem also appears in the asphalt mixtures and should be carefully treated.

Based on the results obtained from this experimental procedure, it is possible to conclude that a longer and higher content of aramid fibers in the asphalt binder improves fatigue and rutting resistance.

Author Contributions: Conceptualization, J.P.; methodology, J.P. and G.N.; formal analysis, J.P.; investigation, G.N.; data curation, G.N., original draft preparation, J.P., G.N. and J.C.; writing—review and editing, J.P., G.N., J.C. and P.P. All authors have read and agreed to the published version of the manuscript.

Funding: This research received no external funding.

Institutional Review Board Statement: Not applicable.

Informed Consent Statement: Not applicable.

Data Availability Statement: Data available by request.

Conflicts of Interest: The authors declare no conflict of interest.

References

1. Mrema, A.H.; Noh, S.-H.; Kwon, O.-S.; Lee, J.-J. Performance of Glass Wool Fibers in Asphalt Concrete Mixtures. *Materials* **2020**, *13*, 4699. [CrossRef] [PubMed]
2. Mohajerani, A.; Hui, S.-Q.; Mirzababaei, M.; Arulrajah, A.; Horpibulsuk, S.; Abdul Kadir, A.; Rahman, M.T.; Maghool, F. Amazing Types, Properties, and Applications of Fibres in Construction Materials. *Materials* **2019**, *12*, 2513. [CrossRef] [PubMed]
3. Poulikakos, L.; Gilani, M.S.; Derome, D.; Jerjen, I.; Vontobel, P. Time resolved analysis of water drainage in porous asphalt concrete using neutron radiography. *Appl. Radiat. Isot.* **2013**, *77*, 5–13. [CrossRef]
4. Jamieson, S.; White, G. Laboratory Evaluation of the Performance of Stone Mastic Asphalt as an Ungrooved Runway Surface. *Materials* **2021**, *14*, 502. [CrossRef] [PubMed]
5. Landi, D.; Marconi, M.; Bocci, E.; Germani, M. Comparative life cycle assessment of standard, cellulose-reinforced and end of life tires fiber-reinforced hot mix asphalt mixtures. *J. Clean. Prod.* **2020**, *248*, 119295. [CrossRef]

6. Decker, D.S. Best practices for crack treatments in asphalt pavements. In Proceedings of the 6th Eurasphalt & Eurobitume Congress, Prague, Czech Republic, 1–3 June 2016.
7. Klinsky, L.; Kaloush, K.; Faria, V.; Bardini, V. Performance characteristics of fiber modified hot mix asphalt. *Constr. Build. Mater.* **2018**, *176*, 747–752. [CrossRef]
8. Asi, I.M. Laboratory comparison study for the use of stone matrix asphalt in hot weather climates. *Constr. Build. Mater.* **2006**, *20*, 982–989. [CrossRef]
9. Putman, B.J.; Amirkhanian, S.N. Utilization of waste fibers in stone matrix asphalt mixtures. *Resour. Conserv. Recycl.* **2004**, *42*, 265–274. [CrossRef]
10. Tang, N.; Huang, W.; Hao, G. Effect of aging on morphology, rheology, and chemical properties of highly polymer modified asphalt binders. *Constr. Build. Mater.* **2021**, *281*, 122595. [CrossRef]
11. Li, H.; Yu, J.; Liu, Q.; Li, Y.; Wu, Y.; Xu, H. Induction Heating and Healing Behaviors of Asphalt Concretes Doped with Different Conductive Additives. *Adv. Mater. Sci. Eng.* **2019**, *2019*, 1–10. [CrossRef]

engineering proceedings

MDPI

Proceeding Paper

Investigation on the Process of Eliminating Abnormal Objects from the Road for the Creation of an AI Program That Can Automatically Detect Potholes †

Moonsup Lee [1], Taehoon Lee [1], Younghan Park [1], Seungyeon Han [1,*], Nuri Lee [2] and Chulki Kim [2]

1 Korea Institute of Civil Engineering and Building Technology, 283, Goyang-daero, Ilsanseo-gu, Goyang-si 10223, Republic of Korea; truepath@kict.re.kr (M.L.); thlee@kict.re.kr (T.L.); hyan@kict.re.kr (Y.P.)
2 Ministry of Land, Infrastructure and Transport, 11 Doum 6-ro, Government Complex-Sejong, Sejong City 30103, Republic of Korea; thisnuri@korea.kr (N.L.); kimck98@korea.kr (C.K.)
* Correspondence: syhan@kict.re.kr; Tel.: 82-31-910-0665
† Presented at the Second International Conference on Maintenance and Rehabilitation of Constructed Infrastructure Facilities, Honolulu, HI, USA, 16–19 August 2023.

Abstract: For effective pothole control on national highways, autonomous pothole identification technology utilizing artificial intelligence was deployed in Korea. There are a number of different objects on the road's surface that resemble potholes. The YOLOv7-E6E model was used to reduce noise, before classifying these objects and potholes. In the algorithm, aberrant objects other than potholes were classified using design and learning techniques. Manhole, automobile, lane-marking, garbage, and shadow elements that are similar to potholes were learned, in order to detect them. "Etc." was used to summarize 15 characteristics, including a broken patch, spalling, crack, ramp, license plate, leaf, and pool. In light of this, learning was conducted using a total of seven classification criteria. The test dataset had a 91% accuracy rate.

Keywords: PMS; AI; pothole; road; automatically detect

Citation: Lee, M.; Lee, T.; Park, Y.; Han, S.; Lee, N.; Kim, C. Investigation on the Process of Eliminating Abnormal Objects from the Road for the Creation of an AI Program That Can Automatically Detect Potholes. *Eng. Proc.* **2023**, *36*, 66. https://doi.org/10.3390/engproc2023036066

Academic Editor: Hosin (David) Lee

Published: 18 September 2023

1. Introduction

Due to global warming, abnormal weather phenomena frequently occur, and the number of potholes on Korean roads has increased rapidly, due to aging roads and an increase in heavy vehicles [1–3]. In Korea, to respond to the increasing occurrence of potholes, two AI pothole detection devices were supplied to 18 land management offices, and the equipment is operated from time to time. To exhaustively detect the pothole, the provided equipment is designed to recognize most abnormal objects on the road pavement as potholes, and then the detection information is transmitted to the general national road pavement data management system (PDMS) [4]. However, there is a disadvantage, in that a vast amount of pothole information is transferred, and excessive human resources are required to select the presence or absence of actual potholes. Therefore, it is necessary to build a web-based AI database technology to automatically and precisely analyze the pothole data received through the detection equipment. This study aims to develop an artificial intelligence model that classifies abnormal objects from the received pothole data.

2. Algorithm to Classify Abnormal Object

2.1. AI Model Selection

AI learning was performed using about 4000 pothole data, using the ResNet, SeNet, Inception v3, and DenseNet models to select an artificial intelligence model. As a result of the learning test, the loss values of SeNet (99.7%), InceptionNet v3 (99.3%), ResNet (99.2%), and DenseNet (99.5%) were obtained. Ultimately, the SeNet model showed the highest accuracy, and was adopted as an artificial intelligence model for anomaly object classification. Figure 1 is a graph comparing the loss values of the artificial intelligence models.

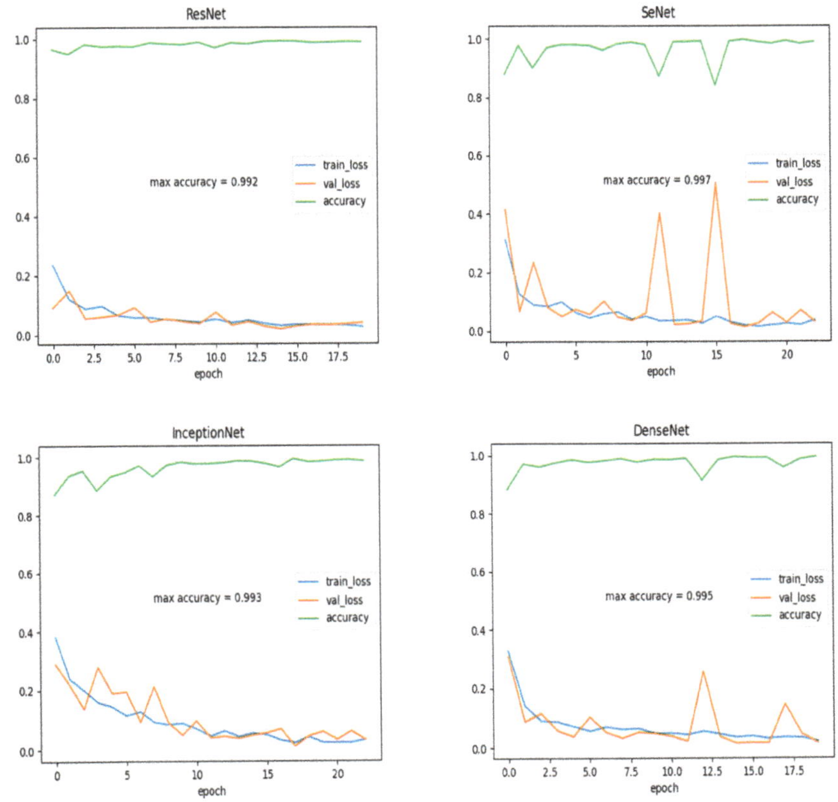

Figure 1. Comparison of the AI model loss values.

2.2. Criteria for Classifying Abnormal Objects

In the transmitted image, objects that could be misinterpreted as potholes were categorized. Potholes were classified as standard objects, and five elements that could be recognized similarly to potholes in the image were, additionally, classified. Six of the elements comprised potholes, manholes, vehicles, lanes, garbage, and shadows, and a further fifteen elements, including cracks and patches, were added together to create the seventh element, "Etc.". These seven elements determined the classification criteria.

2.3. Artificial Intelligence Learning Process

Among the 598,553 data transmitted from the pothole-detection equipment, 1670 images determined to be potholes were used as learning data. The learning was performed using 15,780 images for seven-classification learning, including potholes, manholes, cars, lanes, garbage, shadows, and "Etc.". Table 1 shows the representative images, the number of learnings, and the learning rate of each classification item for abnormal object classification learning.

Table 1. Abnormal object learning contents.

No.	Classification of Learning Data	Learning Data Image	Number of Learning Data	Learning Rate (%)
1	Pothole			10.6
2	Manhole		210	1.3
3	Lane			4.8
4	Car		[illegible]	9.2
	Trash			0.3
	Shadow		120	0.8
	Etc.		11,530	73
	Total		15,780	100

3. Results of Abnormal Object Classification Learning

As a result of inputting existing pothole data for verification into the learned artificial intelligence algorithm, it was found that 100% of potholes were detected among abnormal objects on the road. Furthermore, it showed a 91% accuracy in classifying abnormal objects using the image data. Figure 2 shows the results of pothole detection using images.

Figure 2. Pothole detection results.

4. Conclusions

For efficient pothole management, a study was conducted on developing an abnormal object classification artificial intelligence model for a server database, to detect actual potholes among a wide range of pothole images received from AI automatic pothole-detection equipment. The results are as follows.

1. As a result of carrying out pothole-detection learning with 1,670 images determined to be potholes among 598,553 data transmitted from the pothole detection equipment, potholes among abnormal objects on the road were detected 100% of the time.
2. Twenty-two types of abnormal objects detected on roads and spaces other than roads were classified into seven types. As a result of learning using 15,780 abnormal object image data, the accuracy of the seven types of abnormal objects learned was 91%.
3. The accuracy of the anomaly object classification was improved through its learning about 30,000 additional data stored in the pothole management server. Furthermore, server filtering was implemented, using an artificial intelligence classification model, without requiring manual work in order to advance.

Author Contributions: Conceptualization, M.L.; methodology, M.L.; validation, S.H. and T.L.; investigation, Y.P.; writing-original draft preparation, S.H.; writing-review and editing, M.L., N.L. and C.K.; project administration, M.L. and N.L.; funding acquisition, M.L. and C.K. All authors have read and agreed to the published version of the manuscript.

Funding: This work was supported by consignment business 2023 Pavement Management System (PMS) by the Ministry of Land, Infrastructure and Transport (MOLIT).

Institutional Review Board Statement: Not applicable.

Informed Consent Statement: Not applicable.

Data Availability Statement: All data are new and experimented.

Conflicts of Interest: The authors declare no conflict of interest.

References

1. Hong, S.P. A Study on the Improving Pothole with Snowfall in Winter. Master's Thesis, The University of Seoul, Seoul, Republic of Korea, 2016.
2. Dai, J.; He, K.; Sun, J. Instance-Aware Semantic Segmentation via Multi-Task Network Cascades. In Proceedings of the IEEE Conference on Computer Vision and Pattern Recognition (CVPR), Las Vegas, NV, USA, 27–30 June 2016; pp. 3150–3158.
3. Kim, D.; Sim, K.; Lee, G. Object Detection by Combining Two Different CNN Algorithms and Robotic Grasping Control. *J. Inst. Control Robot. Syst.* **2019**, *25*, 811–817. [CrossRef]
4. Ministry of Land, Infrastructure and Transport. *Final-Report of the National Highway Pavement Management System 2019*; Ministry of Land, Infrastructure and Transport: Sejong City, Republic of Korea, 2019.

engineering proceedings

MDPI

Proceeding Paper

Advancement of a Pavement Management System (PMS) for the Efficient Management of National Highways in Korea †

Seungyeon Han [1], Hyungmog You [1], Myeongill Kim [1], Moonsup Lee [1,*], Nuri Lee [2] and Chulki Kim [2]

1 Korea Institute of Civil Engineering and Building Technology, 283, Goyang-daero, Ilsanseo-gu, Goyang-si 10223, Republic of Korea; syhan@kict.re.kr (S.H.); naekog@kict.re.kr (H.Y.); mill@kict.re.kr (M.K.)
2 Ministry of Land, Infrastructure and Transport, 11 Doum 6-ro, Government Complex-Sejong, Sejong City 30103, Republic of Korea; thisnuri@korea.kr (N.L.); kimck98@korea.kr (C.K.)
* Correspondence: truepath@kict.re.kr; Tel.: 82-31-910-0690
† Presented at the Second International Conference on Maintenance and Rehabilitation of Constructed Infrastructure Facilities, Honolulu, HI, USA, 16–19 August 2023.

Abstract: In order to maintain a suitable road pavement level with limited resources, a management system must be established. In order to achieve this goal, a program using AI (artificial intelligence) was developed to manage and evaluate a sizable volume of survey data. A national highway pavement data management system (PDMS) built on the WEB was also constructed. By connecting several artificial neural networks, the AI crack analysis algorithm was created and taught to automatically recognize cracks in road photos and calculate crack rates. In the PDMS, the current condition of a national highway can be shown on a map, and all the data are updated to allow for verification in increments of 100 m for each lane. The system was also improved to enable the collection of information on the detailed survey section's pavement repair specifics according to survey year.

Keywords: highway; PMS; AI; crack; pavement surface; plastic deformation

Citation: Han, S.; You, H.; Kim, M.; Lee, M.; Lee, N.; Kim, C. Advancement of a Pavement Management System (PMS) for the Efficient Management of National Highways in Korea. *Eng. Proc.* **2023**, *36*, 67. https://doi.org/10.3390/engproc2023036067

Academic Editor: Hosin (David) Lee

Published: 26 September 2023

1. Introduction

Korea is facing budget constraints as the length of managed roads continues to grow. As a result, with limited resources, a system that can maintain road pavement conditions at an appropriate level or higher is required. To this end, a Road Pavement Management System (PMS) was developed to efficiently manage road pavements in a timely and effective manner [1]. Previously, investigations were conducted on road sections, as requested by the Land Management Office. Since 2022, the crack rate, plastic deformation, and longitudinal smoothness of 38,000 km/lane of the national road has been investigated [2]. Due to the increased data generated through this complete inspection, crack analysis performed by existing personnel is ineffective. As an alternative to this, a study on the introduction of an AI crack analysis program was conducted. Furthermore, a study was conducted to advance the pavement data management system (PDMS) for general national highways so as to efficiently manage pavement condition surveys and maintenance performance data.

2. Road Pavement AI Analysis Program

2.1. Program Structure

The developed crack analysis program's artificial intelligence model was designed by combining the crack area detection and classification models. The crack area detection model identifies the cracked area in the road image and binarizes the cracked and non-cracked areas into 0 and 1, respectively. Following that, the crack classification model categorizes the cracks into linear cracks, turtle cracks, and so on. The two models are combined to create a database that can be used to determine the types and risk of cracks.

2.2. Artificial Intelligence Learning Process

By segmenting images and analyzing specifically processed data, additional UNET++ artificial intelligence research was carried out to enhance the functionality of the current experimental crack analysis application. Data scaling and data rotation were performed in an amplification process, along with the learning data.

2.3. Automation of Road Crack Analysis Program

The previously examined road image must be loaded before the program automatically analyzes the road for the automation of crack analysis software. Therefore, a system was created using the automation tool so that the software could trace the path from the folder selected as the road survey image, import the image, and then automatically link and present the analysis results. The program's user interface during automatic analysis is shown in Figure 1.

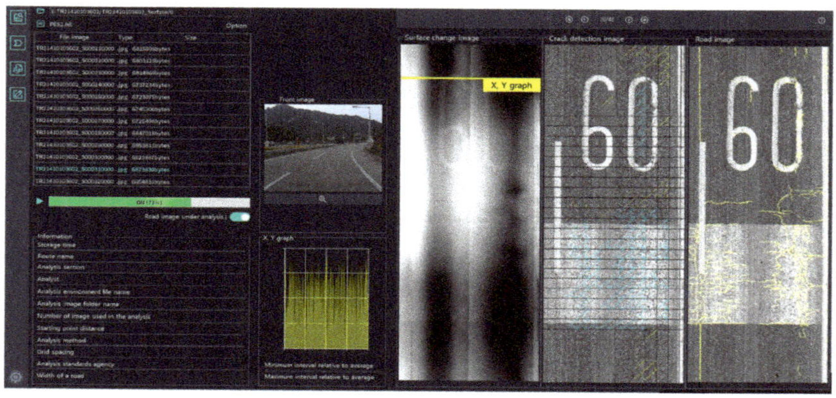

Figure 1. Automated analysis tool user interface.

3. Advanced National Road Pavement Data Management System (PDMS)

3.1. Pdms Database Construction

The database was classified into the route condition management of general national roads, the pavement condition investigation area, the maintenance performance management area, and the general national road pavement condition investigation area. It was established in order to annually collect pavement, bridge, and tunnel condition information and traffic volume data for general national road pavement condition management.

3.2. PDMS Functional Configuration

The route status of general national roads was configured to manage each route's alignment and attribute information based on the geographic information system (GIS). General national road geographic information is used to manage data on the route numbers, road alignments, junctions, the start and end points of bridges and tunnels, and administrative districts for the 51 routes in the country [3]. Attribute information is configured to manage the management agencies, the number of lanes, pavement types, and pavement thickness information. Figure 2a shows the route status of the general national roads as process information and attribute information, respectively.

(**a**) (**b**)

Figure 2. (**a**) Route status of general national roads (geographic information). (**b**) Management of pavement condition survey data.

The general national road pavement condition survey is managed in relation with the data of each section, and the data management function is classified according to the characteristics of the surveying equipment. The PES (Pavement Evaluation Surveyor) takes pictures of the damage on the pavement surface in the survey section and investigates the plastic deformation and longitudinal flatness through sensing. In addition, pavement conditions such as cracks and pothole repairs are evaluated through the analysis system, and the results are uploaded to the PDMS for management.

The visualization function was enhanced, as shown in Figure 2a. As the pavement condition survey of general national roads has been operated as a complete enumeration survey system since 2021, the pavement condition survey and analysis data, regardless of the existing number of lanes, can be realistically verified for each lane.

As shown in Figure 2b, the survey data on the pavement conditions of general national roads across the country were managed by linking the survey area with GIS for the general national roads. For each survey spot, the pavement condition and other status-related information are classified and managed using the designated spot number. By choosing the survey part shown on the map, one is able to check the pavement condition state and other information on current status.

4. Pothole Information Management

4.1. Pothole Information Linkage

Two devices are in operation according to the Land Management Office, and they are mounted on vehicles to detect potholes in the operating section. The detected pothole data were linked to the road work management safety guard app through a related service. In addition, a system was established to link pothole information from among the work information managed by the Road Work Management Safety Keeper app to PDMS.

4.2. Pothole Status Map

The pothole status map is a function that visually provides pothole information using an electronic map, and it is divided into a pothole distribution map and a location map. The pothole status map is built to provide various manipulation tools such as style setting, inquiry options, map control, legend setting, and regional development according to the Land Management Office.

The pothole status map is divided into a pothole distribution map and a location map, as shown in Figure 3a. The pothole distribution map displays the number of potholes according to region as defined by the Land Management Office, with the color according to the legend. The pothole location map is implemented to display the locations of potholes on the electronic map.

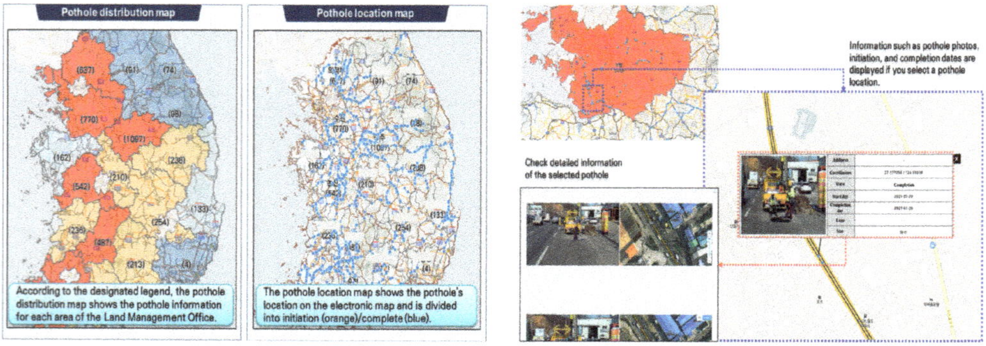

(**a**) (**b**)

Figure 3. (**a**) Types of pothole status map. (**b**) Information verification function according to pothole location.

The blue and orange dots indicate pothole locations on the status map. A blue dot is a pothole that has been repaired, and an orange dot is a pothole in the repair process. When a dot is selected by enlarging the pothole status map, the corresponding pothole information can immediately be checked on the electronic map. When the displayed pop-up of the pothole is clicked, a detailed information page of the corresponding pothole is requested. Figure 3b depicts the information-checking function for each pothole location.

5. Conclusions

In this study, research was conducted to develop a crack analysis program using road survey data, advance the pavement data management system, and establish a pothole information management system to efficiently address the increased management extension. As a result of the study, the following conclusions were obtained:

1. Objective crack analysis results were derived by developing a road pavement crack analysis program using AI. Its accuracy is improved using a designing that excluded anomalous objects on the road surface. In the future, this system's accuracy could be improved by performing additional learning to exclude abnormal items such as manholes and expansion joints. Based on the derived results, an efficient maintenance management system based on consistent standards would be possible.
2. The system was enhanced so that the survey data and analysis results of paved general national roads can be checked in units of 100 m per lane using a map. In addition, improvements were made so that survey agencies, analysis agencies, and road management entities can be used to efficiently manage roads through the visualization of survey and analysis results, maintenance sections, and budgets using the advanced general national road pavement data management system. In the future, the usability of this system could be improved by upgrading the construction data upload function and the automatic budget allocation function.
3. The pothole information management function is used to monitor pothole occurrence and repair status for general national roads and establish a pothole status map and pothole list. The built pothole information management function makes it possible to take immediate action for potholes in general national roads. Furthermore, it would be possible to efficiently conduct pothole repair by analyzing pothole occurrence frequency per region.

Author Contributions: Conceptualization, M.L.; methodology, M.L. and H.Y.; validation, S.H. and M.K.; investigation, M.K. and H.Y.; writing-original draft preparation, S.H.; writing-review and editing, M.L., N.L. and C.K.; project administration, M.L. and N.L.; funding acquisition, M.L. and C.K. All authors have read and agreed to the published version of the manuscript.

Funding: This work was supported by consignment business 2023 Pavement Management System (PMS) by the Ministry of Land, Infrastructure and Transport (MOLIT).

Institutional Review Board Statement: Not applicable.

Informed Consent Statement: Not applicable.

Data Availability Statement: All data are new and experimented.

Conflicts of Interest: The authors declare no conflict of interest.

References

1. Lim, J.K.; Lee, S.W.; Kim, Y.J.; Kim, B.G.; Sin, H.J. A Study of Improving Pavement Data Management System of National Highway. In Proceedings of the 2012 Fall Conference of Korea Information Processing Society, Jeju university, Jeju City, Republic of Korea, 22–23 November 2012; p. 1746.
2. Framework Act on Sustainable Infrastructure Management. Republic of Korea, Act No. 17237. 7 April 2020.
3. Lim, J.K. Development of Condition Change Prediction Model for Asphalt Pavement Using National Highway PMS Database. *J. Korean Road Soc.* **2021**, *23*, 105–106.

engineering proceedings

Proceeding Paper

Application of a Risk Management System of Road Networks Exposed to Volcanic Hazards †

Alondra Chamorro [1,2,3,*], Tomás Echaveguren [2,4], Marta Contreras [2], Manuel Contreras-Jara [1,2], Carlos Pattillo [2], Eduardo Allen [5], Natalia Nieto [1], Joaquín Dagá [1] and Hernán de Solminihac [1,3]

1 Department of Construction Engineering and Management, School of Engineering, Pontificia Universidad Católica de Chile, Santiago 7820436, Chile; mfcontreras3@uc.cl (M.C.-J.); nanieto@uc.cl (N.N.); jadaga@uc.cl (J.D.); hsolmini@ing.puc.cl (H.d.S.)
2 Research Center for Integrated Disaster Risk Management (CIGIDEN), ANID/FONDAP/1522A0005, Santiago 7820436, Chile; techaveg@udec.cl (T.E.); mcontreras3@uc.cl (M.C.); carlos.pattillo@gmail.com (C.P.)
3 Latin American Center for Economic and Social Policies (CLAPES UC), Pontificia Universidad Católica de Chile, Santiago 8331010, Chile
4 Department of Civil Engineering, Engineering Faculty, University of Concepción, Concepción 4070409, Chile
5 Department of Civil and Environmental Engineering, The University of Auckland, Auckland 1023, New Zealand; eall093@aucklanduni.ac.nz
* Correspondence: achamorro@ing.puc.cl
† Presented at the Second International Conference on Maintenance and Rehabilitation of Constructed Infrastructure Facilities, Honolulu, HI, USA, 16–19 August 2023.

Abstract: Risk Management Systems are a valuable tool for estimating the potential losses of natural events, assessing risk reduction strategies, and increasing the resilience of critical infrastructure. The paper discusses the development of SIGeR-RV, a tool for Risk Management of road networks exposed to multi-hazards developed in Chile. The tool was implemented on a web-based Geographic Information System platform. It is able to display hazard maps, calculate risk levels, prioritize mitigation strategies, estimate direct and indirect losses, and assess the social vulnerability of communities exposed to natural hazards. The article includes an application of SIGeR-RV in a road network exposed to the lahar flows of Villarrica volcano in the south of Chile.

Keywords: risk assessment; Risk Management System; natural hazards; volcanic hazard; road networks; resilience; lahar flow

Citation: Chamorro, A.; Echaveguren, T.; Contreras, M.; Contreras-Jara, M.; Pattillo, C.; Allen, E.; Nieto, N.; Dagá, J.; de Solminihac, H. Application of a Risk Management System of Road Networks Exposed to Volcanic Hazards. *Eng. Proc.* **2023**, *36*, 68. https://doi.org/10.3390/engproc2023036068

Academic Editor: Hosin (David) Lee

Published: 12 October 2023

1. Introduction

Because road networks are spatially distributed infrastructures, they are exposed to multiple natural hazards. Roads are critical assets in rural areas, where road networks have little redundancy and the population has little accessibility to services and critical infrastructure [1]. Risk Management Systems (RMS) contribute to assessing the risk of infrastructure exposed to natural hazards and consequently reducing their effects. However, RMS applications require ad-hoc computational tools with comprehensive models, including hazards simulation, fragility modeling, risk assessment, mitigation evaluation, recovery costs, and defining communities' risk tolerance, among others [2]. The paper aims to describe an RMS for road networks developed in Chile called SIGeR-RV [3]. SIGeR-RV estimates risk in terms of traffic disruptions and socioeconomic consequences. A case study illustrates the application of SIGeR-RV to a road network exposed to lahar flows.

2. SIGeR-RV Risk Management System

SIGeR-RV is a web-GIS-based computational tool. This software was developed as part of a research and development project funded by the ANID (National Research and Development Agency of Chile) in collaboration with public and private agencies. The conceptual framework of SIGeR-RV (Figure 1) consists of three modules: input data, risk models, and outputs. A detailed description of the conceptual framework is in Chamorro et al. [3].

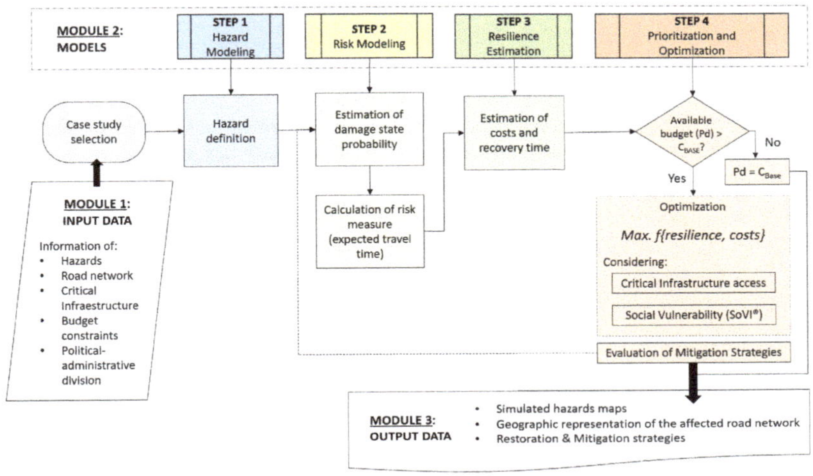

Figure 1. Components and framework of the SIGeR-RV tool. Adapted with permission from Ref. [3]. 2023, Sage Journals.

2.1. Module 1: Input Data

The first data set includes seismic, hydrometeorological, and volcanic hazard simulations. The second data set includes territorial geographic information on coastal lines, borders, and the country's administrative configuration in a separate GIS interface. The third data set is linked to data on road infrastructure. It consists of the georeferenced Chilean road network and its related attributes. The fourth data set includes Critical Infrastructure (CI) information and the geolocation of educational centers, health systems, power stations, rural drinking water systems, emergency facilities, and safety services.

2.2. Module 2: Models

The first step refers to hazard modeling. In this step, the study area exposed to the selected natural hazard is simulated. Seismic hazard maps with 50,000 scenarios are considered for the Chilean subduction zone, whereas hydro-meteorological and volcanic hazards are uploaded as specific hazard maps. The second step considers risk modeling, estimated from calibrated fragility curves of assets exposed to the hazard under study and expressed in terms of an increase in travel time. Fragility curves represent the probability of an expected damage state given a certain hazard intensity [3]. An increase in travel time is estimated using traffic reassignment models based on user equilibrium principles [4]. The third step considers resilience assessment. Based on a user-defined budget, a mitigation and recovery plan is prepared [5]. SiGeR-RV has a pre-loaded library of recovery and mitigation strategies, including direct unit costs, user costs, and the duration of each strategy. Finally, recovery and mitigation strategies are optimized by minimizing the cost-benefit of each mitigation strategy and its effect on the overall resilience of the road network.

2.3. Module 3: Outputs

The software displays the following reports: (1) a graphical GIS representation of the study area with the road network and the simulated hazard; (2) tables with potentially damaged road assets; (3) a GIS representation of the affected road assets; (4) expected travel time for different simulated hazard scenarios; (5) a recovery and mitigation plan with recommended strategies, their costs, and duration; (6) a graphical representation of social vulnerability represented through the developed index; (7) a topological classification of the network; and (8) the most relevant roads that provide access to critical infrastructure.

3. Case Study: Road Network Exposed to Volcanic Hazard

This case study comprises the road network between Villarrica and Pucón cities in the south of Chile that are exposed to laharic flows caused by potential eruptions of Villarrica volcano. Six scenarios were studied considering winter and summer seasons, Hawaiian-strombolian, subplinian, and plinian eruptions [6,7]. Bridge failure probabilities were estimated from fragility models calibrated by Dagá et al. [7].

Figure 2 presents the simulated lahars expressed as volumes of laharic material (in m^3). From the interpolation of the road network, the hazard map, and the fragility curves calibrated by Dagá et al. [7], the failure probability of affected bridges is estimated. Optimal and alternative routes are assessed based on traffic assignment. Figure 2a shows the optimal route (in green) between Villarrica and Pucón after simulating the different volcanic scenarios. Figure 2b shows the alternative route (in red) for the Hawaiian-Strombolian eruption in winter. In some sections, the optimal and alternative routes coincide, whereas the alternative route differs by considering sections on the northern shore of Villarrica Lake. This alternative route, despite being a valid redundancy option, implies a considerable increase in travel time for users compared to the optimal route. This alternative route presents a travel time of 645 min compared to the optimal route of 32 min (see Figure 3).

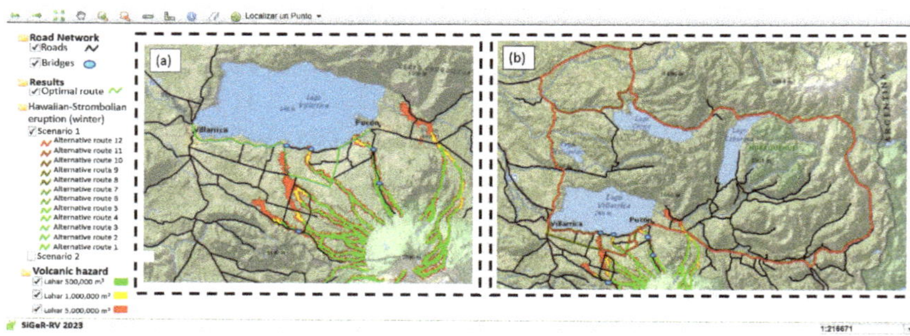

Figure 2. Case study results in map format. (**a**) the optimal route between Villarrica and Pucón cities. (**b**) an alternative route for Hawaiian-Strombolian eruptions in winter.

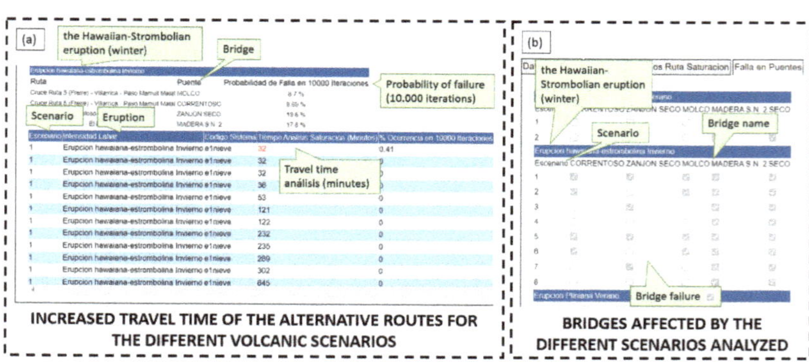

Figure 3. Case study results in table format. (**a**) increased travel time of the alternative routes for the different volcanic scenarios. (**b**) bridges affected by the different scenarios analyzed.

Figure 3a presents detailed information on the probability of failure of affected assets for 10,000 iterations and the travel times for each of these iterations (Column 4). For example, in the scenario presented in Figure 3, 4 assets fail, with a failure probability between 8.7% and 17.8%. Producing an increase in travel times up to 645 min. Figure 3b reports the bridges that have failed in a certain scenario.

The case study evidences that four bridges are mainly affected by laharic flows for the different eruption scenarios that could occur in the Villarica volcano. Knowing these bridges, it is possible for road agencies to allocate resources toward improving the current condition of the bridges, either by implementing some mitigation measure to increase the standard of the asset or by planning in advance restoration measures for possible emergencies.

4. Conclusions

The Risk Management System assists decision-makers in managing the risk of road networks exposed to natural hazards. The article presents a risk management computational tool that allows simulating different natural hazards that affect road networks, territories, and developing communities in Chile. The tool was applied to a road network exposed to the laharic flows of Villarrica volcano.

The results evidence routes and road assets that are more likely to present damages when exposed to certain hazards, as well as alternative routes that provide redundancy to the network. From these results, resources can be efficiently allocated for the recovery and mitigation of road assets when exposed to a specific hazard.

It is important to consider that applying the developed framework and software to other territories with different characteristics and realities requires significant scientific development and model calibration.

Author Contributions: The contribution of the authors to the article is detailed below: study conception and design conceptualization: A.C., T.E., H.d.S. and C.P.; data collection: M.C.-J., E.A., N.N., J.D. and M.C.; analysis and interpretation of results: M.C. and J.D.; draft manuscript preparation: M.C.-J. and M.C.; review and editing, A.C., T.E. and H.d.S.; visualization, M.C. and M.C.-J.; supervision, A.C. and T.E.; project administration, A.C., H.d.S. and T.E.; funding acquisition, A.C. All authors have read and agreed to the published version of the manuscript.

Funding: This research was funded by the National Agency of Research and Development of Chile (ANID), Grant no. ID22I10037/FONDEF/ANID, and the Research Center for Integrated Disaster Risk Management (CIGIDEN) (Grant no. ANID/FONDAP/1522A0005).

Institutional Review Board Statement: Not applicable.

Informed Consent Statement: Not applicable.

Data Availability Statement: SiGeR-RV has models that have been developed from different databases that can be found in greater detail in the references mentioned [1,3,7], among others. In addition, the georeferenced data that serves as a base (for example, political and administrative division, international borders, and critical infrastructures) can be downloaded from the website https://www.ide.cl/ (accessed on 26 September 2023). In turn, georeferenced information about the road network can be obtained from the IDEMOP Platform (https://ide.mop.gob.cl/geomop/) (accessed on 26 September 2023).

Acknowledgments: The authors would like to acknowledge the National Agency of Research and Development of Chile (ANID) for funding the Project Grant no. ID14I20309/FONDEF/ANID.

Conflicts of Interest: The authors declare no conflict of interest.

References

1. Allen, E.; Chamorro, A.; Poulos, A.; Castro, S.; de la Llera, J.C.; Echaveguren, T. Sensitivity analysis and uncertainty quantification of a seismic risk model for road networks. *Comput.-Aided Civ. Infrastruct. Eng.* **2021**, *37*, 516–530. [CrossRef]
2. Free, M.; Anderson, S.; Milloy, C.; Mian, J. Geohazards risk management for infrastructure projects. *Proc. ICE-Civ. Eng.* **2006**, *159*, 28–34. [CrossRef]
3. Chamorro, A.; Echaveguren, T.; Pattillo, C.; Contreras-Jara, M.; Contreras, M.; Allen, E.; Nieto, N.; de Solminihac, H. SIGeR-RV: A Web-Geographic Information System-Based System for Risk Management of Road Networks Exposed to Natural Hazards. *Transp. Res. Rec.* **2023**. [CrossRef]
4. Nie, Y.; Zhang, H.; Lee, D.-H. Models and algorithms for the traffic assignment problem with link capacity constraints. *Transp. Res. Part B Methodol.* **2004**, *38*, 285–312. [CrossRef]

5. Rose, A.; Krausmann, E. An economic framework for the development of a resilience index for business recovery. *Int. J. Disaster Risk Reduct.* **2013**, *5*, 73–83. [CrossRef]
6. INH. *Construcción Mitigación Riesgos Volcánicos y Geológicos Asociados, Comunas de Villarrica, Pucón y Curarrehue, Región de la Araucanía*; Instituto Nacional de Hidráulica: Santiago, Chile, 2013; Available online: http://www.inh.cl/home/documents/proyects/001-09-2014%20Ficha%20Estudio%20Riesgos%20volcanicos,%20regi%C3%B3n%20Araucan%C3%ADa_Estudio.pdf (accessed on 3 May 2023).
7. Dagá, J.; Chamorro, A.; De Solminihac, H.; Echaveguren, T. Development of fragility curves for road bridges exposed to volcanic lahars. *Nat. Hazards Earth Syst. Sci.* **2018**, *18*, 2111–2125. [CrossRef]

engineering proceedings

MDPI

Proceeding Paper

Keynote Presentation: Improving Pavement Sustainability through Integrated Design, Construction, Asset Management, LCA, LCCA, and S-LCA †

John Harvey

University of California Pavement Research Center, University of California, Davis, CA 95616, USA; jtharvey@ucdavis.edu

† Presented at the Second International Conference on Maintenance and Rehabilitation of Constructed Infrastructure Facilities, Honolulu, HI, USA, 16–19 August 2023.

Abstract: Engineers, planners, asset managers, materials suppliers, contractors, and policymakers are focused on improving pavement infrastructure sustainability, and increasingly considering climate change resilience. This focus is often on materials; however, the decisions and practices in design, construction, and asset management are typically more important in achieving the desired environmental, cost, and social outcomes. This presentation discusses the tools of mechanistic–empirical design, asset management, performance-related tests and specifications, construction quality assurance, environmental and social life cycle assessment, and life cycle cost analysis, which can be used together to achieve the desired outcomes, and the data and models of which can be integrated in efficient web-based systems.

Keywords: pavement; design; construction; asset management; life cycle assessment; life cycle cost analysis; social life cycle assessment; integrated data

Citation: Harvey, J. Keynote Presentation: Improving Pavement Sustainability through Integrated Design, Construction, Asset Management, LCA, LCCA, and S-LCA. *Eng. Proc.* **2023**, *36*, 69. https://doi.org/10.3390/engproc2023036069

Academic Editor: Hosin (David) Lee

Published: 29 December 2023

1. Introduction

"Sustainable" in the context of pavements refers to system characteristics that encompass a pavement's ability to [1] achieve the engineering goals for which it was constructed, [2] preserve and (ideally) restore surrounding ecosystems, [3] use financial, human, and environmental resources economically, and [4] meet basic human needs such as health, safety, equity, employment, comfort, and happiness [1]. Sustainability has always been the goal of pavement design, construction, and asset management, with a focus primarily on the engineering goal of handling motor vehicles, and the financial goal of minimizing either initial cost or life cycle cost.

The engineering goals of pavements have also been changing over the years, as knowledge and technology have changed, and as the definition of functionality has changed. As shown in Figure 1, 100 years ago, the focus was on measuring the subgrade bearing capacity and the additional structural capacity of different types of road building materials. The focus was on safely and efficiently carrying cars and trucks. The scope of pavement engineering expanded to pavement types and structures during the deployment of road networks, and then, once networks were deployed, attention turned to efficiently maintaining and rehabilitating the existing pavements considering the whole network, still primarily considering cost (mostly to the agency, but also to the user) in addition to the essential function of safety. Integration with other transportation networks, and a consideration of pavement work zone closures and other pavement-related disruptions on the efficient movement of goods and people, were added afterwards.

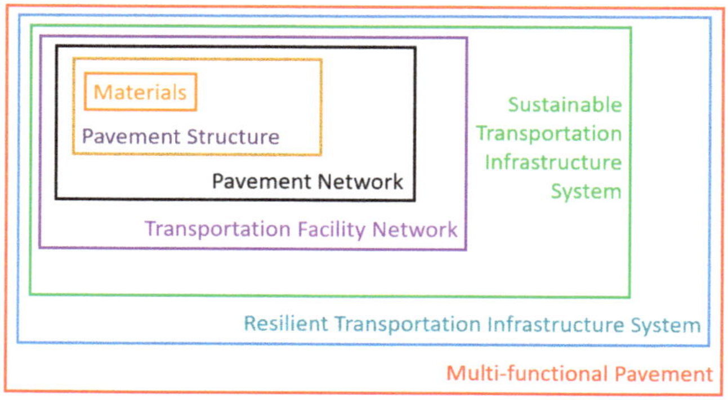

Figure 1. Changes in system boundaries for pavement functionality over last 100 years.

Today, and even more so in the future, networks and the pavements within them will need to achieve environmental goals, and resilience will need to be considered as climate change effects grow and as the requirements for maintaining or quickly restoring functionality during and after extreme precipitation, heat and cold events, and wildfires, for safety, economic, and quality of life reasons, becomes a requirement. In recent decades, it has become apparent that pavements have important environmental impacts and use large quantities of finite resources, which affects quality of life now and will have an increasing impact in the future. Nearly everyone in the world is also now aware of the effects of climate change and there are increasing efforts in every sector of the economy to reduce greenhouse gas emissions and reach net zero emissions, with varying target dates for reaching that goal. There are now market signals for the private sector to compete with in terms of reducing environmental impact, and governments at the local, state, and national levels are similarly sending market signals, setting goals, and using various methods to achieve those goals.

The functionality required of pavements, particularly in urban areas, is also changing to increasingly include active transportation, stormwater pollution mitigation, stormwater and sea level rise flooding risks, human comfort and safety in extreme heat events, and noise.

New tools have been developed to support pavement decision-making through the life cycle and for the wider definitions of the complete system to meet these changing requirements for planners, engineers, asset managers, constructors, and policymakers. These have included tools for pavement materials testing and design, structural design, asset management, life cycle cost analysis, environmental life cycle assessment, and construction quality. New tools are being developed to consider the social effects of access to good pavements, and the social effects of the extraction, processing, and construction of materials. Often, these tools do not communicate with each other in terms of data, models, terminology, and reporting, effectively siloing their use and requiring replication effort to build, support, and use them. Paying attention to the data is essential to using the tools that have not always been well-designed, and insufficient attention, effort, and resources are often applied for maintaining up-to-date and comprehensive data and models.

The pavement is long-lived infrastructure, with very few roads ever abandoned. The pavement life cycle includes planning, design, construction, preservation, maintenance, rehabilitation, reconstruction, and end-of-life, with many loops back into the cycle. Environmental sustainability, cost sustainability, and, increasingly, social sustainability are best and most efficiently achieved by considering the full life cycle and the complete system, which also helps avoid negative unintended consequences.

2. Overview of the Presentation

This presentation reviews the types of tools used in pavements and the requirements for well-planned and -executed processes for building and maintaining the data and models that make them most useful. Figure 2 shows the "pyramid" concept, emphasizing that the majority of the cost, effort, and attention need to be on the data; in the case of a PMS, the models required are for pavement performance prediction and cost which then support life cycle costing and budget optimization. It discusses an approach being implemented in California to develop integrated data definitions and frameworks, databases, models, and web-based tools to support pavement decision-making through its life cycle. This presentation reviews the different stages of the life cycle where environmental sustainability can be improved and their relative importance for different types of roadways. It also reviews how cost and social sustainability can be considered using the integrated tools, and how this integration helps produce better results more efficiently. Several examples are given as follows.

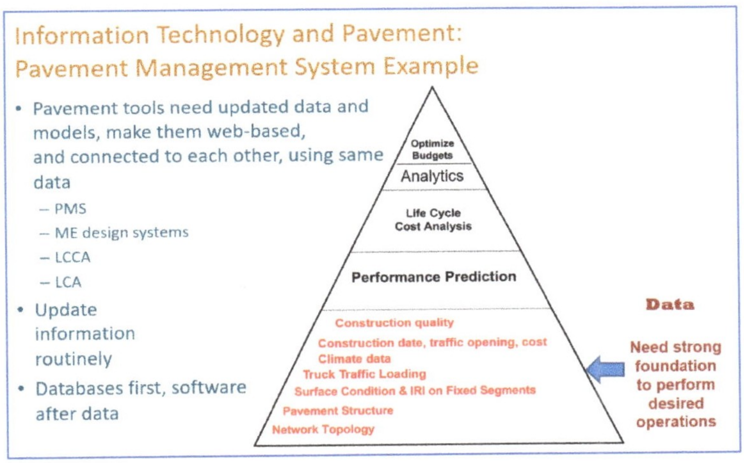

Figure 2. Example of "pyramid" concept for PMS, emphasizing that data, then models, are priorities to support the tool.

3. Calibration of Mechanistic–Empirical Structural Design Methods Using Asset Management Data

This example looks at the recent calibration of the California ME design methods for asphalt and concrete surface pavements using hundreds of thousands of observations from the asset management system, and performance-related materials testing data [2].

4. Improved Life Cycle Cost Analysis Using ME Design and Asset Management Results

This example looks at the recent updating of the California LCCA tool using the results of ME simulations for cracking and roughness performance models of the pavement management system.

5. Environmental Life Cycle Assessment Using Materials Data, ME Design Results, Construction Quality Data, and PMS Models

This part of the presentation presents an overview of life cycle assessment, the California LCA tool eLCAP [3], and the use of materials designs, ME design simulation results, and roughness models from the PMS to quantify environmental impacts over the life cycle, as well as the use of cradle-to-gate LCAs called environmental product declarations (EPD) in construction material procurement. Another example uses PMS models for age-related cracking and materials data to evaluate the effects of asphalt compaction, and a third example compares the use of preservation treatments vs. rehabilitation-only treatments with respect to life cycle cost and life cycle emissions.

6. Environmental Impact Analysis in the PMS and Prioritization of Strategies Using Life Cycle Assessment and Life Cycle Cost Analysis

The implementation of LCA in the pavement management system to calculate global warming potential is illustrated with an example. Another example looks at an approach called the marginal cost of abatement curve, or "supply curve", for assessing the cost effectiveness of different strategies for reducing environmental impacts [4].

7. Integration of Social Vulnerability into Decision Support

The initial steps and applications for the consideration of social vulnerability and environmental impacts in asset management are the final examples shown for integrated tools.

8. Summary

The functional requirements and system boundaries for pavements have increased, and the need for tools considering the full life cycle and complete system in order to support decision making in this more complex environment has also grown. Those tools need to be supported with comprehensive, up-to-date, high quality data and models. This presentation shows examples of the development and use of integrated web-based decision support tools to help meet these challenges.

Funding: Funding for the work described in this paper came from a series of projects sponsored by the California Department of Transportation between 1995 and 2023.

Institutional Review Board Statement: Not applicable.

Informed Consent Statement: Not applicable.

Data Availability Statement: Reports and technical memorandums regarding the research, development and implementation studies leading to the results discussed in this paper can be found at www.ucprc.ucdavis.edu/NewPubPage.aspx (accessed on 10 August 2023).

Conflicts of Interest: The author declares no conflict of interest.

References

1. Van Dam, T.; Harvey, J.; Muench, S.; Smith, K.; Snyder, M.; Al-Qadi, I.; Ozer, H.; Meijer, J.; Ram, P.; Roesler, J.; et al. Towards Sustainable Pavement Systems: A Reference Document. Report FHWA-HIF-15-002. 2015. Available online: https://rosap.ntl.bts.gov/view/dot/38541 (accessed on 10 August 2023).
2. Wu, R.; Harvey, J.; Lea, J.; Jones, D.; Louw, S.; Mateos, A.; Hernandez-Fernandez, N.; Shrestha, R.; Holland, J. Calibration of a Mechanistic-Empirical Cracking Model Using Network-Level Field Data. *Transp. Res. Rec.* **2022**, *2676*, 127–139. [CrossRef]
3. Lea, J.; Harvey, J.; Saboori, A.; Butt, A.A. eLCAP: A Web Application for Environmental Life Cycle Assessment for Pavements. 2022. Available online: https://escholarship.org/uc/item/9f5181j1 (accessed on 10 August 2023).
4. Harvey, J.T.; Butt, A.A.; Lozano, M.T.; Kendall, A.; Saboori, A.; Lea, J.D.; Kim, C.; Basheer, I. Life Cycle Assessment for Transportation Infrastructure Policy Evaluation and Procurement for State and Local Governments. *Sustainability* **2019**, *11*, 6377. [CrossRef]

MDPI
St. Alban-Anlage 66
4052 Basel
Switzerland
www.mdpi.com

Engineering Proceedings Editorial Office
E-mail: engproc@mdpi.com
www.mdpi.com/journal/engproc

www.ingramcontent.com/pod-product-compliance
Lightning Source LLC
Chambersburg PA
CBHW061607220926
51850CB00005B/443